ALSO FROM COLD SPRING HARBOR LABORATORY PRESS

Related Titles

RNA: A Laboratory Manual
RNA Worlds: From Life's Origins to Diversity in Gene Regulation

Other History of Science Titles

A Passion for DNA: Genes, Genomes, and Society by James D. Watson
Davenport's Dream: 21st Century Reflections on Heredity and Eugenics
    edited by Jan A. Witkowski and John R. Inglis
Dorothy Hodgkin: A Life by Georgina Ferry
*George Beadle, An Uncommon Farmer: The Emergence of Genetics in
    the 20th Century* by Paul Berg and Maxine Singer
Mendel's Legacy: The Origin of Classical Genetics by Elof Axel Carlson
*We Can Sleep Later: Alfred D. Hershey and the Origins of Molecular
    Biology* edited by Franklin W. Stahl

RNA

Life's Indispensable Molecule

James Darnell

The Rockefeller University

COLD SPRING HARBOR LABORATORY PRESS
Cold Spring Harbor, New York · www.cshlpress.com

RNA: Life's Indispensable Molecule

Publisher	John Inglis
Acquisition Editor	John Inglis
Director of Development, Marketing, & Sales	Jan Argentine
Developmental Editor	Maria Smit
Project Manager	Inez Sialiano
Permissions Coordinator	Carol Brown
Production Editor	Rena Steuer
Production Manager	Denise Weiss
Sales Account Manager	Elizabeth Powers
Cover Designer	Mike Albano
Compositor	Techset Composition Ltd

Front cover artwork: An RNA molecule morphing into a staircase, representing the historical progress of RNA research, electrifies the dawn sky, representing life. (Credit: Jean-François Podevin; © 2011 Photo Researchers, Inc.)

Library of Congress Cataloging-in-Publication Data

Darnell, James E.
    RNA : life's indispensable molecule / James Darnell.
        p. ; cm.
    Includes bibliographical references and index.
    ISBN 978-1-936113-19-4 (hard cover : alk. paper)
    1. RNA.   I. Title.
    [DNLM: 1. RNA.   2. Molecular Biology--history. QU 58.7]
    QP623.D37 2011
    572.8′8--dc22

                                    2011001258

10   9   8   7   6   5   4   3   2

All Cold Spring Harbor Laboratory Press publications may be ordered directly from Cold Spring Harbor Laboratory Press, 500 Sunnyside Blvd., Woodbury, New York 11797-2924. Phone: 1-800-843-4388 in Continental U.S. and Canada. All other locations: (516) 422-4100. FAX: (516) 422-4097. E-mail: cshpress@cshl.edu. For a complete catalog of all Cold Spring Harbor Laboratory Press publications, visit our website at http://www.cshlpress.com/.

To all my family—both the old and the new

Contents

Preface

A SIZEABLE NUMBER OF COLLEAGUES READ all or portions of this book at various times during its preparation. Most believed, as I did, that a book stressing the history of those phases of molecular biology centered on RNA was a sound and different idea and should produce a useful, even a needed, book. Early discussions with Shai Shaham, Paul Nurse, Sid Strickland, and Jan Breslow helped particularly in shaping the content of what was finally included. My gratitude to each of them.

As the project developed, I received positive encouragement, advice, suggestions, and corrections from a larger group. That list includes David Allis, Jan Breslow, Linda Chaput, Gene Cordes, Bob Darnell, Ford Doolittle, Jeff Friedman, Magda Konarska, Leon Levintow, Peter Model, Tom Muir, Paul Nurse, Lennart Philipson, Bob Roeder, Marjorie Russel, Shai Shaham, Sid Strickland, Jon Warner, and Mike Young.

To all of those friends and colleagues I offer my sincerest gratitude. For the remaining errors and especially for the omitted or neglected references (despite ~1000 included references), the responsibility remains with me.

Initial discussions with John Inglis, the Executive Director of Cold Spring Harbor Laboratory Press, were extremely encouraging and helpful in solidifying the purpose of what is presented. The Cold Spring Harbor Laboratory Press has a staff exceptionally gifted in publishing scientific works, and I am eternally grateful to all of them: in particular, Inez Sialiano, Project Manager, and Rena Steuer, Production Editor, who among many other wise contributions arranged to make figures taken from older articles readable. I especially thank Maria Smit, the Developmental Editor. I've had considerable experience in biology textbook writing and publishing but have never had anything approaching the skillful and thoughtful editorial help I've received on this book. Lois Cousseau, my assistant for more than 30 years, merits very special thanks. Nothing I've written in all this time, all beginning in longhand, would have ever appeared without Lois's cheerfulness, patience, and extraordinary competence. Even the most heartfelt thanks seem insufficient.

Finally, this labor of love was supported with my wife Kristin's unflagging confidence and belief that I could do it.

Author's Note

Pursuing RNA for More Than 50 Years

I F LENGTH OF SERVICE IN PURSUING the mysteries of RNA qualifies one as a reporter of RNA history, I suppose I easily qualify. As luck would have it, my earliest research after a medical internship directed me as a complete naïf toward studying RNA. Some personal events of those long ago years are given here.

I joined the laboratory of Harry Eagle at The National Institutes of Health (NIH) in Bethesda, Maryland, in July 1956, where he had just established the nutritional requirements for the growth of animal cells in culture (Eagle 1955). This landmark advance made growing homogeneous populations of animal cells in some ways comparable to growing uniform populations of bacterial cells and, of equal importance, provided uniformly susceptible cells with which to study animal virus replication.

Eagle had done one collaborative experiment with the virologist Karl Habel on nutritional requirements for poliovirus growth in HeLa cells (Eagle and Habel 1956) but had his hands full with other pursuits, particularly protein turnover. I did not know a virus from a billiard ball, and he could have directed me to work on any number of things. But it was my great good fortune that he suggested I work on virus infection in cultured animal cells, and he assigned me a bench in a room with Robert I. DeMars, Salvador Luria's third graduate student. Luria was, of course, one of the founders of modern bacteriophage research and the molecular biology that grew out of that research. DeMars handed me his copy of Luria's 1953 text *General Virology*, the first successful textbook ever written exclusively about viruses and, incidentally, according to Luria, the first textbook to describe the Watson-Crick structure of DNA (Watson was Luria's first graduate student). DeMars said to me, "Read this and I'll teach you virology." Thus, under DeMars's tutelage, I followed somewhat in lockstep what a phage worker would do, but with an animal virus. I established a plaque assay for the RNA-containing poliovirus (Schaffer and Schwerdt 1956) and showed that every HeLa cell could be infected and yield virus (Darnell 1958).

Together with my colleague, mentor, and close friend Leon Levintow (Levintow and Darnell 1960), we developed a simple purification scheme for poliovirus using CsCl density centrifugation and then, using labeled nucleosides and amino acids, determined the time course of formation of poliovirus protein and poliovirus RNA (Darnell and Levintow 1960; Darnell et al. 1961). The RNA was formed first, as befitted the director of protein synthesis; Heinz Fraenkel-Conrat and colleagues had also shown this to be true for tobacco mosaic virus (Fraenkel-Conrat et al. 1957). Moreover, as had been done by this time with other animal viruses (Colter et al. 1957; Wecker 1959) (and, of course, plant viruses [Gierer and Schramm 1956]), Levintow and I found that the phenol-extracted RNA from poliovirus and from poliovirus-infected cells was infectious (Darnell et al. 1961).

These experiments did not signal any important contribution of ours about how RNA could exert its protein-specifying role. However, this work did earn me a stay in Paris in 1960–1961 as a postdoctoral fellow at the Pasteur Institute with François Jacob in the Service de Physiologie Microbienne directed by André Lwoff. The sojourn in Paris could not have come at a more propitious time for a young researcher interested in the actions of RNA.

When I climbed up to the famous third-floor laboratory in the last week of August 1960, I went into the office and found Sarah Rapkine, Lwoff's assistant, who had rented an apartment for us. She enquired solicitously about our well being (my wife Jane and three little boys were along for our Paris adventure), and seeing that we were at least getting along, she brought me in to meet André Lwoff. After a perfunctory but pleasant conversation and an introduction to Marguerite Lwoff, André's wife and collaborator, André introduced me to François and to Elie Wollman. Within the space of no more than one or two incorrectly worded sentences by me, François announced, smilingly, that we would speak English. Perhaps, for a few milliseconds, my brain thought to be embarrassed at my incompetent unworldliness, but relief instantly overwhelmed pride.

François knew that I had done a little biochemistry with polio-infected HeLa cells, including the use of zonal and density gradient equilibrium sedimentation. We struck up a bargain. He would speak genetics slowly enough for me to get an inkling of what was going on in his experiments, and I would try some "molecular biology" and teach everyone who wanted to know how to make and separate RNA on sucrose gradients. Many geneticists were gearing up to study the newly described messenger RNA (mRNA), as I was to learn.

Although the expression "molecular biology" was still fairly new, I had gotten the drift at NIH that that was what I was doing with polio. But it was clear that I needed badly to catch up on what had been published from Paris. For example, Jacob told me to really understand the experiments that by then were referred to even in Paris as the PaJaMa (Pardee, Jacob, Monod) experiments. (They had been published in 1959 in Vol. 1 of the *Journal of Molecular Biology* [Pardee et al. 1959].) And a few days later (about mid September), François handed me a draft of "the" paper—"Genetic Regulatory Mechanisms in the Synthesis of Proteins" by himself and Jacques Monod (Jacob and Monod 1961). This paper was to become, next to the Watson-Crick papers of 1953, perhaps the most famous paper in molecular biology. He also told me of what he and Sydney Brenner (with Matt Meselson's help) had done at The California Institute of Technology (Caltech) in the spring of 1960 and had not yet published. Later that fall, François showed me the manuscript that he and Sydney had finished that proved the existence of the bacteriophage T4 mRNA. Thus, not only did I get to read before publication the logic behind the idea of a "messenger" (Jacob and Monod 1961) but also the best experiment to establish the existence of mRNA (Brenner et al. 1961). All of the science underlying these conclusions is found in Chapter 2.

Here I was, just turned 30, having done some journeyman work on an RNA-containing animal virus (polio) and having landed in one of only three places on the planet (The Pasteur, the Watson laboratory at Harvard, and Sydney Brenner's laboratory in Cambridge) that knew the secret of information transfer from gene to protein. Now, 50 years later, not a day passes that I don't remember and reflect on my luck as a youngster just getting started.

These early experiences sent me on my way to study RNA for all of the 50 plus years of my scientific career. Whether this long experience equipped and entitled me to be a historian of RNA, we must let the reader decide.

REFERENCES

Brenner S, Jacob F, Meselson M. 1961. An unstable intermediate carrying information from genes to ribosomes for protein synthesis. *Nature* **190:** 576–581.

Colter JS, Bird HH, Moyer AW, Brown RA. 1957. Infectivity of ribonucleic acid polyA from virus infected tissues. *Virology* **4:** 522–532.

Darnell JE. 1958. The adsorption and maturation of poliovirus in singly and multiple infected cells. *J Exp Med* **107:** 633–641.

Darnell JE, Levintow L. 1960. Poliovirus protein: Source of amino acids and time course of synthesis. *J Biol Chem* **235:** 74–77.

Darnell JE Jr, Levintow L, Thoren MM, Hooper JL. 1961. The time course of synthesis of polio-virus RNA. *Virology* **13:** 271–279.

Eagle H. 1955. Nutrition needs of mammalian cells in tissue culture. *Science* **122:** 501–514.

Eagle H, Habel K. 1956. The nutritional requirements for the propagation of poliomyelitis virus by the HeLa cell. *J Exp Med* **104:** 271–187.

Fraenkel-Conrat H, Singer BA, Williams RC. 1957. The nature of the progeny of virus reconsti-tuted from protein and nucleic acid of different strains of tobacco mosaic virus. In *The chemical basis of heredity* (ed. B Glass), pp. 501–517. Johns Hopkins University Press, Baltimore.

Gierer A, Schramm G. 1956. Infectivity of ribonucleic acid from tobacco mosaic virus. *Nature* **177:** 702–703.

Jacob F, Monod J. 1961. Genetic regulatory mechanisms in the synthesis of proteins. *J Mol Biol* **3:** 318–356.

Pardee AB, Jacob F, Monod J. 1959. The genetic control and cytoplasmic expression of "inducibility" in the synthesis of β-galactosidase by *E. coli. J Mol Biol* **1:** 165–178.

Schaffer FL, Schwerdt CE. 1956. Purification of poliomyelitis viruses propagated in tissue culture. *Virology* **2:** 665–678.

Wecker E. 1959. The extraction of infectious virus nucleic acid with hot phenol. *Virology* **7:** 241–243.

Introduction

THIS BOOK WAS WRITTEN FOR SCIENTISTS of all kinds. It is unapologetically historical. But, you say, the world of biology is rocketing ahead at a pace undreamed of even a decade ago. The advancing technological age in biology that began roughly 35 years ago with the "recombinant DNA revolution" now presents a daily mountain of new information. So why be so misguided in the midst of this whirlwind of the new as to turn out a history? And why a history of RNA?

Consider this: How in the next couple of decades are newcomers to biology going to learn, and how and what are established scientists going to teach them? Already, virtually all college-age students have had exposure, often since grade school, to the mantra "DNA makes RNA makes protein." In this computer age, the notion that biology is an information science and that DNA is the library seems a congenial concept to most who are inclined toward an analytical/scientific career. Perhaps the sensible and necessary course to properly prepare declared biology students and analytically trained "transfers" (mathematicians, physicists, engineers) is first to serve up a predigested catechism of settled conclusions achieved in the 20th century by "wet" laboratory experiments. With this concise biological "periodic table" under command, the newcomer then can be efficiently prepared to deal with the rapidly advancing technology both for doing experiments and for collecting and analyzing to a useful purpose the enormous quantity of data that emerges from today's genomic, proteomic, and computationally enhanced microscopic investigations.

It is by no means my intent to deflect teachers/scientists (mostly young, under 40 years of age) who must carry out the indispensable task of getting students ready to enter today's biology world. Rather, my aim in writing this book is to provide a supplement in historical form—both to the younger generation of scientists and teachers and through them to incoming students—that describes how we first learned some of the molecular fundamentals of biology in the days of the "hands-on wet laboratory."

One can legitimately argue whether a 2011 biology student "needs" to know any pre-1990 history. I am not prepared to defend vigorously the affirmative in this debate. But I will argue that many *may choose to know* how we came to know all that we did in the era before commercial kits and genomic sequencing took hold. Many of today's major questions (e.g., about how messenger RNAs [mRNAs] are formed and about how gene control is exercised in eukaryotes) are the same questions that were pursued in 1962–1980. More detailed answers to these questions are arriving today at breathtaking speed, but fundamentally informative and important answers came between 1962 and the early 1980s. How these still-central questions first arose, and how early experiments were structured and answers obtained, it seems to me, ought to be at least available in usable form for teachers and, most of all, the curious students of today.

I have been privileged to listen in on a number of "after hours" (read "faculty club cocktail hour") discussions among physicists. Both elders and youngsters in that community seem able to discuss where ideas, questions, and answers came from, easily back to Maxwell and his equations. Biological science, specifically the role of RNA in current and past life on this planet, also has a history worth knowing, I believe.

This history begins with the following questions: How did macromolecules finally become recognized as the necessary starting place for first learning about biology and now teaching biology? And why was RNA the latecomer in this overall picture?

The bold discovery by James Watson and Francis Crick of the structure of DNA, often told, and well told, by the protagonists themselves, is frequently recited as the "start" of molecular biology. And if one watershed discovery is to be chosen as the "beginning," that discovery is it. But there was a preceding half-century struggle of genetics and physical biochemistry that prepared at first a small group of scientists to grasp what the Watson-Crick structure both predicted and demanded but did not answer. For the discovery of the DNA structure and following discoveries to occur, biology/biochemistry had to take on *macromolecules*. The reign of organic chemistry (and recalcitrant organic chemists) as the main route to understanding life had to be at least momentarily sidestepped so that large molecules, poorly understood and comparatively difficult to study, could become the major research focus. How molecular biology involving macromolecules emerged from these early 20th-century battles is fascinating history.

The double helix discovery instantly revealed, through what Crick in his book *What Mad Pursuit* called "such a beautiful structure" (p. 60), how the molecule worked in inheritance. But the Watson-Crick revelation also lit the fuse that led to uncovering the centrality of RNA to life. The miracle years of 1955–1961—just 50 years ago—finally saw RNA recognized to be not monolithic but a collection of different types of molecules with specific functions. Courtesy of the insight of François Jacob and Jacques Monod, biological specificity among different cells, formally a completely opaque problem, could now at least provisionally be explained by controlling the synthesis of specific mRNAs.

The establishment of RNA function—first by discovering how genetic information is transferred into a readable form and then by proving the intimate roles of RNA in translation—led shortly to deciphering the universal genetic code, the first breakthrough toward which Marshall Nirenberg carried out in 1961. But all of these heady achievements were accomplished (largely) with bacteria and their viruses.

As biologists took these ideas to eukaryotic cells, first with cultured human cells, RNA remained the major focus. Throughout the 1960s and 1970s, a new world of macromolecular genetics was unearthed through studies of eukaryotic RNA. Unknown at the time, storage of information in the DNA of eukaryotes was very different from that of bacteria. Simply copying DNA into RNA did not suffice for genetic function. Primary RNA transcripts required molecular carpentry of various kinds—generically termed *RNA processing*—to produce functional RNAs. This era culminated in 1977 with the discovery of pre-mRNA splicing to produce functional mRNA. Both the complicated machinery for digging out the primary transcript as well as the processing to make a specific mRNA opened our eyes to additional points at which regulation of mRNA might occur. All of this was well established *before* facile genomic sequencing confirmed these conclusions.

Soon thereafter (1979–1981), a second bombshell burst. Chemical catalysis can be performed by pure RNAs, most often held in the proper tertiary structure inside cells by protein scaffolds. These major new concepts largely dealing with making and controlling functional RNAs also preceded the era of rapid DNA sequencing.

The young student of today or their youngish mentors can hardly be blamed for knowing very few details of this era, which ended before they were born (in case of students) or before they had finished their first decade or had begun their college years (in the case of young professors). The intellectual sweep of these many achievements before the early 1980s would, of course, be available by reading a selected sample of the hundreds of original papers from 1960 to 1980. But a relatively abbreviated historical discussion told from the point of view of a long-interested RNA biochemist has been unavailable. This is what compelled me to assemble the material in this book. I note here that I began working with the RNA of poliovirus in the late 1950s in the laboratory of Harry Eagle at the National Institutes of Health. This was followed by a year (1960–1961) with François Jacob at L'Institut Pasteur at the time when the concept and proof of mRNA were just being described (although I had absolutely nothing to do with these landmark experiments). However, my own laboratory work was, and still is, directed by these early very fortuitous training experiences and perhaps will help the reader to forgive the personal voice that appears in various spots throughout the book.

Chapter 1 presents early discoveries that were not fitted into an understandable fabric of cell function for decades. For example, more than 100 years ago, organic chemists were able to identify all of the nucleobases, even placing uracil only in "yeast" nucleic acid (aka RNA) and thymine only in "nuclein," later "thymus" nucleic acid (aka DNA). However, only in 1920 was deoxyribose finally identified as the sugar in DNA.

The peptide bond was described and accepted as the most probable link among amino acids by 1902. But a long disputatious history of the molecular nature of proteins followed, literally until after World War II. How could scientists of 1950 begin to think of cells making proteins by uniting amino acids in the correct order (step by step and therefore uncovering RNA functions) until after 1951, which brought Linus Pauling's models of the α-helix and Fred Sanger's sequencing of the first chain of insulin?

The monumental accomplishments of George Beadle and Edward Tatum in showing that genes were responsible for the function of individual proteins (enzymes) and the discovery of DNA as the genetic material by Oswald Avery, Colin MacLeod, and Maclyn McCarty are stories that preceded Watson and Crick and are known by many at least in outline. But a recitation of exactly what experiments these heroes performed does not trip lightly off the tongue of the majority of today's biologists, young or old.

Therefore, in diplomatic language, after "frank discussions" with my editors, and with the support of many colleagues who have read early versions of the book, Chapter 1 presents some of this history of the centrality of macromolecules, with my hope that, at the very least, it will be entertaining.

Chapter 2 needs no such defense. If we were going to have a history of RNA, it was obligatory to recount the signal achievements of the 1950s–1960s that finally brought RNA out of the shadows. On reflection, viruses with only RNA as a genome were obvious candidates to first catch attention for the genetic/biochemical importance of RNA. This proved to be the case, with TMV (tobacco mosaic virus) and the RNA formed after T-even (T2, T4) bacteriophage infection of *Escherichia coli* leading the way.

Although the gene/protein connection was made in the 1940s, it took in vitro protein synthesis by rat liver extracts, largely carried out by Paul Zamecnik and colleagues, to begin to truly connect proteins and RNA in 1953–1958. These years saw the discovery of transfer RNA (tRNA) and established a role for ribosomes (and presumably their RNA) in making proteins. Given these advances, it remains something of a puzzle as to why it took so many remarkably gifted scientists so long after the Watson-Crick structure (~7 years) to hit intellectual and molecular pay dirt with the idea of and discovery of mRNA. This is one of the most intriguing stories in the history of molecular biology. The secret lay in closer attention to the genetics of gene regulation that explained switches in the proteins that the cell made. Making mRNA in a controlled fashion in the test tube was then accomplished with bacterial systems in the late 1960s and early 1970s. Discussion of all these accomplishments constitutes Chapter 2.

Chapter 3 guides the reader stepwise through achievements that unlocked the universal genetic code. Virtually all biology students and many from other disciplines will know the conclusions of this era. The aim, however, is to put some experimental meat on the bones of the catechism symbolized by "DNA makes RNA makes protein." The clarity of these conclusions led Gunther Stent, a physicist turned biologist, to title a 1968 paper in *Science* "That Was the Molecular Biology That Was," in which he seemed to argue it was all over but the shouting.

The stunning achievements on bacterial gene functions *were* the logical takeoff point for beginning to work toward understanding how eukaryotic cells controlled their genes and performed such tricks as differentiation. Chapter 4 picks up this problem that began, however, with several years of inability to define mRNA in eukaryotic cells. Kinetic studies following incorporation of labeled RNA precursors into properly separated classes of RNA in cultured animal cells and in animal cells infected with DNA viruses finally wrestled this problem to the ground. Processing of preribosomal RNA (pre-rRNA) was uncovered by 1962–1963 and processing of pre-tRNA by 1968, but RNA processing to make mRNA was not completely understood until the final details of splicing of adenovirus pre-mRNA into mRNA were discovered in 1977. This ~15-year period (1962–1977) is recounted in detail in Chapter 4. Also described in this chapter are the initial discoveries of the biochemistry of the three eukaryotic RNA polymerases and the original illustration of the complexity of nuclear factors required to initiate RNA synthesis correctly in cultured human cells and cell-free systems.

Between the discovery of RNA splicing in 1977 and the early 1980s, additional astonishing discoveries occurred. The capacity of RNA to perform enzymatic functions was recognized. Also, the involvement of previously unknown small ribonucleoprotein particles (in particular, their RNA) in carrying out splicing to make mRNA was discovered. These topics are also introduced in their appropriate historical frame in Chapter 4.

Finally, complementary DNA (cDNA) cloning and pulse-labeled nuclear RNA allowed the measurement of the rate of synthesis of individual genes, which clinched the previously widely assumed, but not yet proven, primary control of gene expression at the level of transcription.

The end of Chapter 4 marks a dividing point in the book. A reasonably comprehensive historical accounting of important events in which RNA is the chief actor ends.

Chapter 5 is an attempt to provide a useful summary of important events in work on RNA after the early 1980s. Because the regulation of mRNA is the central event in all biological specificity, a discussion of the arcane array of proteins that control regulated transcription of chromatin is the first order of business. The positively required factors for the initiation and manufacture of an mRNA from pre-mRNA had first to be understood. This allowed more recent proofs of the wide variety of negative-acting proteins and protein complexes in preventing initiation and of the details of differential pre-mRNA processing, also a regulated process.

The most recent stunning advances in regulation of mRNA translation efficiency and lifetime have come from discoveries of yet more and different RNA molecules, both short and long *noncoding* RNAs. How these were uncovered and initial insight into how they function are products of research in the last ∼15 years, and new discoveries and insights into noncoding RNAs continue with each new journal issue. A running summary, necessarily incomplete, of all this experimental activity brings up the rear of Chapter 5. No attempt is made (even if it could have been) to be comprehensive in the material of Chapter 5. Rather, important areas are included with discussion of some key discoveries, and up-to-date references are provided. The shelf life of all of these new findings makes discussions about them admittedly problematic. But no attempt to describe a history of RNA could fail to include a digest of this recent material.

Chapter 6 is brief and highlights, first, a research area that looks back on 3 billion plus years to how RNA likely had an indispensable role in initiating life on the planet. Organic chemical and geochemical advances are being made that may enlighten us about events in the Archaean era. This is possibly the most difficult of all areas in biology, but it no longer seems the impossible field that it did a couple of decades ago.

Perhaps of equal difficulty is the tangled problem of the origin of cells. Challenges to the conventional wisdom of prokaryote → eukaryote evolution began with Carl Woese's discovery of archaea more than 30 years ago and remains a fascinating, unsettled area today despite hundreds of genomic sequences of microorganisms. One of the most challenging unsolved problems in this area also centers on RNA. How and when did splicing of RNA arise—before or after cells arose? If we had an unambiguous answer to this question, might we not also better trace how the three cellular kingdoms arose and persisted? The evolution of cells is intimately tied to the idea of an initial *RNA world*, a hypothetical but increasingly probable time in the evolution of life. Given this likely history and considering the many functions of RNA in the cells of today, shouldn't RNA share a wedge of the spotlight with DNA?

1

The Dawn of Molecular Biology
History of Macromolecules before RNA

THE STUDY OF RNA SITS AT THE VERY CENTER of contemporary molecular
biology. This was hardly always the case. It was just over 50 years ago that
different, discrete RNA molecules with different functions were first
recognized. What took so long?

First of all, the organic chemists who dominated the early decades of
20th century biochemistry were accustomed to dealing with simple organic
molecules for which they could obtain a chemical structure and often could
painstakingly synthesize. They simply did not believe in any kind of biological
macromolecules. Second, the most powerful quantitative biological science—
genetics—did not turn in a molecular direction until the 1950s, even though
specific genes were proved to control specific enzyme functions in the 1940s
(Beadle and Tatum 1941; Beadle 1945).

Not until 1951, with Linus Pauling and Robert Corey's description of
the basic structural elements of proteins (α helices and β-pleated sheets),
were proteins widely accepted to be long strings of amino acids that folded
into three-dimensional (3D) shapes. Only with Fred Sanger's sequencing of
insulin, first reported in 1951 (Sanger and Tuppy 1951a), were proteins finally
recognized to be long strings of amino acids *in a specific order*.

In retrospect, it seems that only when scientists actually began to grapple
with the fact that cells must have specific machinery to place amino acids in
a specified order in proteins was the inexorable progress that led to RNA
assured.

Jean Brachet, a Belgian embryologist, and Torbjörn Caspersson, a Swedish biochemist, hinted at the importance of RNA before 1950 (Brachet 1941;
Caspersson 1947). They both commented that tissues thought to make a large

9

amount of protein had a large amount of cytoplasmic basophilic staining material—presumably RNA (or PNA, for pentose nucleic acid, as it was often called). But they offered essentially no explanation for this correlation. An excellent summary of all of the early thoughts and experiments connecting high RNA content with presumed high growth rate or high protein production was given in an article by Brachet (1955) in 1955. But the article also illustrates the muddled thinking, at the time, about protein synthesis.

Finally, the Watson-Crick structure of DNA came in 1953, revealing the simple solution as to how a long linear polymer, DNA, could contain and replicate genetic information from generation to generation. The era of a long informational biopolymer providing directions to construct a specific long polymer of amino acids was born. Five more years passed, during which the initial understanding of the aminoacylation of transfer RNA (tRNA) by "activated" amino acids followed by amino acid incorporation into proteins by RNA-rich ribosomes was accomplished. By this point (1958), RNA was definitely connected to protein synthesis. Finally, messenger RNA (mRNA) was hypothesized by François Jacob and Jacques Monod in 1959–1960, and bacterial mRNA was identified in 1961.

The world of experimental biology changed forever with these discoveries: From that time on, it was accepted that the specific abilities of a cell *depended on its mRNA content.* For, although DNA contains the blueprint for life, it is the control of mRNA formation and its subsequent use that underlies all biological specificity.

Since the 1960s, extraordinary effort has been devoted to and success achieved in broadening our knowledge about gene control. Control of mRNA synthesis, especially most recently in all types of eukaryotic cells—from single cells to specific cells in animals and plants—has been intensively studied.

Among the many advances the years have brought was the discovery in 1977—astonishing at the time—that mRNA in human cells is spliced together out of long primary nuclear RNA transcripts. These primary transcripts had first been identified in 1962 and were studied intensively between 1962 and 1977. This most surprising event of RNA splicing is necessary because the protein-specifying information in most eukaryotic genes, in distinction to bacterial genes, is encoded in separated stretches of DNA, so the RNA copies are useless until the protein-coding pieces are brought together.

Almost immediately came another surprise. In the early 1980s, RNA was found to have chemical capabilities that were formerly thought to be the exclusive property of proteins. RNA is capable not only of both encoding and interpreting genetic information, but also of acting catalytically in a test tube on its own and as part of protein/RNA machines in cells.

Finally, in the past 15 years, new excitement has resulted from recognizing many new types of RNA. All of this frenzied activity has placed RNA at the center of retrieving and using in a regulated manner all of the information stored in DNA. Because of the evident importance of RNA in present-day cells, an ever-deepening interest has been kindled in RNA as the likely first informational molecule, potentially even before cells existed.

So if the aim of this book is to "tell the story of RNA," the first response might be "Fine, the story can begin only a relatively short time ago—i.e., 50 or 60 years ago." But the rich history of discoveries that finally led to accepting two macromolecules—protein and DNA—begins even earlier, and these discoveries instructed the thinking and experiments that led directly to RNA. How and why scientists finally turned to RNA to explain *genetics in action* is fundamental to understanding the history of modern biology. This chapter reviews some highlights of the decades before the arrival of the Watson-Crick structure of DNA, when RNA was still shrouded in mystery and macromolecules were just about to become the starting point for the modern teaching of and learning about biology.

PROTEINS WERE THE CENTRAL FOCUS OF EARLY BIOCHEMISTRY

Early in the 19th century, chemists began to grind up both plant and animal tissues and to analyze substances in milk, egg whites, blood serum, and the gelatinous material formed after boiled meats cooled. Heat or acid treatment caused "coagulation" in all of these materials, suggesting that a common substance might exist in them. These coagulates were known generically as fibrins, albumins, or gelatins.

The main scientifically relevant analysis at this early time was determining the presence and relative concentrations of the $\sim15-20$ elements that could be measured. The Dutch physician and chemist Gerrit Jan Mulder (1802–1880) was adept at performing elemental analyses of coagulated substances. He found a relatively constant ratio of carbon, hydrogen, oxygen, and nitrogen in different samples. Mulder first used the name *protein* to describe the various coagulated substances (Mulder 1838), following a suggestion in a letter from the great Swedish chemist Jöns Jacob Berzelius: "The word *protein* that I propose to you...I would wish to derive from *proteios* [Greek for first, foremost, primary]..." (Mulder 1844, p. 171; Vickery 1950). In his 1844 textbook, Mulder wrote "In plants as well as in animals there is present a substance which is produced in the former, constitutes a part of the food of the latter, and plays an important role in both. ...It is without doubt the most important of all the known substances of the organic kingdom, and without it life on our planet would probably not exist" (Mulder 1844; Vickery 1950).

Breakdown of Proteins by Stomach Enzymes

Progress in discovering the constituents of proteins depended on organic chemists and on scientists who would later be called physiologists. Theodor Schwann, an early proponent of the cell theory, found that stomach contents, known to contain hydrochloric acid, must contain something else that helped to digest meat: Hydrochloric acid alone did not work nearly as well as the stomach contents. He isolated a fraction of stomach contents that could digest meat and called this *pepsin*. (Whether he actually partially purified pepsin is questionable. In 1930, John Northrop [1930] purified and crystallized pepsin; see p. 23.)

A colorful contribution to the understanding of the digestive power of stomach contents came in the 1820s and 1830s. In 1822, William Beaumont, an American Army physician, was brought a patient, Alexis St. Martin, who had suffered a gunshot wound to his upper abdomen. Beaumont treated the man with the remedies of the time. Miraculously, the patient survived, and he lived with Beaumont for some years thereafter. Scar tissue formed around a hole in St. Martin's abdominal wall; this hole remained open to his stomach. St. Martin allowed Beaumont to study the ability of his stomach to digest various kinds of food by dangling the food into his stomach cavity on a string. Among the conclusions Beaumont reached were that the acid gastric juices did a particularly good job of digesting meat—much better than vegetables—and that fatty food was hardly digested at all. These early physiological observations were recorded by Beaumont (1833). Incidentally, Beaumont died in 1853 and St. Martin, the patient, lived until 1880.

The digestion of proteins by stomach enzymes produces products from proteins that are much more readily soluble (even in alcohol); these products could no longer be coagulated by heat, the main distinguishing characteristic of proteins at the time. These products, later named *peptides*, had a key role in understanding protein structure after 1900.

Discovery of Amino Acids as Part of Proteins

Throughout the 19th century, organic chemists had begun to discover amino acids initially as a related family of organic molecules and later as components released by acid hydrolysis of proteins. Glycine was clearly identified as aminoacetic acid and alanine as aminopropionic acid by 1850. Some compounds such as asparagine were first isolated as free substances, in this case from asparagus juice, in the first decade of the 19th century. All of these and aspartic acid, leucine, and valine were known for some time before being found in acid hydrolysates of protein. The reactive sulfur in cysteine was found very early, but cysteine was characterized as a constituent of protein only in the latter part of the 19th century.

All 20 of the common amino acids in proteins, however, were not discovered until the early 20th century. For example, the structure of the second sulfur-containing amino acid methionine was finally determined in 1928 (Barger and Coyne 1928). Because of their labile amide groups, the proof that glutamine and asparagine were part of proteins was not uncovered until the 1930s. And the final member of the list of 20 common amino acids was identified only in 1935 (Table 1-1).

In 1935, William C. Rose, an American biochemist, reported the dietary necessity of individual amino acids in humans by feeding a synthetic diet with the already known 19 amino acids to volunteer students. He fed newborn rats the same diet (McCoy et al. 1935; Womack and Rose 1935). The students lost weight and were in negative N_2 balance until he added threonine, the final of the common 20 amino acids in protein to be identified.

For balanced protein synthesis, of course, all 20 amino acids are required. Humans can make limited quantities of some amino acids and quite adequate quantities of others. But Rose found that humans and rats must get eight "essential" amino acids from the outside to maintain metabolic balance. These eight are leucine, isoleucine, valine, threonine, methionine, phenylalanine, tryptophan, and lysine (summarized in Rose et al. 1955).

It was left to Francis Crick and Jim Watson in their considerations about a genetic code to propose a list in 1954 of the 20 common amino acids that

Table 1-1. Dates of recognition of amino acids as protein constituents, based on isolation from protein hydrolysates

1819	Leucine	1889	Lysine
1820	Glycine	1890	Cystine/Cysteine
1846	Tyrosine	1895	Arginine
1865	Serine	1896	Histidine
1866	Glutamic acid	1901	Valine
1869	Aspartic acid	1901	Proline
(1873	Asparagine, see text)	1901	Tryptophan
(1873	Glutamine, see text)	1903	Isoleucine
1875	Alanine	1922	Methionine
1881	Phenylalanine	1935	Threonine

Reproduced from Tanford and Reynolds 2001 by permission of Oxford University Press (using dates of Vickery 1972); Vickery and Schmidt 1931; threonine date corrected to 1935 based on Womack and Rose 1935.

are actually found in protein. They omitted amino acids such as hydroxypro-line that, although clearly present in protein, were likely to have been modified after protein synthesis. This list of 20 was obviously a necessary step in thinking about how DNA might encode proteins (Crick 1988, pp. 91–92).

Emil Fischer: Peptide Linkage and Organic Synthesis of Peptides

Toward the end of the 19th century, organic chemists and physiologists generally agreed that amino acids (in fact, α amino acids of the L, not D, configuration) were at least the major constituents of proteins. But how amino acids were linked chemically in proteins was unsettled. Thoughts of the way in which cells could assemble amino acids into protein were still many decades away.

A requisite first step to understanding what was actually happening when enzymes or acid hydrolysis broke down protein was to determine how the amino acids in proteins were linked. Solving this crucial question was largely the work of the brilliant organic chemist Emil Fischer (1852–1919). Although Fischer's Nobel Prize in Chemistry in 1902 (the second Nobel Prize in Chemistry to be given) was for his earlier synthetic organic work on sugars and purines, his work on peptides may well be his greatest legacy. Fischer assumed that the most likely link between amino acids in proteins was an amide link between the carbonyl carbon of one amino acid and the amino group of its neighbor, splitting out H_2O and forming an anhydride linkage (Fig. 1-1). By 1901, he had achieved the first synthetic peptide linkage, glycyl–glycine. (The R_1 and R_2 in Fig. 1-1 were simply H atoms in glycyl–glycine.) Fischer first used the term *peptide* while describing the bond in a lecture in 1902

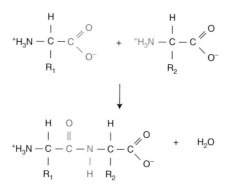

Figure 1-1. The peptide bond, named by Emil Fischer in 1902, is an anhydride linkage. R_1 and R_2 each refer to one of the 20 common side chains.

(Fischer 1902; Goodman et al. 2003). Franz Hofmeister, one of the earliest "physiologists" to do serious biochemistry, was at the same meeting and had reached the same conclusion about the amino acid linkage. Hofmeister favored the peptide linkage mainly because of the unlikelihood of enzymatic cleavage of a C–C or =C–N–C= bond in protein degradation (Hofmeister 1902).

Because the enzymatic digestion of meat yielded a hopeless mixture of peptides of varying composition, Fischer decided to study the problem of specificity in protein linkage by using a reverse strategy. He would use known synthetic chemistry to synthesize peptides of known composition and then compare the chemical behavior of the synthetic peptides with the digestion products of meat.

By 1907, Fischer's laboratory had prepared hundreds of specific peptides, some reaching a length of 18 amino acids. There was no way to tell whether natural peptides had been formed. However, the peptides that he made could be digested by enzymes in extracts of the stomach or pancreas, clear evidence that those peptides were like those that existed in actual proteins (Fischer 1907).

Fischer believed that there was an upper limit to polypeptide length in proteins—that proteins were composed of 50 or fewer amino acids—and he hoped that such proteins could be synthesized chemically (Fischer 1923). He wrote in a 1905 letter to his fellow organic chemist Adolf Baeyer "My entire yearning is directed toward the first synthetic enzyme. If its preparation falls into my lap...I will consider my mission fulfilled" (quoted from ref. 85 in Fruton 1999, p. 187). But at this point, perhaps only 15 of the 20 amino acids used in proteins were known, and nothing at all was known of actual sequences, so there was little prospect of success.

Successful Ordered Peptide Synthesis in the 20th Century

The first organic synthesis of a full-length natural peptide did not occur until the 1950s. The organic chemist Vincent du Vigneaud (1952) first determined the sequence of oxytocin and then chemically synthesized this nonapeptide hormone, an effort for which he received the Nobel Prize in Chemistry in 1955.

Finally, a means of chemically synthesizing long peptides (short proteins) was devised in the 1960s and 1970s by Bruce Merrifield (1986) at The Rockefeller Institute in New York. By this time, the exact sequences of many proteins were known, so that they or selected fragments of them could be prepared. The trick was to devise a method in which the last amino acid in the final desired chain was anchored chemically to a fixed substrate; each additional amino acid was added, chemically linked by techniques of organic chemistry, and the system was washed free before the next amino acid was added. Merrifield's chemical synthesis proceeded from the COOH (carboxyl) end to the NH$_2$ (amino) end, which is the opposite of natural protein synthesis, as we discuss in Chapter 2. Merrifield received the 1984 Nobel Prize in Chemistry for his synthetic work. The basic Merrifield technique is still in wide use today to make specified peptides and small proteins up to ~100 amino acids in length (Fig. 1-2) (Merrifield 1993).

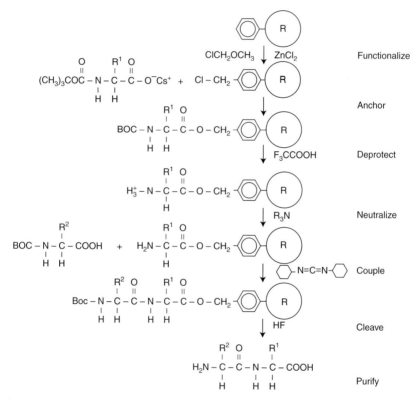

Figure 1-2. Merrifield's solid-phase peptide synthesis. The procedure starts with the desired carboxy-terminal amino acid (R^1) blocked on its amino terminus with an organic group (BOC shown in image is the *tert*-butyloxycarbonyl group); this blocked amino acid is then bound to a chloromethyl group on the resin (R). The blocking group is then removed to receive the second amino acid (R^2), which is blocked on its amino terminus. The second amino acid is coupled to the resin-bound R^1 residue via dicyclohexylcarbodi-imide. This procedure can be repeated as desired. The diagram shows cleavage of a new dipeptide after one round of synthesis. (Redrawn, with permission, from Merrifield 1986 [©AAAS]; article based on Merrifield's 1984 Nobel Lecture.)

EARLY ADVANCES IN PROTEIN CHEMISTRY

After the recognition in the late 19th and early 20th centuries that proteins contained chemically linked amino acids, several disputatious decades followed. Two interconnected questions arose repeatedly:

1. Were proteins long, uninterrupted polypeptide chains?

2. Were smaller chemically active organic molecules somehow associated with proteins to act as the actual catalytic agents responsible for promoting chemical reactions in cells?

To the modern reader, it very likely seems surprising that these questions took so long to be answered completely. And it is perhaps even more surprising to learn that despite progress toward recognition of the large size of proteins as far back as the late 19th century and quite conclusive experiments taking place between 1920 and 1930, the questions persisted throughout the 1930s.

The Question of Chain Length: "Molecular Weight"

That hemoglobin, already "purified" physiologically in red blood cells, could form crystals in "almost dried blood" had been recognized by 1840 (Hunefeld 1840). In the 1880s, quite pure globins had been studied for some time, and elemental analysis was used in an attempt to determine an accurate molecular weight. The minimal number of a single type of atom in any molecule is obviously one, and from the iron and sulfur content of globins measured by several different investigators, molecular weight estimates of hemoglobin were in the range of 12,000–16,000. This was further confirmed in the 1890s by the O_2-binding ability of hemoglobin (Zinoffsky 1886; for review, see Edsall 1962; Tanford and Reynolds 2001, Chapter 4). These estimates were accurate, because $\sim$16,700 is the correct mass estimate of a single globin chain from many different animals.

The work of one of the most careful early investigators of proteins perhaps deserves special mention. Thomas Osborne, working at the Connecticut Agricultural Research Station, extracted proteins from the seeds of many plants. The high protein concentrations of the water solutions from some of these seeds led to crystal formation, which is not a guarantee but a strong suggestion of purity. His elemental analyses of these crystalline proteins for sulfur, amides (glutamine and asparagine), and free amino groups (lysine, arginine) allowed him to estimate a number of molecular weights in the 10,000–20,000 range for many of these proteins, which were later proved to be accurate (Osborne 1902, 1909; Osborne and Harris 1903).

Colloid Chemistry: Obstacle to Progress in Protein Chemistry

So, why were such obviously sound measurements not sufficient to convince the biochemists/organic chemists of $\sim$1900–1920 that proteins were very large molecules?

The major stumbling block came from the field of study termed *colloid chemistry*. Thomas Graham, a Scotsman, was essentially the founder of this field in the 1860s. Graham (1861) described two classes of molecules: (1) Simple salts and sugars diffused rapidly and could pass through a semipermeable membrane. (2) Other materials such as starch and dissolved proteins diffused slowly and could not pass through the membranes that he used. He called the first class *crystalloids* and the second *colloids*. The nature of the bonds that held substances together—chemical (covalent) or physical (noncovalent)—had not been completely clarified at this time. The structure of atoms was, of course, not yet understood. The constant "combining power" or *valenz* (valence), as it came to be called for molecules in organic compounds, was just being defined.

Later work on colloids proved that complex noncovalently associated molecules, both organic and inorganic, clearly exist. Richard Dickerson gives a particularly clear discussion of ideas about colloids (Dickerson 2005, pp. 2–4) and how they led to confusion about molecular weights of proteins (also see Eisenberg 1996). Dickerson points out that soap molecules in water form micelles that are colloidal. The long chains of fatty acids in the soap molecules are directed inward, and the carboxyl residues of the acid head groups interface with water. The contents of the interior of each micelle are determined by other surrounding substances—grease, for example, which can be solubilized by soap and carried away inside the micelle. The sizes and contents of such micelles, of course, are not constant. Therefore, there is no meaningful "molecular weight" of a colloid. But the grip of these ideas on biochemists is clearly illustrated in a review of the second edition of the influential textbook by Ross Gortner published in 1938. In a generally positive review, George R. Burr writes "The new book is divided into the same general parts found in the first edition: I The colloid state of matter, 348 pages, II proteins 226 pages..." (Burr 1939).

One important reason that reticence persisted to accept long polypeptide chains as the primary protein structure was that Emil Fischer himself doubted the existence of large polypeptides. As he wrote in 1916, "In my opinion, however, the methods applied to the determination of the molecular weight of the hemoglobins are less certain than had been assumed previously. Although they crystallize beautifully, no guarantee of homogeneity is given, and even if one concedes this and accepts the validity of a molecular weight of 15,000–17,000 for several hemoglobins, it should always be remembered that the hematin, from all that we know of its structure, can bind several globin units. . . .On the other hand, I gladly concur in the view of Hofmeister and many other physiologists that proteins of molecular weight 4,000–5,000 are not rare. If one assumes an average molecular weight of 142 for the amino acids, this would correspond to a content of 30–40 amino acids" (Fischer 1923, p. 40; quoted by Fruton 1999, p. 199).

As the eminent protein chemist and historian Joseph Fruton concludes, ". . . the clear verdict of the most famous organic chemist of his time who had worked on proteins was that values above 5,000 for the molecular weights were not acceptable. Nor can there be any doubt, in my opinion, that Fischer's disappointment with the outcome of his efforts to synthesize a protein (perhaps even an enzyme) affected his attitude to the views of people like Franz Hofmeister, who he termed 'only physiologists.' Small wonder that many of Fisher's biochemical contemporaries found more merit in the physical-chemical approach to the problems of protein chemistry!" (Fruton 1999, p. 200).

Physical Evidence of the Long Polypeptide Nature of Proteins

Instead of organic chemists, scientists trained in physics who became interested in biological materials led the way to the convincing proof of the high-molecular-weight nature of proteins. Their advances ultimately subdued the reluctance of colloid chemists to believe in large molecules.

An important line of physical research was pioneered by S.P.L. Sorenson and M. Hoyrup (1915–1917), who applied classical Gibbs thermodynamics equations to proteins. Using a sensitive osmometer, they measured the proportional changes in osmotic pressure created by various concentrations of crystalline ovalbumin. From such changes, the molarity and hence molecular weight could be determined. They estimated a molecular weight for ovalbumin of 34,000 (close to the actual molecular weight of >40,000 measured later).

The Sorenson initiative was continued to great effect in the 1920s by Gilbert Adair at Cambridge University. By osmometry, he showed in 1925 that oxyhemoglobin had a molecular weight of 67,000, indicating that it was a tetramer of single chains, each of which had a molecular weight of 16,700 (Adair 1925; for review, see Johnson and Perutz 1981). Adair's experiments were performed in a scientific atmosphere in which physics and biology were in the process of merging under the influence of Sir Lawrence Bragg, the crystallographer and son of W.J. Bragg. The two shared the 1915 Nobel Prize in Physics. The younger Bragg's influence on biology became particularly important in the 1930s and afterward as leader of the Cavendish laboratory, where both the DNA structure and the first atomic structure of proteins were accomplished.

Theodor (The) Svedberg and the Ultracentrifuge

One of the most convincing physical measurements arguing for the high molecular weight of proteins (later, macromolecules in general) was made by the Swedish physicist Theodor (The) Svedberg. Ironically, Svedberg started out by studying the properties of colloidal suspensions of inorganic materials. *The Formation of Colloids* (Svedberg 1921) was an amplified version of his Ph.D. thesis of 1909. This early work impelled him to develop the ultracentrifuge to try to understand the interspersed phases of colloids.

After much trial and error in building a machine that could exert a force of $100,000 \times$ gravity (g) in a centrifugal field and also in solving the physics of particles moving through such a centrifugal field, he was ready to examine various objects. When he finally got the machine working effectively, he anticipated that like other colloids, proteins, under some disaggregation conditions at least, would show multiple sedimenting species characteristic of a mixture of colloidal components. One of the early pure proteins he examined was hemoglobin, and it sedimented as a single sharp-moving boundary. The molecular weight was calculated to be 67,000—almost exactly what Adair published in 1925 on the basis of osmotic pressure measurements. Svedberg eventually examined a long list of proteins, even hemocyanins, one of the largest easily purified proteins from insects. The molecular weights all were large, with hemocyanin reaching above 6,000,000 (Svedberg 1937).

However, some of the proteins, notably the hemocyanins, although highly purified, showed some evidence of smaller components as if disaggregation was occurring. Although his evidence convinced many others that proteins could be large molecules, eventually Svedberg reverted to the idea that perhaps all proteins were composed of some sort of repeating unit that has a molecular weight of only a few thousand. Strangely enough, Sorenson, Svedberg's Scandinavian predecessor, also backtracked on his findings about the molecular weight of ovalbumin, again turning to the idea of colloids. However, by the late 1930s, Svedberg's work had largely convinced many physical chemists interested in proteins that proteins were high-molecular-weight polypeptides.

Some Organic Chemists Accept Large Polymeric Molecules

Organic chemists continued in the 1920s and 1930s to resist the idea of large natural molecules as well as large covalently linked polymers of synthetic material. However, there were notable exceptions. By 1930, one of the pioneers of synthetic polymers, Wallace Carothers, the discoverer of nylon, took a very clear stand in an instructive 1931 review entitled "Polymerization" (Carothers 1931). The portion of the review pertinent to natural products includes a discussion of cellulose and rubber. Among those organic chemists contributing to his conclusions, Carothers pays particular homage to Hermann Staudinger, whose group contributed to the structure of both cellulose and rubber (Staudinger 1953).

By 1930, massive evidence—both chemical and physical—had demonstrated that "a single bivalent radical derived from the removal of water from glycose (glycopyranose)" was the structural unit in the long-chain polymer that is cellulose. Carothers called attention to the likely fact that such polymer chains are not all of equal length and therefore will not have uniform physical behavior.

Rubber was a particular refuge of the colloid school because raw rubber in organic solvents formed "sols" or "gels" that were thought to be evidence of noncovalent linkages in the assembly of the basic isoprene units that comprise rubber. Again, Carothers pointed out that differing chain lengths cause differing solubility properties. Finally, a product similar to rubber could be assembled synthetically by chemically linking isoprene units. Developing this process to an industrial level as a source of synthetic rubber was very important in World War II. Carothers steered clear of any discussion of proteins, but the language of the review makes it clear that long-chain covalently linked biologically important polymers—not mysterious, loosely interactive smaller "organic" units—ought to have been expected by chemists by 1931.

Staudinger invented the word *macromolecule* in 1922 to describe all of the very large synthetic and giant natural molecules. In 1953, after the conclusion of World War II, Staudinger received a belated Nobel Prize for Chemistry.

THE BIRTH OF ENZYMOLOGY

As noted earlier (p. 17), there were at least two main obstacles regarding the nature of proteins that needed to be overcome at the turn of the 19th century: (1) the chemical and physical basis for their high molecular weight and (2) the role of the protein molecule itself—as opposed to some adsorbed, chemically active organic molecule—in promoting chemical reactions.

By the end of the 19th century, there was a long history of secreted proteins that had the ability to chemically change various substrates. But all of these proteins acted outside cells. In addition to proteases in gastric contents, extracts of barley (malt) and human saliva were found in the 1830s to be capable of breaking down starch. Changes of this sort seemed to occur without the disappearance of the agent that caused the change. The Swedish chemist Berzelius, in commenting on this phenomenon, suggested that such agents of change be called *catalysts*. He wrote that "[their] cause will perhaps be discovered in the future in the catalytic power of the organic tissues of which the organs of the living body consist" (from Berzelius' 1836 paper; quoted by Fruton 1999, p. 145). In 1876, the organic chemist Willy Kuhne suggested the name *enzyme* for these apparently catalytic activities (Fruton 1999, p. 148).

Eduard Büchner and Cell-Free Fermentation

One of the most time-honored molecular interconversions was the fermentation by yeast of the disaccharide sucrose (glucose + fructose), the principal sugar in grape juice, to produce the two-carbon ethyl alcohol. This conversion was thought to require several chemical steps by the "organized ferments" within the cell as opposed to the "unorganized enzymes" in tissue extracts.

In 1897, Eduard Büchner and his brother Hans, a microbiologist, were experimenting with broken cell preparations of the microorganisms (yeasts) that fermented grape juice to produce wine. A paste of broken yeast cells was prepared by passage through a hydraulic press. By the time they finished, there was no fermentative action. In an effort to preserve activity during the preparation, they suspended the yeast cells in high concentrations of sucrose, which makes the suspension more viscous. Following incubation of these extracts, the sucrose disappeared and alcohol was produced.

What had previously been thought to be a property only of the whole intact cell was found to be performed by the protein(s) of properly prepared cell extracts. This remarkable discovery, rewarded with a Nobel Prize in 1906 (Büchner 1966), was thought by Büchner to be caused by a single protein that he named *zymase*. It is actually several proteins. Many writers date the real beginning of enzymology to these experiments; cells could be broken and individual enzymatic activities could be purified from the extracts. The ever-present naysayers claimed that in Büchner's experiments, bits of "protoplasm" actually performed the fermentation.

Protein Crystals: Proof that Enzymes Are Proteins

The first two enzymes to be highly purified and crystallized, however, were proteins that were secreted outside of cells. At Cornell University, in his first independent laboratory, James Sumner began to purify urease, an enzyme in plant seeds and bacteria that splits urea. Six years after he began purifying the enzyme from jack bean seeds, he obtained crystals in his purest preparation (Sumner 1926). No other cell constituent for which he tested was present except that protein enzyme. Despite this success, organic chemists who had become interested in enzymes such as the famous Richard Willstätter, a subsequent Nobelist, refused to believe that the protein itself rather than some active chemical bound to the protein was the catalyst for enzymatic action (Fruton 1999, p. 208).

By 1930, John Northrop had not only purified and crystallized the protease pepsin, but he also proceeded to characterize his final product so thoroughly that it had to be accepted as nothing but protein. His conclusion in his famous paper is classic: "The preceding experiments have shown that no evidence for the existence of a mixture of active and inactive material in the crystals could be obtained by re-crystallization, solubility determinations in a series of solvents, inactivation by either heat or alkali, or by the rate of diffusion. It is reasonable to conclude therefore that the material is either a pure substance or a solid solution of two very closely related substances. If it is a solid solution of two or more substances it must be further assumed that these substances have about he same degree of solubility in the various solvents used, as well as the same diffusion coefficient and rate of inactivation or denaturization by heat. It must also be assumed that both substances are changed by alkali at the same rate and to the same extent. This could hardly be true with the possible exception of two closely related proteins. It is conceivable that two proteins might be indistinguishable by any of the tests applied in this work. But in this case it would follow that the enzyme itself was a protein and this, after all, is the main point" (Northrop 1930, p. 763).

Northrop went on to also crystallize trypsin and chymotrypsin (Northrop 1935). These experiments left no doubt that at least some enzymes were single proteins. As we note again in Chapter 2, Sumner and Northrop (together with Wendell Stanley) shared a Nobel Prize in Chemistry in 1946.

EARLY X-RAY ANALYSIS OF PROTEINS, PEPTIDES, AND AMINO ACIDS

The 1930s saw the earliest effective use of X-ray analysis of proteins, which by the 1940s had convinced many physical biochemists that proteins were regular, long, linear polypeptides. This early history is admirably summarized by R.E. Dickerson in *Present at the Flood* (Dickerson 2005, pp. 2–9 and 58–65). Some of the early findings, based on X-ray analysis and principles of physical chemistry that contributed prominently to the acceptance of proteins as long, linear polypeptides, are mentioned here.

Perhaps the most influential of these early studies were those of William T. Astbury, a former student of William Henry Bragg, the founder (with his son, William Lawrence Bragg) of X-ray crystallography to study matter. Astbury worked in the Textile Physics Laboratory in Leeds in Northern England. He was naturally interested in fibers of all kinds—hair, wool, silk, etc. Using X-ray analysis, he examined these fibers in both a stretched and natural state. All gave repetitious diffraction patterns, signifying a regular and probably linear molecular structure. Moreover, a single wool fiber stretched (by steam) produced a pattern resembling that of a silk fiber and then after cooling returned to the same repetitious pattern as before. This stretching and refolding is now known to be due to the breaking and reforming of hydrogen bonds. Astbury concluded that the proteins were long chains of peptide-linked amino acids with the long axis of their chains disposed in parallel along the long axis of the fiber. So fibers such as wool and silk, at least, were most likely long polypeptides (Astbury and Woods 1933).

In 1934, J.D. Bernal and Dorothy Crowfoot (later Crowfoot-Hodgkin) at Cambridge produced the first clear X-ray-diffraction pictures of a crystallized globular protein (pepsin) (Bernal and Crowfoot 1934). They commented cautiously on the fact that they found a very discrete pattern: From "the intensity of the spots. . .it can be inferred that the arrangement of atoms inside the molecule is of a perfectly definite kind" (Bernal and Crowfoot 1934). From the data reported by Bernal and Crowfoot, Astbury commented in an accompanying interpretive paper that "The two chief rings (in the photos) have spacings of about 11.5 Å and 4.6 Å corresponding to the 'side chain spacing' and the 'backbone spacing' respectively of an *extended polypeptide*" (Astbury and Lomax 1934; my emphasis). These photographs were the first indications that a globular protein might be composed partially of "neatly" folded stretches of linear polypeptides.

Cyclol Theory: Role of Dorothy Wrinch

What might be interpreted as a last dying gasp of resistance to the idea of long polypeptide chains as the basic primary structure of proteins was cyclol theory. In this theory, proteins were composed of short, covalently closed rings stacked on one another.

The cyclol idea reached its zenith in the mid to late 1930s. The main protagonist was an English mathematician named Dorothy Wrinch, who while visiting Oxford wrote a flurry of papers describing a hypothetical cyclic ring structure in proteins (Wrinch 1936, 1937, 1938). Apparently a persuasive and by all accounts determined and forceful individual, she convinced Irving

Langmuir, the distinguished chemist (Langmuir received the 1932 Nobel Prize in Chemistry), of her theory (Langmuir 1939).

By this time, Linus Pauling, already very prominent among chemists for his contributions to the nature of the chemical bond and to resonance in unsaturated organic compounds, had become interested in proteins (Pauling 1993). Earlier, he had invited Alfred Mirsky of The Rockefeller Institute to California Institute of Technology (Caltech) for a sabbatical visit. Mirsky and others had shown that some proteins by careful denaturation would, under the proper conditions, renature. This meant to Pauling that there must be both covalent and noncovalent bonds that held the polypeptide together before denaturation and that the polypeptide, thus held together, could then refold, reforming the original structure. Pauling and Mirsky published a paper in 1936 in which they suggested that hydrogen bonds between the nitrogen and oxygen of the *physically* (not necessarily *linearly*) adjacent residues in the 3D native structure as well as hydrogen bonds between side-chain groups might contribute to protein stability (Mirsky and Pauling 1936).

So Pauling was primed to combat cyclol theory. Both J.D. Bernal (Bernal 1939) and Pauling, along with Pauling's Caltech colleague Carl Nieman (Pauling and Niemann 1939), wrote devastating articles undercutting the Wrinch cyclol proposals. They describe how the Wrinch proposals used the rudimentary X-ray data of Bernal and Crowfoot very selectively and point out a bit acerbically that the hypothetical proposed cyclol structures could not possibly exist because there was no space for the amino acid side chains in the cyclol models. Perhaps the item of lasting importance to note in all of this furor is the, by then, obvious deep interest in protein structure aroused in Linus Pauling.

Pauling and Corey: The α-Helix and β-Pleated Sheets

Pauling was unconvinced that Astbury's proposed linear structures for the fibrous proteins were precisely correct. He believed that, from his chemical triumphs of the late 1920s and early 1930s on bond angles and atomic distances in organic molecules, he could arrive at a better proposal. To accumulate substantial data to this end, he invited the experienced crystallographer Robert B. Corey to Caltech to join him for what was originally to be 1 year. Corey and Caltech students solved the 3D crystal structure of glycine, the first crystal structure of an amino acid, and followed this up with other structures before World War II intervened. Pauling and Corey returned to these projects after the war and by 1948 had solved the atomic structures of a dozen amino acids, a diketopiperazine, a number of amides, and even two peptides (Corey 1945; Pauling 1993). These data were crucial in several ways for providing the parameters that would be used in thinking about polypeptide structures. Not only were precise bonding angles and atomic distances of amide bonds noted, but perhaps most importantly, the structures showed the planar nature of the atoms in a peptide bond.

With all of these data under his belt, Pauling returned to the question of the pathway of the linked amino acids in a long polypeptide, and particularly to his earlier thoughts about Astbury's early X-ray data on the proteins in hair and wool.

Pauling tells the story (Pauling 1993), repeated by several authors (e.g., Dickerson 2005, p. 60), of how he came to the conclusion that the amino acids were tracking a helical path in space with a regular periodicity. The solution rested on knowing the exact size of amino acids and the distances between the peptide bonds in linked amino acids. While he was in an Oxford hotel room recuperating from a bad cold, he drew out a chain of peptide linkages on a sheet of paper and folded the paper cylindrically so that amino acids would lie in a helical fashion on the surface of the paper. He adjusted the folding conditions so that the amino nitrogen of one amino acid could make a hydrogen bond with the carboxyl oxygen of another amino acid several residues away in the chain. With this crude modeling, he arrived at the conclusion that each amino acid could form hydrogen bonds with a residue some distance away on the polypeptide chain. But the number of residues in each turn around the surface of the helical structure was not a whole number. Following their return to Caltech, Pauling and Corey refined the α-helix polypeptide model using correct 3D models of amino acids. They settled on two acceptable structures, one with a repeat of 3.7 residues per helical turn and another with 5.1 residues per turn, and all of the peptide bonds were planar (Fig. 1-3). In 1951, Pauling and Corey published an astonishing *eight* papers on "The Structure of Proteins" (together with H.R. Bronson on the first paper). These papers describing their findings appeared in two consecutive issues of the *Proceedings of the National Academy of Sciences*. (The papers with perhaps the most lasting impact are Pauling and Corey 1951a, 1951b, and Pauling et al. 1951.) These not only described the α-helix in proteins but proposed a second major structure, the β-pleated sheet. These two fundamental structural motifs have been proved by crystallographers to be the major structural elements in all proteins. After the Pauling and Corey papers were published, there was apparently very little objection to believing that proteins were composed of long polypeptide chains with most likely a limited number of structural motifs.

There are many other important and fascinating aspects of the maturation of protein chemistry from the 1930–1950 decades, covered economically but with deep insight by Charles Tanford and Jacqueline Reynolds (2001).

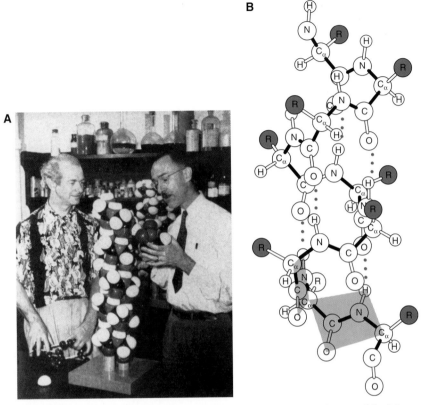

Figure 1-3. (*A*) Linus Pauling (*left*) and Robert Corey (*right*) with their model of the α-helix in 1951. (Courtesy of the Archives, California Institute of Technology.) (*B*) The Pauling-Corey protein α-helix ball-and-stick model. The amino acid side chains (R, green shading) do not determine the course of the helix in space. The peptide bonds are planar (gray shading) as the helix proceeds. Hydrogen bonds (red dots) between residues on the same face of the helix contribute stability. There is not an even number of amino acids per turn (∼3.6 or 3.7). (Redrawn, with permission, from Darnell et al. 1986.)

FRED SANGER: PROTEINS HAVE A SPECIFIC SEQUENCE

One crucially important advance in protein chemistry had begun before and was completed after the Pauling-Corey conclusions of 1951. Fred Sanger, an organic chemist, began just after World War II to take on the problem of the *primary* structure of proteins, i.e., the sequence of amino acids in a small protein. The pancreatic hormone insulin was available in pure form, and its molecular weight was thought to be between 6,000 and 12,000 and therefore to contain ∼100 amino acids. Sanger's experimental design was to split insulin into overlapping peptides, each of which could be sequenced, and then to piece together the entire sequence. Few believed that he could do this. Even those who thought proteins might be polypeptides didn't think that the sequence would be necessarily fixed across all copies of the protein.

Sanger's first success, reported in 1945, used dinitrofluorobenzene, which reacts with the *free* α amino groups in a protein or peptide. In a continuous polypeptide chain, there would only be a single free amino group on one end of the chain. Using this reagent, Sanger found that insulin had two such free amino groups: glycine and phenylalanine (Sanger 1945). Sanger's senior colleague Albert Charles Chibnall, head of the Cambridge Biochemical Laboratory, had determined the total amino acid composition of insulin and discovered a quite high concentration of cysteine. Sanger suggested that insulin was composed of two cysteine–cross-linked (S–S, a disulfide bond), but independent, polypeptide chains to account for the two free amino groups. He used oxidative cleavage using performic acid to separate an A chain with glycine at the amino terminus and a B chain with phenylalanine at the amino terminus (Sanger 1949). (It would be almost another two decades before it was proven that insulin is produced as a single polypeptide and that, after folding and S–S cross-linking, undergoes a proteolytic clip to result in the two linked chains that Sanger sequenced [Steiner and Over 1967].)

The next stage, obtaining reproducible and eventually overlapping peptides, was accomplished first with the B chain. Sanger broke the polypeptide chain into peptides by using controlled mild acid hydrolysis or enzymatic cleavage followed by isolation with paper chromatography of the resulting overlapping peptides. The yellow color imparted to the amino-terminal peptide by dinitrofluorobenzene was helpful in providing an anchor for the amino terminus (Sanger 1945).

The Edman protein degradation technique that removes amino acids one at a time from the amino terminus was established by 1950 (Edman 1950) and greatly aided the sequencing of isolated peptides. By 1951, Sanger and his coworker Hans Tuppy reported the sequence of the 30 amino acids in the B chain (Sanger and Tuppy 1951a,b). Sanger's opening to an influential review in 1952 illustrates the lingering effect of the 1930s–1940s disputes over the basic structure of proteins. He wrote "...as an initial working hypothesis it will be assumed that the peptide theory is valid, in other words, that a protein molecule is built up only of chains of α-amino (and α-imino) acids bound together by peptide bonds. ...While this peptide theory is almost certainly valid..., it should be remembered that it is still a hypothesis. ...Probably the best evidence in support of it is that since its enunciation in 1902 no facts have been found to contradict it" (Sanger 1952, p. 3).

By the time this review was published, Sanger and E.O.P. Thompson had virtually completed the sequence of the 21 amino acids of the A chain. The

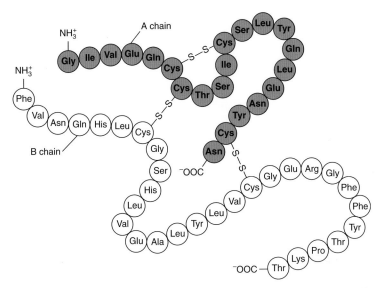

Figure 1-4. Human insulin, showing two disulfide bridges (S–S) between cysteine residues of the A and B chains and one bridge connecting two residues of the A chain. (Redrawn, with permission, from Darnell et al. 1986.)

A-chain sequence was published in 1953 (Sanger and Thompson 1953). In 1955, the cysteine residues involved in the cross-links between the B and A chains had been located (Fig. 1-4) (Ryle et al. 1955).

Using Sanger's strategy, the 39 amino acids of the anterior pituitary hormone ACTH (adrenocorticotrophic hormone) had also been sequenced by 1956 (Shepherd et al. 1956), and other proteins followed soon after.

This monumental work forced most chemists and biochemists to recognize that proteins were long arrays of peptide-linked amino acids *with a definite order*. The question now posed for proteins was, How could they possibly be made so precisely so that such an order could be achieved? In little more than a decade, that too would be answered, *but not until RNA entered the picture.*

THE SECOND LONG LINEAR POLYMER: DNA

In this section, we continue to explore the underpinnings of the now well-known linear relationship of the three long polymers—DNA, RNA, and proteins. Historically, DNA was next in line, but early studies melding chemistry and biology, through genetics, required more than 70 years to establish the nature of DNA. Although DNA was discovered as a chemical substance by Friedrich Miescher ~1870, even its basic chemical composition was not understood for another four decades, and its connection to genes was not established until 1944. But genetics, originating with Mendel in 1865 and rediscovered in 1900–1902, made unambiguous progress after 1900. Chromosomes, discovered in the late 19th century as long, microscopically visible structures stained by basic dyes, were first proposed to be the carriers of Mendel's genes by 1902. By the 1930s, the concept of chromosomes as long strings of genes was thoroughly understood. Moreover, by the 1930s, chromosomes were known to contain DNA as well as protein. By 1940, geneticists with a "chemical impulse" were raising the question, What is the chemical nature of the gene? This question was barely answered when Watson and Crick, feeling certain that DNA was the genetic material, solved its structure. The Watson-Crick discovery was the completion of the era that recognized DNA as a long informational polymer.

A sound conception of two of the three long interrelated polymers was thus in place by 1953, setting the stage for seeking a molecular connection between these first two that finally resulted in recognizing the central role of RNA (Chapter 2).

Miescher Discovers DNA

As we have seen, proteins, as the major constituent in cells, claimed almost all of the attention in 19th century chemistry/biochemistry. But DNA (not called that at the time) was discovered as a separate, nonprotein entity by Friedrich Miescher in the late 1860s. It was known that dead white blood cells comprise the majority of pus from wounds. Miescher recovered these cells from bandages and discovered that the cells consisted largely of shrunken nuclei. With a sulfate solution, he was able to extract a material that did not behave as protein and had a high concentration of phosphate. The material was not destroyed by gastric proteases (pepsin). Miescher's paper was finally published in 1871 by the famous chemist Felix Hoppe-Seyler (in his private journal) only after confirmation by Hoppe-Seyler.

Believing that the material came from nuclei, Miescher (1871) called it *nuclein*. He then also studied salmon sperm and later bull sperm because they were almost devoid of cytoplasm. By 1874, Miescher settled on some defining characteristics of nuclein: swelling in salt solutions, solubility in alkali, and resistance to pepsin.

Discovery of Nucleic Acid Bases: Levene's Tetranucleotide Hypothesis

The pyrimidines—cytosine and thymine—were characterized by Albrecht Kossel and his associates as part of "nuclein." By 1894, Kossel had identified all four nucleic acid bases in nuclein and confirmed its very high phosphorous content. Only after Emil Fischer's synthetic work on guanine and adenine was it clear that both were also authentic constituents of nuclein (Fischer 1899).

In addition, in Kossel's lab, uracil was found in "yeast nucleic acid" in the 1890s. Phoebus Levene pointed out early in the 1900s that yeast nucleic acid contains little or no thymine (Kossel and Neumann 1893, 1894; Levene and Bass 1931). But it was only after another several decades that yeast nucleic acid was recognized to be almost exclusively RNA.

Thus, all of the nucleic acid bases found in DNA and RNA had been recognized quite early. Chemists also found sugars associated with the nucleic acid bases. By 1909, D-ribose had been identified in yeast nucleic acids, but deoxyribose was not proven to be part of "thymus nucleic acid" until much later (for review, see Levene and Bass 1931). (Thymus cells are almost all lymphocytes with very little cytoplasm, and DNA was referred to well into the 1930s as "thymus nucleic acid.")

The discovery of the presence of phosphate in association with the nucleobases also came quite early. The high concentration of ATP in muscle led to the discovery of inosinic acid (deaminated AMP) in 1847 (for review, see Baddiley 1955, p. 160). "Yeast nucleic acid," almost completely RNA that was extracted by and essentially completely degraded by alkali, yielded the four nucleotides—AMP, CMP, GMP, and UMP. All were characterized in Phoebus Levene's laboratory (Levene 1918, 1919).

Levene formalized an idea that has left a negative legacy, diminishing the memory of all of the fine chemistry to have come from his group. He concluded that the nucleic acid bases plus their attached sugars and phosphates most likely existed in cells as tetranucleotides. He proposed that units of these four "nondescript building blocks" served some structural purpose (Levene and Bass 1931). This tetranucleotide hypothesis was used over and over again as an excuse to exempt nucleic acids from specific roles in the cell. Proteins, with their many more numerous and chemically active amino acid side chains, were thought to be the only repositories of specificity (for review, see Hunter 1999).

As discussed in the next section, *chromosomes* had been recognized in the 1880s, and by 1902–1903, they were thought to be the controlling elements in heredity. The great Columbia University biologist Edmund B. Wilson would write in 1895, "Now chromatin [the staining material in chromosomes] is known to be closely similar if not identical with a substance known as nuclein—a tolerably definite chemical compound of nucleic acid and albumin [Miescher's preparations contained some protein]. ...We reach the remarkable conclusion that inheritance may, perhaps, be effected by the physical transmission of a particular compound from parent to offspring" (Wilson 1895; quoted in Fruton 1999, p. 396).

It was another 50 years before this prescient remark was validated.

GENETICS: QUANTITATIVE AND PRECISE
FROM THE BEGINNING

The difficulty imposed on 19th and early 20th century biochemists and chemists in dealing with macromolecules is obvious to us today. Practical and reproducible techniques for purifying and characterizing unbroken large nucleic acid molecules are modern inventions. The organic chemists of an earlier time only felt secure about specific pure smaller compounds that they could identify and in most cases synthesize and crystallize.

On the other hand, classical genetics has a robust early history that is as valid today as it was in 1900. Plant breeding was already understood in the 18th century. Flowers were recognized to contain both pollen (male) and ovules (female), elements that allowed for self-fertilization. Furthermore, pollen from one plant could be used to pollinate (cross-fertilize) another. Seed formation followed, with which to begin a new generation of plants.

A contemporary of the great naturalist Carl Linnaeus, Joseph Kölreuter published results of a series of cross-pollination experiments on related tobacco plants in the 1760s (see Carlson 2004). Extensive plant hybridization, even on garden peas, had been performed before the Austrian monk Gregor Mendel performed his famous experiments with garden peas in the 1850s and 1860s. Mendel's discoveries, reported in 1865, were possible because a great variety of seeds that reproducibly produced specific types of plants were available from plant hybridizers. Out of a collection of more than 20 varieties of one species, *Pisum sativum*, he chose two strains, already inbred and therefore genetically pure. When these were self-fertilized, they exhibited seven traits in an either/or fashion. All genetics textbooks illustrate these—green *or* yellow seeds; round *or* wrinkled seeds; purple *or* white flowers, etc. When he cross-fertilized these marked strains, he kept meticulous records, counted large numbers of offspring, and organized his findings according to the generation in which particular characters appeared. Thus, his observations were statistically conclusive. Many writers have commented on Mendel's determination to study the events of plant hybridization quantitatively. He was, after all, a trained mathematician.

All genetics students are familiar with Mendel's discoveries after crossing the two strains: the observance of dominance in the first generation (F_1) and the recovery of the *recessive* allele in a 1:3 ratio in the second (F_2) generation. Although Mendel both presented his work publicly and published the work in a scientific journal (Mendel 1865; reprinted in Peters 1959), it failed to catch the attention of mainstream 19th century biologists. In 1900, three different plant biologists—Hugo de Vries, Carl Correns, and Erich Tschermak-Seysenegg—rediscovered Mendel's work (for details, see Carlson 2004, pp. 99–107). Each had done at least part of Mendel's experiments again, but before the publication of their own work, they each either came across or were made aware of Mendel's thorough previous experiments. Thus, all of a sudden, the biological world rediscovered Mendel.

A relevant point for this discussion is that genetics is exact. The original fundamental conclusions that were already in place by 1901–1902 could have been widely recognized at the time of Mendel's publication in 1865, and they have not changed since. Mendel is reputed to have said shortly before his death in 1884, "My day will come." And so it did.

Discovery of Chromosomes in Heredity

The connection between Mendel's genetics and anything recognizable in the cell came only after another 19th century current of research had set the stage. With the discovery of improved microscopes and methods for fixing, staining, and preparing tissues for microscopic examination, the cell theory strongly promulgated by Theodor Schwann and Matthias Schleiden in the late 1830s was widely adopted.

Walther Flemming studied rapid cell division in regrowing tissue after amputation of extremities in salamanders. He used basophilic aniline dyes that stain DNA and noted the thread-like "colored bodies" that were partitioned equally following cell division. By 1880, Flemming had described mitosis; he named the colored bodies *chromosomes* (Flemming 1880; for translation, see Voeller 1948, pp. 43–77).

Both August Weissman and Oscar Hertwig had described quite early the union of sperm and egg nuclei during fertilization of sea urchin and frog eggs. Weissman concluded that the nuclear content of germ cells was the carrier of information between generations—this was his germ plasma theory (Weissman 1893; Hertwig 1885; Carlson 2004, p. 26)—but the fate of chromosomes during the fusion of gametes was not understood.

The connection among chromosomes, nuclear fusion, and Mendel's laws of heredity came independently from experiments by Theodor Boveri in Europe and Walter S. Sutton, a graduate student at Columbia University in New York. Both closely observed spermatogonia, the somatic cells that, after mitotic divisions and a final meiotic division, give rise to sperm. Boveri worked with *Ascaris*, a worm, and Sutton with grasshoppers. The different chromosomes in each species could be distinguished microscopically by their

characteristic shape and length. There are only two morphologically distinguishable types of chromosomes in *Ascaris* and 11 in grasshoppers. Both
Boveri and Sutton observed that the morphological types existed in pairs
in the *dividing* cells of the testis. However, after the final meiotic division,
only a single chromosome of each morphological type was delivered to
the sperm. Both also recognized that when germ cells fuse to start a new
individual, the chromosome number (two pairs in *Ascaris*, 11 in grasshopper)
was again restored. This "reduction" division that occurred to produce germ
cells had been noticed earlier but had not yet been connected to Mendel's
work.

Sutton recognized that the loss of one of the two chromosomes of each
morphological type in the final germ cell division paralleled the independent
distribution of Mendel's "factors" in the different strains of garden peas.
He proposed correctly in two very thorough papers (Sutton 1902, 1903)
that Mendel's factors resided on the chromosomes. Because each parent
contributes one chromosome of each morphological type, each parent was
contributing one form of each character (gene). Sutton was only 25 at the
time, and his papers were published alone. He did his work in the laboratory
of the famous biologist E.B. Wilson, but to Wilson's consternation, Sutton did
not finish his graduate thesis. Instead, he went for a year or two into a business venture, then into medicine, became a surgeon, and tragically died of
appendicitis at age 39.

Genetics in the Early 20th Century: The Morgan School and the Columbia "Fly Lab"

The word *gene* was first used by Wilhelm Ludwig Johannsen, who studied the
size of beans (*Phaseolus vulgaris*) and established the concept of genetically
"pure lines" by inbreeding. He took the root of *pangenesis* to coin a relevant
compact word, gene (Johannsen 1903). William Bateson further clarified the
concept in 1905 to describe those determinants in an organism that allowed
specific traits to be inherited from parent to offspring in a regular fashion.
Bateson had discovered Mendel's work at about the same time as de Vries
and Correns and was responsible for translating Mendel's 1865 paper into
English (for review, see Bateson 1909).

By ~1910, Thomas Hunt Morgan, who was originally interested in
embryology, turned his attention to the genetics of the fruit fly *Drosophila
melanogaster*. He obtained his first batch of flies from Fernandus Payne, to
whom Morgan had suggested raising flies in the dark to test Jean-Baptiste
Lamarck's ideas of adaptive inheritance. Payne had originally collected the
flies on some bananas in Schermerhorn Hall at Columbia to begin his

experiments. From Payne's donated cultures, Morgan began his work in 1908 (Carlson 2004, p. 169).

At first, Morgan was uncertain of the Sutton/Boveri idea that chromosomes carried genes. After several years of working successfully with fruit flies by himself, he was joined in 1911–1913 by one of the most famous and productive collections of scientists in the history of biology. Alfred Sturtevant, Calvin Bridges, and Hermann Müller were all undergraduate students and then Ph.D. students at Columbia. The impact of the students in advancing fly genetics work is indicated by what happened in 1915: the publication of an influential monograph (entitled *The Mechanism of Mendelian Heredity*) on which future genetics textbooks were based (Morgan et al. 1915). It had four authors: Morgan, Sturtevant, Müller, and Bridges. A fascinating and compact history of the operation of the famous fly lab and the monumental accomplishments of the group appear in *Mendel's Legacy* by E.A. Carlson (Carlson 2004, pp. 167–224).

Whereas Morgan may have been uncertain before 1910 of the chromosome theory of inheritance, the importance of the chromosome in genetics was ensured by the work of this group. Suffice it to say here that Morgan and his students perfected the selection of mutant fruit flies and, through many dozens of crosses, established without doubt the theory of chromosomal inheritance. They discovered among many things the chromosomal determination of sex (Morgan 1910) and the regular arrangement of genes along the length of the *Drosophila* chromosomes as revealed through meiotic "crossing-over." The Morgan group correctly interpreted their genetic results as recombination between paired chromosomes that could be observed at meiosis. These results established the first genetic maps (Fig. 1-5) (Sturtevant 1913; Morgan 1926).

Among the many interesting and useful cytological observations that enhanced the idea of a connection between chromosome location and definite genetic function was the discovery of expanded chromosomes in salivary glands. The discovery came from the cytologist Theophilus S. Painter (1934), who had established a second small *Drosophila* genetics outpost at the University of Texas, where he was joined by, among others, H.J. Muller, who showed the power of X irradiation to induce mutations. Painter observed segments of salivary chromosomes that had been copied repeatedly so that when they were stained, the regularity in the underlying structure of the amplified regions became obvious. In these enlarged structures, it was possible to directly visualize the occurrence of rare instances of large-scale breakage and rejoining between parts of the three autosomes.

Later, Calvin Bridges used salivary gland chromosomes to make very precise maps of the chromosomal bands. Subsequently, these allowed

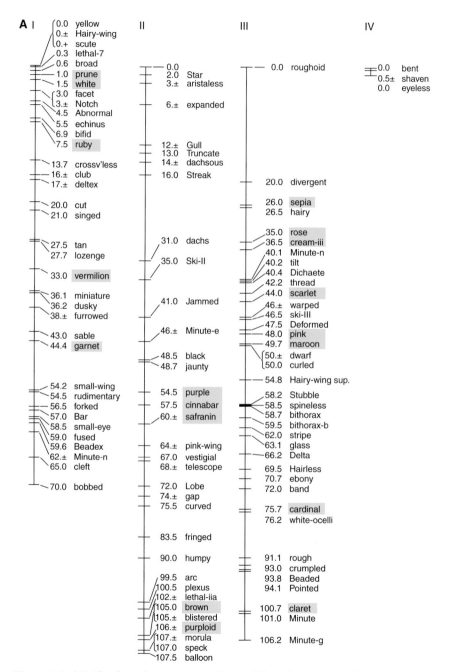

A

I
- 0.0 yellow
- 0.± Hairy-wing
- 0.+ scute
- 0.3 lethal-7
- 0.6 broad
- 1.0 prune
- 1.5 white
- 3.0 facet
- 3.± Notch
- 4.5 Abnormal
- 5.5 echinus
- 6.9 bifid
- 7.5 ruby
- 13.7 crossv'less
- 16.± club
- 17.± deltex
- 20.0 cut
- 21.0 singed
- 27.5 tan
- 27.7 lozenge
- 33.0 vermilion
- 36.1 miniature
- 36.2 dusky
- 38.± furrowed
- 43.0 sable
- 44.4 garnet
- 54.2 small-wing
- 54.5 rudimentary
- 56.5 forked
- 57.0 Bar
- 58.5 small-eye
- 59.0 fused
- 59.6 Beadex
- 62.± Minute-n
- 65.0 cleft
- 70.0 bobbed

II
- 0.0
- 2.0 Star
- 3.± aristaless
- 6.± expanded
- 12.± Gull
- 13.0 Truncate
- 14.± dachsous
- 16.0 Streak
- 31.0 dachs
- 35.0 Ski-II
- 41.0 Jammed
- 46.± Minute-e
- 48.5 black
- 48.7 jaunty
- 54.5 purple
- 57.5 cinnabar
- 60.± safranin
- 64.± pink-wing
- 67.0 vestigial
- 68.± telescope
- 72.0 Lobe
- 74.± gap
- 75.5 curved
- 83.5 fringed
- 90.0 humpy
- 99.5 arc
- 100.5 plexus
- 102.± lethal-iia
- 105.0 brown
- 105.± blistered
- 106.± purploid
- 107.± morula
- 107.0 speck
- 107.5 balloon

III
- 0.0 roughoid
- 20.0 divergent
- 26.0 sepia
- 26.5 hairy
- 35.0 rose
- 36.5 cream-iii
- 40.1 Minute-n
- 40.2 tilt
- 40.4 Dichaete
- 42.2 thread
- 44.0 scarlet
- 46.± warped
- 46.5 ski-III
- 47.5 Deformed
- 48.0 pink
- 49.7 maroon
- 50.± dwarf
- 50.0 curled
- 54.8 Hairy-wing sup.
- 58.2 Stubble
- 58.5 spineless
- 58.7 bithorax
- 59.5 bithorax-b
- 62.0 stripe
- 63.1 glass
- 66.2 Delta
- 69.5 Hairless
- 70.7 ebony
- 72.0 band
- 75.7 cardinal
- 76.2 white-ocelli
- 91.1 rough
- 93.0 crumpled
- 93.8 Beaded
- 94.1 Pointed
- 100.7 claret
- 101.0 Minute
- 106.2 Minute-g

IV
- 0.0 bent
- 0.5± shaven
- 0.0 eyeless

Figure 1-5. (*A*) The four chromosomes of *Drosophila melanogaster*, with the map positions of the approximately 110 mutants identified by T.H. Morgan's group by 1926. Eighteen separate gene loci, each affecting eye color (Morgan 1926), are indicated with yellow highlighting. (*Continued on facing page.*)

Figure 1-5. (*Continued*) (*B*) Diagram of the four chromosomes of *Drosophila ampelophila*, a close relative of *D. melanogaster*, drawn by Calvin Bridges. The male has two unequal chromosomes (the X and the Y; *bottom*), whereas the female has an equal pair (two Xs) (Bridges 1916). (Reprinted from Darnell et al. 1986.)

breakpoints to be mapped precisely (Bridges 1935). It is safe to say that the Morgan School led the way to the ultimate speculation about the chemical nature of the gene (Morgan 1919, 1926).

Early Chemical Analysis of Chromosomes

Despite all of the progress in establishing chromosomes as the bearers of genes, biochemists and biologists in general were not quick to recognize the central position of genes in controlling life. As late as the 1941 Cold Spring Harbor Symposium on Quantitative Biology, the famous geneticist Hermann Muller (both a Morgan student and famous on his own for the discovery of, among other things, radiation-induced mutations [Muller 1927]) said in his Symposium summary, "...only recently, however, has the long *non-recognition* of the gene by physiologists and biochemists in general at last given way...and its [the gene's] startling characteristics brook no further pretense of its *non-existence*" (Muller 1941; my emphasis).

Once some scientists accepted the importance of chromosomes as carriers of the gene, the main question became "What is the chemical nature of the chromosome and of the gene?" The most substantial early work (1930s) on the chemical nature of chromosomes came from the cytological approaches of Torbjörn Caspersson and Einar Hammarsten in Sweden (Caspersson 1936). Caspersson constructed microscopes that gave high resolution of cells not only in chemically stained tissues but also in tissues that were illuminated with ultraviolet (UV) light. Nucleic acids have a strong absorbance in the UV range, and Caspersson observed that chromosomes strongly absorbed UV light, indicating a high nucleic acid content. Using chemical tests, the nucleic acid seemed to be composed mostly of DNA. In fact, in 1938,

Hammarsten, more the biochemist of the two, had prepared protein-free DNA that by ultracentrifugation had a molecular weight of 800,000–1,000,000 (Signer et al. 1938). Chromosomes obviously had in them very high-molecular-weight DNA. Still, in 1936, Caspersson wrote, "If the genes are considered to be chemical substances there can be only one known class of substances to which they may be reckoned to belong, namely the proteins..." (Caspersson 1936; translated and quoted by Fruton 1999, p. 429). This was still the prevalent attitude at the 1941 Cold Spring Harbor conference referred to above.

But there were some hints of things to come. At that same 1941 conference, Jack Schultz, a Morgan trainee who later worked with Caspersson (Caspersson and Schultz 1951), was an early aspirant to learn the important ingredient in the genes residing on chromosomes. Schultz, at least, was keeping an open mind. His paper first bowed in the direction of the high degree of specificity that might be achieved by proteins but then added that "much new data is necessary before we can exclude the possibility of specificities in the nucleic acids themselves" (Schultz 1941).

FROM GENETICS TO MOLECULAR BIOLOGY: "ONE GENE, ONE ENZYME"

Moving beyond the early era of genetics, cytochemistry, and biochemistry, giant steps leading toward the molecular biology of the 1950s came in the 1940s. One giant step was the connection of individual genes to individual protein functions.

Connecting Genes and Proteins: Eye Color in *Drosophila*

The first connection of a specific gene to a specific protein came from studies with the fruit fly *Drosophila*. Normal flies have a deep-red eye color due to a pigment that is made from the amino acid tryptophan through a series of enzymatic reactions. Mutant flies that lack the deep-red eye color are found rarely (one in several thousand); the first was found by T.H. Morgan himself in 1910 (Morgan 1910). George Beadle and Boris Ephrussi (1936) used two such genetically different eye color mutants called *vermillion* and *cinnabar* from the original fly collection of the T.H. Morgan laboratory (see Fig. 1-5). After grafting early embryonic eye bud tissue from either of the two types into the body cavity of a wild-type fly, the mutant tissue developed into supernumerary normal red eyes. Some chemical substance apparently diffused from the normal cells and corrected a defect in the eye cells of either mutant. Beadle and Ephrussi proposed that multiple enzyme steps were required for formation of the red pigment. Each single gene, such as *vermillion* and *cinnabar*, might be responsible for one of the pigment-forming enzymes. Furthermore, because the *vermillion* mutant eye tissue could be "cured" when placed on developing *cinnabar* eye tissue but the reverse (*cinnabar* on *vermillion*) did not work, they suggested that the enzymatic events in the pathway were ordered

vermillion (enzyme 1) → *cinnabar* (enzyme 2) → pigment

(Beadle and Ephrussi 1936, 1937).

Although they did not make this conclusion clearly in their original papers, the inference in Beadle's later accounts was clear (Beadle 1945). The product of enzyme 1 existed in *cinnabar* tissue and could diffuse out to cure *vermillion*, but no product of enzyme 2 existed in *vermillion* tissue to diffuse out and cure *cinnabar*.

Enzymes in the Bread Mold *Neurospora crassa*

Beadle then paired with the younger microbiologist Edward L. Tatum to show that a large number of biochemical events, each performed by a single enzyme, could be damaged individually by mutations (Beadle and Tatum 1941; Beadle 1945). Originally, Tatum, a biochemist, pursued the nature of the diffusible substance that affected *Drosophila* eye pigment development. But then the Beadle and Tatum combination switched to the pink bread mold *Neurospora crassa*. This approach arose after Beadle attended a class lecture on *Neurospora* genetics given by Tatum during which he (Beadle) realized that the genetics of this mold offered the great advantage of furnishing many genes that probably affected growth. *Neurospora* can grow on a simple minimal medium containing a carbon source and ammonium-containing compounds to furnish the "fixed" nitrogen that allows the synthesis of amino acids. They used ammonium tartrate because, as they pointed out (Beadle and Tatum 1941), the carbon in tartrate is not liberated in a form usable as a carbon source. *Neurospora* can make all necessary constituents on this minimal medium.

Muller had established in late 1927 that the X irradiation of *Drosophila* led to mutations in the offspring of these flies (Muller 1927). So Beadle and Tatum used X irradiation of *Neurospora* spores to induce mutations. They took a collection of the irradiated spores and diluted them sufficiently in rich medium to grow colonies from single irradiated spores. Individual colonies were then tested for growth on minimal medium. Individual colonies unable to grow on minimal medium were then tested for growth on minimal medium plus specific nutrients (first in groups and then single nutrients).

In this way, they isolated dozens of mutants lacking the ability to make specific compounds (Beadle 1945). Some of the necessary chemical steps needed to make these compounds were known to be performed by individual enzymes. As this work progressed, it was noted that typically the growth-deficient mutant strains had lost only one enzyme in the pathway to make a necessary compound.

Beadle and Tatum proposed that "one gene was responsible for one enzyme," a largely successful prediction. The idea had to be modified later to "one gene, one polypeptide" because, of course, some enzymes are composed of two or more different polypeptide chains. The experiments of Beadle and Tatum were hugely important in convincing scientists that genes, whatever they were, determined the functioning of proteins. Still, it was too early to emphasize that the order of bases in DNA controlled the order of amino acids in proteins. (For example, Sanger's sequence of insulin was still a decade away.) Furthermore, how the gene might possibly direct protein synthesis was clearly still a deep mystery. It seems that RNA was not even an afterthought at this time.

Escherichia coli Has Genes

Tatum and Joshua Lederberg then found in 1947 that *Escherichia coli*, the modern world's most-used laboratory organism, could also be used in genetic studies of biochemical processes and other types of cellular events such as susceptibility or resistance to bacteriophage infection (Lederberg and Tatum 1946).

Lederberg mixed two *E. coli* strains differing in each of three traits $(a^-b^-c^- \times a^+b^+c^+)$ and showed that all of the different traits could be transferred from one cell to another at one time, ruling against induced mutations, each of which would have been very rare. He had proved that two *E. coli* cells can stick to each other and genes can be transferred between cells. He called this *bacterial conjugation* (Lederberg 1947). Later, additional conjugating strains were found, first by William (Bill) Hayes; some of these strains (Hfr strains) had a higher frequency of gene transfer (Hayes 1953). By this time, many dozens of genes had been identified in *E. coli*, and in these high-frequency strains, dozens of genes could be transferred during one "mating"—one more proof that genes were lined up on chromosomes. Each of the different Hfr strains always started its gene transfer with the same gene, but between strains, the starting point differed.

Using one of these special strains, Elie Wollman and François Jacob mapped many dozens of the genes on the *E. coli* chromosome (see Chapter 2). Not only did the gene transfer start in the same place, but the transfer of the genes also obeyed a strict order. Jacob and Jacques Monod in the

1955–1960 era used the conjugation procedure to discover how genes control protein synthesis. However, the question of the physical nature of DNA in the bacterial chromosome was still a mystery. Jacob drew diagrams depicting each chromosome as long "spaghetti." But the idea of the bacterial chromosome as one long DNA molecule was not proposed directly.

One practical reason for the lack of appreciation of the great length of DNA was the fragility of long DNA molecules (Levinthal and Davison 1961). It was not until 1963 that the entire bacterial chromosome ($\sim 3 \times 10^6$ base pairs [bp]) was seen by John Cairns using autoradiography and by A.K. Kleinschmidt in the electron microscope after gentle lysis of cells and spreading on electron microscope grids (Fig. 1-6; for review, see Cairns 1972).

In addition to bacterial conjugation, some strains were found to transfer a limited but regular smaller group of genes. These cells were transferring small circular DNA molecules that Lederberg termed *plasmids* (Lederberg 1951). In later decades, the use of Lederberg's techniques established *E. coli* as a major tool in biotechnology. Stanley Cohen and Herb Boyer used purified plasmids as vectors for genes of interest that had been inserted into the plasmids (Cohen et al. 1973). This was the necessary founding technique for the recombinant DNA revolution of the late 1970s.

In 1958, Beadle and Tatum, together with Lederberg, shared a Nobel Prize for their immense contributions to genetics and the recognition of the "one gene, one enzyme" idea.

"TRANSFORMING PRINCIPLE": DNA IS THE GENETIC MATERIAL

The story of the "transforming principle (TP)," which was so important to many of the founders of molecular biology, provides deep insight into the mystery that the chemical nature of genes still presented in the early 1940s as well as insight into the stubbornness of human nature.

The story began with the British microbiologist Fred Griffith. In the late 1920s, he observed that when mice were infected with a *rough* nonvirulent variant (say, of type-II pneumococci) (Fig. 1-7), nothing happened. Likewise, *heat-killed smooth* virulent (type I) pneumococci were harmless. But if the mice were treated with both heat-killed type-I smooth pneumococci and with living rough type-II organisms, the mice died of pneumonia. Their lungs were full of living pathogenic smooth type-I organisms. The smooth bacteria have a gelatinous coat that provides protection by inhibiting phagocytosis. Somehow, the living cells had acquired the type-I coat character from the "dead" type-I organisms. This phenomenon, called *transformation*, interested Oswald Avery at The Rockefeller Institute for Medical Research because his laboratory had found that the coat on pathogenic strains was mainly

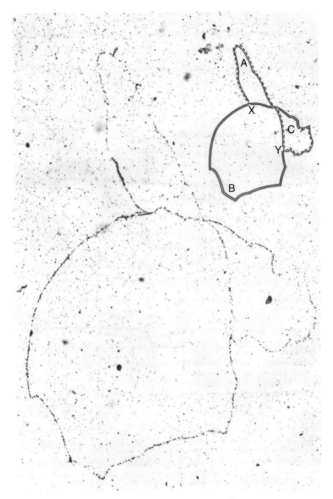

Figure 1-6. Autoradiograph of *E. coli* DNA labeled with ³H thymidine. Structure is also shown diagrammatically (*top right*), with the two replication forks arising from bidirectional replication, labeled X and Y. (Reprinted, with permission, from Cairns 1963.)

polysaccharide. Each different polysaccharide was specifically antigenic, so what led to its synthesis was of great importance. (In the 1970s, a vaccine devised by Robert Austrian consisting of a mixture of the different pneumococcal polysaccharides proved to be protective and is now widely used to protect the elderly as well as infants.)

Characterization of Pneumococcal Extracts

Avery had a long series of collaborators who studied the transformation problem with him. In the 1930s, J. Lionel Alloway and Martin Dawson

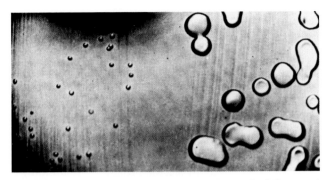

Figure 1-7. Colonies of *Streptococcus pneumoniae*. (*Left*) Small, dull-appearing, "rough" colonies; (*right*) large, glistening, "smooth" colonies. Transformation by DNA was proved with such organisms. (Reprinted from Avery et al. 1944.)

showed that bacteria-free extracts of pneumococci, when mixed with a suspension of pneumococci of a different type, could cause transformation (Dawson and Sia 1931; Alloway 1933). Identifying the chemical nature of the TP became the goal. By 1941, Avery and Canadian associate Colin MacLeod had isolated a rough strain, R36, that was absolutely stable but regularly responded to the transforming extracts. MacLeod had adopted an extraction technique devised by M.G. Sevag (still in use today) that left most protein behind but the TP intact (Sevag 1934). The extract is shaken with a mixture of organic solvents (chloroform and an alcohol), and the great majority of the protein is denatured. Following centrifugation, the protein forms a "wafer" at the organic–aqueous interface. Although it was not appreciated at the time, when the aqueous phase was removed, all nucleic acid was left still in solution, including the TP.

A young physician, Maclyn McCarty, joined the team in 1941 in the same month that MacLeod was leaving. (MacLeod became Chairman of New York University Department of Microbiology, where late in his career he helped start Robert Austrian on his way to developing the pneumococcal vaccine.)

McCarty made a number of key advances in preparing TP and in the fractionation of extracts, beginning with using the correct concentration of ethyl alcohol after the chloroform step. This ethyl alcohol step produced a stringy precipitate that could be collected by wrapping it around a stirring rod. This trick of collecting the precipitate may have been pointed out to McCarty and Avery by Alfred E. Mirsky, a senior Rockefeller scientist who had been extracting nucleic acid from the nuclei of animal cells. (This is the same Mirsky with whom Linus Pauling had had a useful 1936 collaboration; see p. 25.) The irony of the possibility that Mirsky lent a helping hand (if, in fact, he did) to the Avery/McCarty enterprise will become clear a bit later.[1]

McCarty measured the chemical contents of his most purified extracts and found very little protein. There was a positive reaction with orcinol for RNA (still called "yeast RNA" at this time) and a weak reaction with diphenylamine, a reaction that had only recently been recognized to be for *deoxyribose*. Thus, it appeared that there was DNA in the extracts, although it was not widely appreciated at the time that bacteria even contained DNA.

McCarty then secured some assistance from Moses Kunitz, a Rockefeller scientist and an acknowledged master of protein crystallization procedures, who had crystallized pancreatic ribonuclease (Kunitz 1940) and who had also obtained crystalline pepsin from his mentor John Northrop. Kunitz gave McCarty both crystalline ribonuclease and pepsin. Treatment of the most purified TP preparations with either enzyme had no effect on the TP. McCarty took some of the cruder tissue (pancreatic) extracts from which pepsin and ribonuclease had been purified and showed that TP was easily destroyed in such extracts. Thus, some active substance, presumably an enzyme, could destroy TP, although the activity did not seem to be directed toward a protein or RNA.

For help in characterizing the TP, McCarty and Avery turned in April 1943 to Alexandre Rothen, a physical chemist who had built one of the first ultracentrifuges in the United States. From these analyses, it quickly became clear that the active substance in the best TP preparations was a very high-molecular-weight material. It was so large that all of the activity in the TP could be collected at the bottom of an ultracentrifuge tube as a clear pellet. When redissolved, the pellet retained all of the TP activity in the original solution. The preparation contained DNA but no measurable RNA or protein and again was undamaged by crystalline proteases and ribonuclease.

Avery, MacLeod, and McCarty Knew What They Had

As we discuss below, some backbiting commentators would later say that the Rockefeller scientists did not know what they had found. This is clearly not so. Oswald Avery, by this time 65 years old, not in the best of health, and by

---

[1] It was my great good fortune to know McCarty quite well. In my medical school lab experience in Robert Glaser's laboratory, I had become familiar with streptococci and received from Glaser a "letter of introduction" to Mac (as he was universally known). In December 1954, I visited him and we talked about rhamnose, which he had just identified in strep cell walls. I had virtually no idea what rhamnose was but did my best to keep that to myself. After I moved to The Rockefeller University in 1974, I had the pleasure of getting to know Mac quite well and had many useful conversations with him including listening to him reminisce about the "old days."

nature extremely judicious and cautious, was nevertheless convinced that the TP was DNA and therefore that DNA was capable of changing the enzymatic functioning of cells. In the spring of 1943, he was prepared to speak out, at least for internal consumption. Avery wrote in his yearly report to the Board of Scientific Advisors of The Rockefeller Institute, "If the present studies are confirmed and the biologically active substance isolated in highly purified form as the sodium salt of deoxyribonucleic acid actually proves to be the transforming principle, as the available evidence now suggests, then nucleic acids of this type must be regarded not merely as structurally important but as functionally active in determining the biochemical activities and specific characteristics of pneumococcal cells" (quoted in *The Transforming Principle* [McCarty 1986, pp. 154–155]).

A second proof that Avery knew the importance of his discovery comes in a long letter he wrote to his brother Roy Avery, a Professor of Microbiology at Vanderbilt University in Nashville, Tennessee. In this letter of May 1943, he summarizes all of the work and writes that the problem "touches the biochemistry of the thymus type of nucleic acids which are known to constitute the major part of the chromosomes but have been thought to be alike regardless of origin and species. It touches genetics, enzyme chemistry, cell metabolism & carbohydrate synthesis, etc. But today it takes a lot of well documented evidence to convince anyone that the sodium salt of deoxyribonucleic acid, protein free, could possibly be endowed with such biologically active & specific properties..." (McCarty 1986, p. 159). (Of considerable interest in connection with this letter is a conversation between Horace Freeland Judson and Max Delbrück [Judson 1996, p. 40]. Delbrück in 1943 was at Vanderbilt, and Roy Avery showed the letter to him. In fact, Roy Avery and his wife had to dig through papers to unearth the letter at Delbrück's request many years later so that Delbrück could use it at a symposium honoring Avery.)

In contrast to Avery's approach, McCarty wrote, McCarty and MacLeod were "much less inclined to be cautious." But the famous paper that the trio then prepared to write was a model of care and caution. They each worked on drafts of the paper during the summer of 1943, and Avery, on November 1, 1943, delivered the final copy by hand to the office of *The Journal of Experimental Medicine*, a journal published by The Rockefeller Institute. The paper, appearing on February 1, 1944, was entitled "Studies on the Chemical Nature of the Substance Inducing Transformation of Pneumococcal Types. Induction of Transformation by a Deoxyribonucleic Acid Fraction Isolated from Pneumococcus Type III" (Avery et al. 1944). The last sentence of the paper pulled no punches: "The evidence presented supports the belief that a nucleic acid of the deoxyribose type is the fundamental unit of the transforming principle of Pneumococcus Type III."

It is important to insert here one last scientific note before dealing with the social/scientific aftermath of the paper. McCarty continued the work by purifying deoxyribonuclease, the pancreatic enzyme capable of destroying DNA. Although he did not get it to crystallize, Kunitz did a bit later (Kunitz 1950). By 1946, McCarty had purified DNase to the point where his preparation lacked any RNase or proteolytic activity. This DNase instantly killed the TP. McCarty and Avery published these results in 1946 (McCarty and Avery 1946). In the Discussion section of this later paper, the authors make it explicitly clear that "transformation depends upon the presence of a highly polymerized and specific form of desoxyribose nucleic acid." They go further in the Discussion section to say "The objection can be raised that the nucleic acid may merely serve as a carrier for some hypothetical substance, presumably protein, which possesses the transforming activity. . . .[T]here is no evidence in favor of such a possibility, and it is supported chiefly by the traditional view that nucleic acids are devoid of biologic specificity." In virtually all of the disputatious discussions that followed the reports from the Avery laboratory that McCarty and Avery did not know what they had discovered, reference to the Discussion section of this second 1946 paper is seldom made.

Reception of the Avery et al. 1944 Paper

The aftermath of the publication of the epochal 1944 *Journal of Experimental Medicine* (*JEM*) paper is interesting. We have seen already how the first monumental paper on genetics—that of Gregor Mendel—was not appreciated by the scientific culture of the 1860s. The Avery, MacLeod, and McCarty (1944) paper also did not create an immediate sensation. First of all, it was wartime—the end of World War II in Europe was still 15 months off and VJ (victory over Japan) day was 18 months away. In addition, *JEM* was not a place where either genetics or much biochemistry was published, and perhaps few in either discipline read the papers until later.

But perhaps the most imposing obstacle to acceptance of DNA as the active agent in TP was the failure at the time to credit any cellular macromolecules other than protein with sufficient magic to embody the specificity required of the biochemical events of which a cell was capable. Nucleic acids were still regarded by many to be nonspecific, a sentiment not peculiar to the Rockefeller community and widely shared. As noted earlier (p. 31), Phoebus Levene, the well-known Rockefeller Institute chemist, had in his 1931 treatise proposed the tetranucleotide hypothesis, which implied that nucleic acids could not embody specificity (Levene and Bass 1931). However, obstructionism to accepting the obvious conclusions stated by the Avery group was evinced most strongly by A.E. Mirsky, the Rockefeller biochemist and personal acquaintance of Levene. Mirsky told McCarty in 1947 that DNA could not possibly be the transforming principle because "nucleic acids are all alike."

In time, Mirsky became a campaigner against the belief that DNA could be the active ingredient in TP. As illogical as was Mirsky's stance, it could not be refuted: No preparation from a cell extract as of 1944 could be shown to be *totally* free of protein. Mirsky's objection, probably based on jealousy more than reason, was cloaked under the claim that the small amount of potential protein contamination (<0.5%) only allowed Avery, MacLeod, and McCarty to conclude that a *nucleoprotein* containing DNA was the transforming agent. In the end, Mirsky's vehement and frequent objections to an exclusive role of DNA as the bearer of genetic information probably led to the failure of an immediate wider acceptance of DNA as the genetic material and of a Nobel Prize for Avery's group.

In 1972, 17 years after Avery's death, the book *Nobel: The Man and His Prizes* (Nobel Foundation and Odelberg 1972) was published. It contained the following statement: "The discovery, because of its far-reaching implications, aroused much interest, and Avery was proposed for a Nobel Prize. But doubts were also expressed, and the Nobel Committee found it desirable to postpone an award." McCarty, who died in 2005, wryly notes in his masterful account, *The Transforming Principle* (McCarty 1986, p. 219), that this statement of 1952 was made 3 years *before* Avery died.

One of the main evaluators of the Nobel Committee was Einar Hammarsten, the chemist who, ironically, had in 1938 prepared the first well-authenticated high-molecular-weight DNA (Signer et al. 1938). He was negative toward Avery's work. He believed, as did many others, that only proteins could be complicated enough to do the work of genes. Peter Reichard, a professor at the Karolinska Institute in Stockholm who was a student of Hammarsten, has written a very informative account of these events that reveals much about the thinking of some scientists in the 1945–1955 era (Reichard 2005).

Many Scientists "Got It"

The record is clear that not all scientists were oblivious to the importance of the Avery, MacLeod, and McCarty findings. The 1944 paper had an immediate effect on one pioneer, Joshua Lederberg, the discoverer of bacterial genes. Lederberg wrote in 1987, "When biologists of that era (∼1944–45) used terms like protein, nucleic acid, or nucleoprotein, it can hardly be assumed that the words had today's crisp connotations of defined chemical structure. Sleepwalking, we were all groping to discover just what was important about the chemical basis of biological specificity. It was clear to the circle I frequented at Columbia [this had been Morgan's department] that Avery's work was the most exciting key to that insight" (Lederberg 1987).

Erwin Chargaff's seminal biochemistry identified the A = T and C = G equalities in the DNA from many organisms, a finding that provided a central biochemical fact about DNA that contributed to the Watson-Crick structure (see his summary lecture published in *Federation Proceedings* [Chargaff 1951]). Chargaff was an organic chemist originally trained in Austria who came to the United States in the 1930s. At Columbia University, he and his colleagues set up procedures, including the newly discovered techniques of paper chromatography for separating and quantitating the nucleic acid base composition of both DNA and RNA. (As acknowledged by Chargaff, the use of crystallized DNase and RNase provided by Moses Kunitz was greatly beneficial to making reasonably pure preparations of DNA and RNA.) The result of the greatest importance to come from these analyses was, of course, the "Chargaff rules" of equality of A and T and of G and C. This was achieved by analysis of many different organisms—bacteria, yeast, animals, and plants (Table 1-2). However, the importance of Chargaff's results at the time was to instantly refute the "tetranucleotide hypothesis" of Phoebus Levene. Variations in the base content in different organisms proved that equimolar mixtures of the four bases acting as a tetranucleotide could not be true.

Chargaff repeatedly described the importance of the Avery, MacLeod, and McCarty (1944) paper. A 1971 review presents a good summary of Chargaff's

Table 1-2. Molar ratios of bases in DNA preparations of different origins

Source	Adenine to guanine	Thymine to cytosine	Adenine to thymine	Guanine to cytosine	Purines to pyrimidines
Ox	1.29	1.43	1.04	1.00	1.1
Man	1.56	1.75	1.00	1.00	1.0
Hen	1.45	1.29	1.06	0.91	0.99
Salmon	1.43	1.43	1.02	1.02	1.02
Wheat	1.22	1.18	1.00	0.97	0.99
Yeast	1.67	1.92	1.03	1.20	1.0
Hemophilus influenzae, type c	1.74	1.54	1.07	0.91	1.0
Escherichia coli K–12	1.05	0.95	1.09	0.99	1.0
Avian tubercle bacillus	0.4	0.4	1.09	1.08	1.1
Serratia marcescens	0.7	0.7	0.95	0.86	0.9
Hydrogen organism *Bacillus Schatz*	0.7	0.6	1.12	0.89	1.0

The equality of A and T and G and C are shaded. (Reproduced, with permission, from Chargaff 1951.)

thoughts: "This discovery [the 1944 paper] almost abruptly appeared to *fore-shadow a chemistry of heredity and moreover made probable the nucleic acid character of the gene.* It certainly made an impression on a few, not on many, but probably on nobody a more profound one than on me" (Chargaff 1971; my emphasis). The 1944 paper led Chargaff to turn all of the efforts of his laboratory to the study of the base composition of different DNA samples.

Unfortunately, Chargaff, whose meticulous biochemistry was so important, developed an antipathy to the fast-moving, and in his mind, slovenly rush of molecular biology that followed the Watson-Crick structure (Chargaff 1974, 1978). His objections and a refutation of his claim that he contributed in an essential manner to the understanding of the *structure* of DNA is extremely well chronicled in Horace Freeland Judson's text *The Eighth Day of Creation* (Judson 1996, pp. 631–637). Although Chargaff definitely recognized the universal A = T, G = C rule, he did not at all fit his results into a three-dimensional, two-stranded idea.

James Watson, during his graduate work with bacteriophages in Salvador Luria's laboratory, was completely aware of the Avery lab's work. He writes in *DNA, The Secret of Life* that "when I arrived at Indiana University in the fall of 1947 [at the age of 19, incidentally] with plans to pursue the gene for my PhD thesis, Avery's paper came up over and over in conversations" (Watson 2003, p. 40).

McCarty quotes a passage from Watson's popular account of the discovery of the DNA structure, *The Double Helix* (Watson 1968, p. 143): "Given the fact that DNA was known to occur in the chromosomes of all cells, Avery's experiments strongly suggested that future experiments would show that all genes were composed of DNA. If true, this meant to Francis [Crick] that proteins would not be the Rosetta stone for unraveling the true secret of life. Instead, DNA would have to provide the key to enable us to find out how the genes determined, among other characteristics, the color of our hair, our eyes, most likely our comparative intelligence, and maybe even our potential to amuse others. Of course there were scientists who thought the evidence favoring DNA was inconclusive and preferred to believe that genes were protein molecules. Francis, however, did not worry about these skeptics" (McCarty 1986, p. 223). And obviously, neither did Jim Watson.

Between the mid 1940s and 1951, considerable progress on bacterial transformation occurred. Several additional scientists—Rollin Hotchkiss, Hattie Alexander, and Harriet Taylor—showed that DNA preparations not only from pneumococci but also from other bacteria (*Haemophilus influenzae*) could transform cells for a variety of traits (Hotchkiss 1957). Obstinacy and stridency persisted, however, and still not all biologists had assimilated the conclusion that "DNA is the genetic material" until later in the 1950s.

THE PHAGE SCHOOL FINDS DNA: ON TO WATSON AND CRICK

According to some historical accounts of thoughts at the time, Avery, MacLeod, and McCarty's findings were "premature" and that is why they were not immediately universally accepted (Stent 1972). According to these accounts, not until experiments by Al Hershey and Martha Chase in 1952 (Hershey and Chase 1952) was the importance of DNA as the genetic material widely accepted.

The T-even series bacteriophages enter *E. coli* and cause production of more phages in 20–30 min. These phages, of course, exhibit an array of specific genetic functions, and they contain DNA enclosed in a protein shell. Hershey and Chase used radioactive bacteriophage (labeled $^{32}PO_4$ in DNA and ^{35}S in proteins) and followed the fate of the labeled molecules in the infected cells (Hershey and Chase 1952). The great majority of the phage DNA entered the cell; *however, 3% of the protein also entered the cell.* Of course, 3% protein was far more than the trace contaminant of protein in McCarty's purest "transforming principle." But the world, especially the phage world, was ready now to believe that DNA, not protein, carried the genetic information.

The most likely reason that the 1952 Hershey/Chase experiment was accorded such an important place in settling the fact that DNA was the genetic material is that by the 1950s, the "phage school" of biology had collected such a close-knit and brilliant group of workers, many of whom were originally physicists following the physicist Max Delbrück into quantitative biology (see Cairns et al. 1966, which was dedicated to Max Delbrück). These gifted scientists were not nearly as likely to read or know "older" literature (all the way back to 1944!), especially that published in the *Journal of Experimental Medicine*, as to know and exhibit more respect for the new bacterial and phage genetics that Delbrück, Luria, and Hershey had pioneered. It also bears noting that Delbrück himself, who knew of the Avery work, did not originally view it as groundbreaking (see p. 45). But it was clear that many in the experimental biology community of the pre-1953 era, i.e., the pre-Watson-Crick era, knew that genes controlled proteins and that genes were DNA.

Watson has written extensively, not only in the popular book *The Double Helix* (Watson 1968), but, e.g., in the more recent *DNA, The Secret of Life* (Watson 2003), about his early career leading up to the discovery of the DNA structure. After his training at the University of Indiana (he received his Ph.D. in 1950), he was completely convinced of the importance of DNA as the genetic material (Watson 2003, pp. 38–43). The Pauling-Corey papers, on the discovery of the α-helix in proteins based on model building with limited but accurate data about the structure of amino acids and peptides, greatly impressed him. Furthermore, at a conference in Italy in 1951, Watson heard Maurice Wilkins give a talk on preliminary studies of X-ray analysis of DNA fibers. Watson importuned Luria, his Ph.D. advisor, to help him to get his postdoctoral fellowship transferred to Cambridge, where X-ray analysis of macromolecules was concentrated. He describes arriving in 1951, being assigned to a room together with a genial physicist still working for a graduate degree, named Francis Crick. The progress from 1951 to 1953 in the building of a correct model structure of DNA has been repeatedly and absorbingly related (Watson 1968, 2003; Crick 1988). That thrilling outcome marks the end of an era of ignorance about long polymeric molecules. In 1953, both proteins as long polypeptides and DNA as long, double-stranded molecules were settled entities. In Chapter 2, we can finally turn to RNA—the other long, functional biopolymer indispensable for genetic expression.

REFERENCES

Adair GS. 1925. A critical study of the direct method of measuring the osmotic pressure of haemoglobin. *Proc R Soc Lond A Math Phys Sci* **108:** 627–637.

Alloway JL. 1933. Further observations on the use of pneumococcal extracts in effecting transformation of type in vitro. *J Exp Med* **57:** 165–278.

Astbury WT, Lomax R. 1934. X-ray photographs of crystalline pepsin. *Nature* **133:** 795.

Astbury WT, Woods HJ. 1933. X-ray studies of the structure of hair, wool and related fibers. II. The molecular structure and elastic properties of hair keratin. *Philos Trans R Soc Lond A* **232:** 333–394.

Avery OT, MacLeod CM, McCarty M. 1944. Studies on the chemical nature of the substance inducing transformation of pneumococcal types. Induction of transformation by a desoxyribonucleic acid fraction isolated from pneumococcus type III. *J Exp Med* **79:** 137–158.

Baddiley J. 1955. Chemistry of nucleosides and nucleotides. In *The nucleic acids* (ed. Chargaff E, Davidson JN), Vol I, pp. 137–188. Academic Press, New York.

Barger G, Coyne FP. 1928. The amino acid methionine: Constitution and synthesis. *Biochem J* **22:** 1417–1425.

Bateson W. 1909. *Mendel's principles of heredity.* Cambridge University Press, Cambridge.

Beadle GW. 1945. Biochemical genetics. *Chem Rev* **37:** 15–96.

Beadle GW, Ephrussi B. 1936. The differentiation of eye pigments in *Drosophila* as studied by transplantation. *Proc Natl Acad Sci* **21:** 225–247.

Beadle GW, Ephrussi O. 1937. Development of eye colors in *Drosophila*: Diffusible substances and their interrelations. *Genetics* **22:** 76–86.

Beadle GW, Tatum EL. 1941. Genetic control of biochemical reactions in *Neurospora*. *Proc Natl Acad Sci* **27:** 499–506.

Beaumont W. 1833. *Experiments and observations on the gastric juice and the physiology of digestion.* Allen, Plattsburgh, NY [reprinted by Andre Combe, available online under William Beaumont].

Bernal J D. 1939. Vector maps and the cyclol hypothesis. *Nature* **143:** 74–75.

Bernal JD, Crowfoot D. 1934. X-ray photographs of crystalline pepsin. *Nature* **133:** 794–795.

Brachet J. 1941. La détection histochimique et le microdosage des acides pentosenucléiques. *Enzymolgia* **10:** 87–96.

Brachet J. 1955. The biological role of the pentose nucleic acids. In *The nucleic acids* (ed. E Chargaff, JN Davidson), Vol II, pp. 475–519. Academic Press, New York.

Bridges CB. 1916. Non-disjunction as proof of the chromosome theory of heredity. *Genetics* **1:** 1–52.

Bridges CB. 1935. Salivary chromosome maps with a key to banding the chromosomes of *Drosophila melanogaster. J Hered* **26:** 60–64.

Büchner E. 1966. Cell-free fermentation. In *Nobel lectures in chemistry 1901–1921*, pp. 103–120. Elsevier, Amsterdam.

Burr GO. 1939. Outlines of biochemistry. By Ross Aiken Gortner. *J Phys Chem* **43:** 392–393

Cairns J. 1963. The chromosome of *Escherichia coli. Cold Spring Harbor Symp Quant Biol* **28:** 43–46.

Cairns J. 1972. DNA synthesis. *Harvey Lect* **66:** 1–18.

Cairns J, Stent GS, Watson JD, eds. 1966. *Phage and the origins of molecular biology.* Cold Spring Harbor Laboratory, Cold Spring Harbor, NY.

Carlson EA. 2004. *Mendel's legacy*, p. 43. Cold Spring Harbor Laboratory Press, Cold Spring Harbor, NY.

Carothers WH. 1931. Polymerization. *Chem Rev* **8:** 353–428.

Caspersson T. 1936. Veber den chemishen aufbau der structure des zellkerns. *Skand Arch Physiol* (suppl. 8) **23:** 1–151.

Caspersson T. 1947. The relation between nucleic acid and protein synthesis. *Symp Soc Exp Biol* **1:** 129–151.

Caspersson T, Schultz J. 1951. Cytochemical studies in the study of the gene. In *Genetics in the 20th century* (ed. LC Dunn), pp. 155–171. Macmillan, New York.

Chargaff E. 1951. Structure and function of nucleic acids as cell constituents. *Fed Proc* **10:** 654–659.

Chargaff E. 1971. Preface to a grammar of biology. A hundred years of nucleic acid research. *Science* **172:** 637–642.

Chargaff E. 1974. Building the tower of babble. *Nature* **248:** 776–779.

Chargaff E. 1978. *Heraclitean fire: Sketches from a life before nature.* Rockefeller University Press, New York.

Cohen SN, Chang AC, Boyer HW, Helling RB. 1973. Construction of biologically functional bacterial plasmids in vitro. *Proc Natl Acad Sci* **70:** 3240–3244.

Corey RB. 1945. X-ray studies of amino acids and peptides. *Adv Protein Chem* **4:** 385–406.

Crick F. 1988. *What mad pursuit: A personal view of scientific discovery.* Basic Books, New York.

Darnell JE Jr, Lodish HF, Baltimore D. 1986. *Molecular cell biology.* Freeman, New York.

Dawson MH, Sia RHP. 1931. In vitro transformation of pneumococcal types: I. A technique for inducing transformation of pneumococcal types in vitro. *J Exp Med* **54:** 681–700.

Dickerson RE. 2005. *Present at the flood: How structural molecular biology came about.* Sinaur, Sunderland, MA.

du Vigneaud V. 1952. *A trail of research.* Cornell University Press, Ithaca, NY.

Edman P. 1950. Method for the determination of the amino acid sequence of peptides. *Acta Chem Scand* **4:** 283–293.

Edsall JT. 1962. Proteins as macromolecules: An essay on the development of the macromolecule concept and some of its vicissitudes. *Arch Biochem Biophys* (Suppl.) **1:** 12–20.

Eisenberg H. 1996. Birth of the macromolecule. *Biophys Chem* **59:** 247–257.

Fischer E. 1899. Synthesen in der Puringruppe. *Ber Dtsch Chem Ges* **32:** 455–504.

Fischer E. 1902. Veber die Hydrolyse der Proteinstoffe. *Chem Zeitschrift* **26**: 939–940.

Fischer E. 1907. Synthetical chemistry in relation to biology. *J Chem Soc* **91**: 1749–1765.

Fischer E. 1923. *Untersuchungen über Aminosäuren, Polypeptide und Protein. II (1907–1919)*. Springer, Berlin.

Flemming W. 1880. Beiträge zur Kenntniss der Zelle und ihrer Lebenser-scheinungen, I. *Arch Mikrosk Anat* **16**: 302–306.

Fruton JS. 1999. *Proteins, enzymes, genes: The interplay of chemistry and biology.* Yale University Press, New Haven, CT.

Goodman M, Cai W, Smith ND. 2003. The bold legacy of Emil Fisher. *J Pept Sci* **9**: 594–603.

Gortner RA. 1938. *Outlines of biochemistry: The organic chemistry and the physiochemical reactions of biologically important compounds and systems.* Wiley, New York.

Graham T. 1861. Liquid diffusion applied to analysis. *Philos Trans R Soc Lond* **151**: 183–224.

Hayes W. 1953. The mechanism of genetic recombination in *Escherichia coli. Cold Spring Harb Symp Quant Biol* **18**: 75–93.

Hershey AD, Chase M. 1952. Independent functions of viral protein and nucleic acid in growth of bacteriophage. *J Gen Physiol* **36**: 39–56.

Hertwig O. 1885. Das Problem der Befructung und der Isotropie des Eis, eine Theorie der Bererbung. *Jenaische Z Medizin Naturwiss* **18**: 276–318.

Hofmeister F. 1902. Ueber Bau und Gruppering der Eiweisskurper. *Natwiss Rundsch* **17**: 529–533, 545–549.

Hotchkiss RD. 1957. Criteria for quantitative transformation of bacteria. In *A symposium on the chemical basis of heredity* (ed. WD McElroy, B Glass), pp. 321–335. Johns Hopkins Press, Baltimore.

Hunefeld FL. 1840. *Der Chemismus in der Thierischen Organization.* Brockhaus, Leipzig.

Hunter GK. 1999. Phoebus Levene and the tetranucleotide structure of nucleic acids. *Ambix* **46**: 73–103.

Johannson W. 1903. In *Uber Erblichkeit in Populationonen und in Reinen Linien G. Fischer Jena* [English translation *Selected readings in biology for natural sciences* (ed. Gall H, Putschan E)], Vol. 3, pp. 172–215. Chicago University Press, Chicago.

Johnson O, Perutz M. 1981. Gilbert Smithson Adair. *Biog Mem FRS* **27**: 1–17.

Judson HF. 1996. *The eighth day of creation: Makers of the revolution in biology.* Cold Spring Harbor Laboratory Press, Cold Spring Harbor, NY.

Kossel A, Neumann A. 1893. Ueber das Thymin, ein Spaltungsproduct der Nucleinsäure [About the thymide, a cleavage product of the nucleic acid]. *Ber Dtsch Chem Ges* **26**: 2753–2756.

Kossel A, Neumann A. 1894. Darstellung und Spaltungsprodukte der Nucleinsäure (Adenylsäure). *Ber Dtsch Chem Ges* **27**: 2215–2222.

Kunitz M. 1940. Crystalline ribonuclease. *J Gen Physiol* **24**: 15–32.

Kunitz M. 1950. Crystal desoxyribonuclease. *J Gen Physiol* **35**: 349–377.

Langmuir I. 1939. The structure of proteins. *Proc Phys Soc* **51**: 592–612.

Lederberg J. 1947. Gene recombination and linked segregation in *Escherichia coli. Genetics* **32**: 505–525.

Lederberg J. 1951. Genetic studies with bacteria. In *Genetics in the twentieth century* (ed. Dunn LC), pp. 263–289. Macmillan, New York.

Lederberg J. 1987. Genetic recombination in bacteria: A discovery account. *Annu Rev Genet* **21**: 23–46.

Lederberg J, Tatum EL. 1946. Gene recombination in *Escherichia coli. Nature* **158**: 558.

Levene PA. 1918. The structure of yeast nucleic acid: III. Ammonia hydrolysis. *J Biol Chem* **33**: 425–428.

Levene PA. 1919. The structure of yeast nucleic acid: IV. Ammonia hydrolysis. *J Biol Chem* **40:** 415–424.

Levene P, Bass LW. 1931. *Nucleic acids.* Chemical Catalogue, New York.

Levinthal C, Davison PF. 1961. Degradation of desoxyribonucleic acid under hydrodynamic shear. *J Mol Biol* **3:** 674–683.

McCarty M. 1986. *The transforming principle: Discovering that genes are made of DNA.* Norton, New York.

McCarty M, Avery OT. 1946. Studies on the chemical nature of the substance inducing transformation of pneumococcal types. II. Effect of desoxyribonuclease on the biological activity of the transforming substance. *J Exp Med* **83:** 89–96.

McCoy AL, Meyer CE, Rose WC. 1935. Feeding experiments with mixtures of highly purified amino acids. VIII. Isolation and identification of a new essential amino acid. *J Biol Chem* **112:** 283–302.

Mendel G. 1865. Versuche uber Pflanzen. [Translated by WA Bateson as "Experiments in plant hybridization."] *Verh Naturforshung Ver Brunn* **4:** 3–47.

Merrifield B. 1986. Solid phase synthesis. *Science* **232:** 341–347.

Merrifield RB. 1993. *Life during a golden age of peptide chemistry.* American Chemical Society, Washington, DC.

Miescher F. 1871. Uber die chemische Zusammersetzung des Eiters. *Medicinische-chemische Untersuchungen* **4:** 441–460.

Mirsky AE, Pauling L. 1936. On the structure of native, denatured and coagulated proteins. *Proc Natl Acad Sci* **22:** 439–442.

Morgan TH. 1910. Sex limited inheritance in *Drosophila*. *Science* **32:** 120–122.

Morgan TH. 1919. *The physical basis of heredity.* Lippincott, Philadelphia.

Morgan TH. 1926. *The theory of the gene.* Yale University Press, New Haven, CT.

Morgan TH, Sturtevan AH, Muller HJ, Bridges CB. 1915. *The mechanism of Mendelian heredity.* Henry Holt, New York.

Mulder GJ. 1838. Zusammensetzung von Fibrin, Albumin, Leimzucker, Leucin U.S.W. *Ann Pharm* **28:** 73–82.

Mulder GJ. 1844. *Versuch einer allegemeinen Physiologische chemie.* Vieweg, Brannschweig, Germany.

Muller HJ. 1927. Artificial transmutation of the gene. *Science* **66:** 84–87.

Muller HJ. 1941. Résumé and perspectives of the Symposium on Genes and Chromosomes. *Cold Spring Harbor Symp Quant Biol* **9:** 290–308.

Nobel Foundation Odelberg W. 1972. *Nobel: The man and his prizes.* American Elsevier, New York.

Northrop JH. 1930. Crystalline pepsin. *J Gen Phys* **13:** 739–766.

Northrop JH. 1935. The chemistry of pepsin and trypsin. *Biol Rev* **10:** 263–282.

Osborne TB. 1902. Sulphur in protein bodies. *J Am Chem Soc* **24:** 140–167.

Osborne TB. 1909. *The vegetable proteins.* Longmans, London.

Osborne TB, Harris IF. 1903. Nitrogen in protein bodies. *J Am Chem Soc* **25:** 323–353.

Painter TS. 1934. Salivary chromosomes that attach on the gene. *J Hered* **19:** 465–476.

Pauling L. 1993. How my interest in proteins developed. *Protein Sci* **2:** 1060–1063.

Pauling L, Corey RB. 1951a. Atomic coordinates and structure factors for two helical configurations of polypeptide chains. *Proc Natl Acad Sci* **37:** 235–240.

Pauling L, Corey RB. 1951b. The pleated sheet, a layer configuration of polypeptide chains. *Proc Natl Acad Sci* **37:** 251–256.

Pauling L, Niemann C. 1939. The structure of proteins. *J Am Chem Soc* **61:** 1860–1867.

Pauling L, Corey RB, Branson HR. 1951. The structure of proteins: Two hydrogen bonded helical configurations of the polypeptide chain. *Proc Natl Acad Sci* **37:** 205–216.

Peters JA. 1959. *Classic papers in genetics.* Prentice-Hall, Englewood Cliffs, NJ.

Reichard P. 2005. Oswald T. Avery and the Nobel Prize in Medicine. *J Biol Chem* **227:** 13355–13302.

Rose WC, Wixom RL, Lockhart HB, Lambert GF. 1955. The amino acid requirements of man. XV. The valine requirement, summary and final observations. *J Biol Chem* **217:** 987–996.

Ryle AP, Sanger F, Smith LF, Kitai R. 1955. The disulfide bonds of insulin. *Biochem J* **60:** 541–556.

Sanger F. 1945. The free amino acids of insulin. *Biochem J* **39:** 507–515.

Sanger F. 1949. Fractionation of oxidized insulin. *Biochem J* **44:** 126–128.

Sanger F. 1952. The arrangement of amino acids in proteins. *Adv Protein Chem* **7:** 1–62.

Sanger F, Thompson EOP. 1953. The amino acid sequence in the glycyl chain of insulin. *Biochem J* **53:** 353–356, 366–374.

Sanger F, Tuppy H. 1951a. The amino-acid sequence in the phenylalanyl chain of insulin. I. The identification of lower peptides from partial hydrolysates. *Biochem J* **49:** 463–481.

Sanger F, Tuppy H. 1951b. The amino-acid sequence in the phenylalanyl chain of insulin. II. The investigation of peptides from enzymic hydrolysates. *Biochem J* **49:** 481–490.

Schultz J. 1941. The evidence of the nucleoprotein nature of the gene. *Cold Spring Harbor Symp Quant Biol* **9:** 151–167.

Sevag MG. 1934. Eine neue physikalische enteiweissungemethode zur darstellung biologisch wirksamer substanzen. *Biochem Z* **273:** 419.

Shepherd RG, Willison SD, Howard KS, Bell PH, Davies DS, Davis SB, Eigner EA, Shakespeare NE. 1956. Studies with corticotropin. III. Determination of the structure of β-corticotropin and its active degradation products. *J Am Chem Soc* **78:** 5067–5076.

Signer R, Caspersson T, Hammersten E. 1938. Molecular shape and size of thymonucleic acid. *Nature* **141:** 122. doi: 10.1038/141122a0.

Sorenson SPL, Hoyrup M. 1915–1917. Studies on proteins. *C R Trav Lab Carlsberg* **12:** 1–372.

Staudinger H. 1953. "Macromolecular chemistry: Nobel Lecture, December 11, 1953." http://nobelprize.org/nobel_prizes/chemistry/laureates/1953/staudinger-lecture.pdf.

Steiner DF, Over PE. 1967. The biosynthesis of insulin and a probable precursor of insulin by a human islet cell adenoma. *Proc Natl Acad Sci* **57:** 473–480.

Stent GS. 1972. Prematurity and uniqueness in scientific discovery. *Sci Am* **227:** 84–93.

Sturtevant AH. 1913. The linear arrangement of six sex-linked factors in *Drosophila* as shown by their mode of association. *J Exp Zool* **14:** 43–59.

Sumner JB. 1926. The isolation and crystalization of the enzyme urease. *J Biol Chem* **69:** 435–441.

Sutton WS. 1902. On the morphology of the chromosome group in *Brachystola magna*. *Biol Bull* **4:** 24–39.

Sutton WS. 1903. The chromosomes in heredity. *Biol Bull* **4:** 231–251.

Svedberg T. 1921. *The formation of colloids.* JA Churchill, London.

Svedberg T. 1937. The ultracentrifuge and the study of high-molecular weight compounds. *Nature* **139:** 1051–1062.

Tanford C, Reynolds J. 2001. *Nature's robots: A history of proteins.* Oxford University Press, New York.

Vickery HB. 1950. The origins of the word "protein." *Yale J Biol Med* **22:** 387–393.

Vickery HB. 1972. The history of the discovery of the amino acids. II. A review of the amino acids described since 1931 as components of native proteins. *Adv Protein Chem* **26:** 81–171.

Vickery HB, Schmidt CLA. 1931. The history of the discovery of the amino acids. *Chem Rev* **9:** 169–319.

Voeller B. 1948. *The chromosome theory of inheritance.* Appleton-Century-Crofts, New York.

Watson JD. 1968. *The double helix.* Atheneum, New York.

Watson JD. 2003. *DNA: The secret of life.* AA Knopf, New York.

Weissman A. 1893. *The germ plasma: A theory of heredity* [translation by Parker WN, Rönnfeldt H]. Scribners, New York.

Wilson EB. 1895. *An atlas of the fertilization and karyogeneis of the ovum.* Macmillan, New York.

Womack M, Rose WS. 1935. Feeding experiments with mixtures of highly purified amino acids. VIII. Isolation and identification of a new essential amino acid. *J Biol Chem* **112:** 275–282.

Wrinch DM. 1936. The pattern of proteins. *Nature* **137:** 411–412.

Wrinch DM. 1937. The cyclol theory and the "globular" proteins. *Nature* **139:** 972–973.

Wrinch DM. 1938. The structure of the insulin molecule. *Science* **88:** 148–149.

Zinoffsky O. 1886. Ueber die Grösse des Hämoglobinmoleküls. *Hoppe-Seylers Z Physiol Chem* **10:** 16–34.

2

RNA Connects Genes and Proteins

Ribosomes, tRNA, and Messenger RNA

THE WATSON-CRICK DEDUCTION IN 1953 of the correct structure of DNA (deoxyribonucleic acid, already identified as the genetic material) is deservedly revered as a watershed moment in biology. The two famous Watson-Crick papers (Watson and Crick 1953a,b) ushered in an era in biological research in which only precise biochemical/biophysical answers would now suffice to explain the inner workings of cells. It seemed clear that the long "trackless wastes" of Cs (cytosines), As (adenines), Gs (guanines), and Ts (thymines) somehow contained all of the information needed to make a new organism.

What James D. Watson and Francis Crick first emphasized, and the scientific world immediately embraced, was that the structure explained how DNA could be copied correctly from generation to generation, because each strand could direct the copying of its complementary partner. Of course, major problems remained about the details of how new DNA strands would be produced: How long were DNA molecules anyway? Could an enzyme be smart enough to take apart and copy very long DNA chains without getting them tangled up? Because information was passed from generation to generation with few if any mistakes, the copying machine somehow had to be error free (or nearly so). These and other problems of DNA replication and chemistry promptly attracted biochemists and geneticists, with much ensuing progress. But that is not our story.

In many ways, the biggest puzzle in 1953 was how the information in DNA could be recovered and then used—translated—to make perfectly ordered proteins in the right amount, in the right cell, at the right time. It is instructive to reflect on how little was known in 1953 regarding how information might be stored in DNA. By the early 1940s, George Beadle, Boris Ephrussi, and Edward L. Tatum had produced strong evidence that genes controlled the function of enzymes and presumably other proteins (reviewed in Chapter 1). The genetic link between genes and proteins was established, but the biochemical link was completely opaque.

Only 2 years had passed since Linus Pauling and Robert B. Corey had proposed in their *eight* papers in *The Proceedings of the National Academy of Sciences* that proteins were linear chains of amino acids that folded into regular shapes in their native state (see Chapter 1). And Fred Sanger, working at Cambridge (as were, of course, Watson and Crick), was just completing the amino acid sequence of the A chain of insulin (Sanger and Thompson 1953); the B-chain sequence had been reported in 1951 (Sanger and Tuppy 1951a). This was the very first evidence that proteins even had a fixed *primary* sequence of amino acids.

In 1953, knowledge of the molecular nature of RNA (ribonucleic acid) was scarce. In the early 1950s, different types of RNA molecules had not been seriously discussed, much less demonstrated. RNA was simply a cell constituent, a nucleic acid, with ribose instead of deoxyribose. Even the "ribose" was apparently not completely accepted, because prominent reviews written in the early 1950s spoke of "pentose nucleic acid" (see Chargaff and Davidson 1955). Histological stains (basophilic dyes) identified RNA in eukaryotic cells as mainly cytoplasmic with some presence also in the nucleus. The four nucleotides in RNA (C, A, G, and U [instead of T]) were known, but the repeating 5',3'-phosphate linkages between ribose moieties had only been settled in the early 1950s (Fig. 2-1; for review, see Brown and Todd 1955). The 1920s idea of Phoebus A. Levene (Levene and Bass 1932) that perhaps RNA was only a repeating tetranucleotide that served a "scaffolding" function was still current in some quarters. It was not even known whether RNA was strictly linear or might contain branched molecules. Daniel M. Brown and Sir Alexander R. Todd (1955) mention the possibility that such branchpoints could not be discounted. It was a quarter century later that such molecules were found when splicing was discovered in eukaryotes (see Chapter 4).

The path to uncovering different types of RNA that connected genes to proteins was due to the confluence of the following several disparate lines of research:

1. The earliest contribution came from virology. Purified plant viruses were found in 1937 (Bawden and Pirie 1937) to have ~3% RNA and no DNA. The presence of RNA in viruses provided a source of a pure RNA. Techniques for extracting RNA were very crude. However, Seymour Cohen and Wendell Stanley reported that tobacco mosaic virus (TMV) RNA

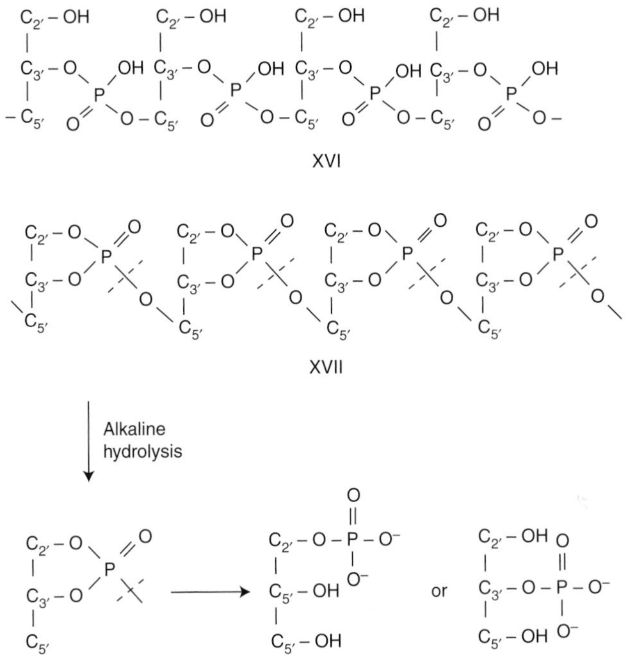

$C_{2'}$–OH $C_{2'}$–OH $C_{2'}$–OH $C_{2'}$–OH

$C_{3'}$–O, OH $C_{3'}$–O, OH $C_{3'}$–O, OH $C_{3'}$–O, OH
  P   P   P   P
–$C_{5'}$ O O–$C_{5'}$ O O–$C_{5'}$ O O–$C_{5'}$ O O–

XVI

$C_{2'}$–O, O $C_{2'}$–O, O $C_{2'}$–O, O $C_{2'}$–O, O
  P   P   P   P
$C_{3'}$–O, O $C_{3'}$–O, O $C_{3'}$–O, O $C_{3'}$–O, O
$C_{5'}$ $C_{5'}$ $C_{5'}$ $C_{5'}$

XVII

Alkaline hydrolysis

$C_{2'}$–O, O $C_{2'}$–O–P–O⁻ $C_{2'}$–OH O
  P | ||
$C_{3'}$–O $C_{5'}$–OH O⁻   or $C_{3'}$–O–P–O⁻
$C_{5'}$ $C_{5'}$–OH $C_{5'}$–OH O⁻

Figure 2-1. The 3′,5′ linkage of ribonucleotides in RNA, as shown in structure XVI, was proved in Sir Alexander Todd's laboratory (reviewed in Brown and Todd 1955). $C_{2'}$, $C_{3'}$, and $C_{5'}$ represent the carbons in ribose that have OH groups to which phosphate linkages occurred. Alkaline cleavage of RNA involves the formation of a cyclic intermediate, as depicted in structure XVII; resolution of this cyclic intermediate yields a mixture of 2′ and 3′ mononucleotides. However, the alkali-produced cyclic intermediate would be formed whether the linkage before cleavage was 5′–2′ or 5′–3′, and it required further chemistry by the Todd group to settle on the 5′–3′ linkage. The fact that DNA was already proven to be 3′–5′-linked was an important part of Todd's argument for the 3′–5′ linkage in RNA. Enzymatic degradation to mononucleotides, all 3′ phosphate, proved the 5′–3′ linkage (for review, see Heppel and Rabinowitz 1958). (Modified, with permission, from Brown and Todd 1955.)

preparations free of protein contained molecules as large as 300,000 Da, making clear that at least some RNA was very long (Cohen and Stanley 1942). In 1956–1957, TMV RNA was proven to contain the genetic information of the virus.

2. The discipline that came to be known as cell biology originally was limited to light microscopy and chemical stains to trace molecules in cells. After the end of World War II, cell fractionation techniques on animal tissues such as liver and pancreas and the use of the electron microscope vastly improved the knowledge about the location and biochemical

activities of cell constituents, including those related to protein synthesis. With better knowledge of cell structure, cytoplasmic fractions of liver cells containing different kinds of RNA molecules (transfer RNA [tRNA] and the RNA contained in ribosomes) were proven capable of incorporating amino acids into proteins.

3. Bacterial studies supported conclusions about tRNA and ribosomes in protein synthesis. But even more important, experiments concerning the selection of which proteins a bacterial cell produced—the question of *gene control*—captured the interest of the world of bacterial genetics. From these experiments came the first ideas about gene regulation that led to the proposal of the existence and then the discovery of messenger RNA (mRNA).

These several topics that brought RNA to prominence in the world of biochemistry and genetics are explored in this chapter.

RNA VIRUSES: RNA AS GENETIC MATERIAL

Viruses multiply by binding to, entering, and using the synthetic machinery of the host cells, a fact not fully appreciated in the mid 1950s. However, viruses clearly have "genetic" capacity because many different viruses of plants, animals, and bacteria always cause the production of more virus particles exactly like themselves with the same potential for causing symptoms. Plant virologists knew in the early 20th century that different strains (mutants) of the same type of virus could faithfully cause the production of the particular strain that originally caused the infection (Fig. 2-2).

The first viruses to be highly purified in the 1930s and 1940s were plant viruses, and they consisted of protein and RNA with no detectable DNA. The presence of the RNA in these viruses was initially ignored because it was still believed that only proteins were complicated enough to have "genetic" function.

In 1935, Wendell Stanley purified TMV, a rod-shaped plant virus that was only later found to contain RNA (Stanley 1935). The pure TMV formed crystals. As we noted in Chapter 1, James B. Sumner had earlier crystallized the first enzymatic protein, urease, in 1926 (Sumner 1926), and John Northrop crystallized pepsin in 1930 (Northrop 1930). Stanley's feat in purifying and crystallizing TMV was taken to mean that the virus was entirely protein. In fact, Stanley's paper was entitled "Isolation of a Crystalline Protein Possessing the Properties of Tobacco Mosaic Virus" (Stanley 1935). Stanley, Northrop, and Sumner shared the 1946 Nobel Prize in Chemistry for these purifications and crystallizations. It was not until 2 years after Stanley's report of TMV

2519 R

Ni 2519

2519 R

A14

Figure 2-2. Two different strains of TMV regularly produce different-sized lesions. The viruses used in the experiment were the wild-type strain A14; the mutant strain Ni2519, which grows poorly; and the revertant 2519R, which grows more like wild type. Leaves were inoculated separately on each half-leaf. The stain inoculated in each case is shown in the figure. (*Bottom*) Half-leaf inoculated with A14 and half-leaf inoculated with the revertant 2519R are both seen to form fairly large lesions on each half of the leaf. (*Top*) Infection on the lower half of the leaf with the mutant Ni2519 did not form visible lesions. All of these strains, like many others, always cause distinctive characteristic lesions when transferred from plant to plant. (Reprinted, with permission, from Zimmern and Hunter 1983 [©Macmillan].)

crystals that F.C. Bawden and N.W. Pirie reported that RNA made up 3% of TMV (Bawden and Pirie 1937). This crucial fact was still overlooked for some time.

Genetically different strains of TMV were recognized because they produced different lesions on tobacco leaves (Fig. 2-2). The Stanley group cultivated and purified these different strains. Injection of different strains into rabbits produced antibodies that could distinguish among the different strains. It was possible (but laborious) to analyze the proportion of each amino acid in the proteins of different strains. C. Arthur Knight, a collaborator of Stanley, found small differences in the total amino acid distribution of the TMV coat protein from different strains (Black and Knight 1953). However, he found no differences in the average content of RNA nucleotides (Knight 1952). Stanley held firmly to the notion that the "specific" part of the virus, the protein, had the genetic potential of the virus.

RNA Is the Genetic Material in Tobacco Mosaic Virus

In the mid 1950s, the importance of RNA in viruses came to the fore. Two groups of investigators proved the genetic role of TMV RNA in specifying TMV protein. By implication, these results raised the possibility of an equivalent role of RNA in cellular protein synthesis, although the idea did not immediately take hold.

Heinz Fraenkel-Conrat, a refugee German chemist, had joined the Stanley group in 1952 at the Virus Laboratory at the University of California at Berkeley. Fraenkel-Conrat's extensive experience included 1 year (1951–1952) in Fred Sanger's laboratory in England, where he learned to work with proteins. By 1955, Fraenkel-Conrat had succeeded in separating nondenatured protein from TMV, and with the use of the detergent sodium dodecyl sulfate (SDS), he obtained fairly high-molecular-weight RNA from TMV. Together with Robley Williams, a senior member of the Virus Laboratory and an expert electron microscopist, it was found that the separated TMV protein and RNA would spontaneously re-form long, rod-like particles with the same diameter as the virus (Fraenkel-Conrat and Williams 1955). Most of the particles were not as long as the longest virus particles, but a few were (Fig. 2-3). Remarkably, a small fraction of the infectivity of the virus could be restored. The protein by itself formed rods of imprecise length that were not infectious. Thus, the RNA was at least *required* for infectivity.

A group of virologists in Germany headed by Gerhard Schramm had been working on plant viruses, including TMV, before and even during World War II. After returning from military service (reportedly as a member of the SS [Schutzstaffel, the police wing of the Nazi party]) (Deichmann 1999), Schramm resumed work on viruses with young collaborators. Schramm and Alfred Gierer, one of these collaborators, succeeded in obtaining TMV RNA completely free of protein by exposing a suspension of the virus in water to phenol. The virus protein was denatured at the phenol-water interface. Centrifugation separated the phenol and water phases. The pure RNA remained in the aqueous phase and could be collected by precipitation with ethanol.

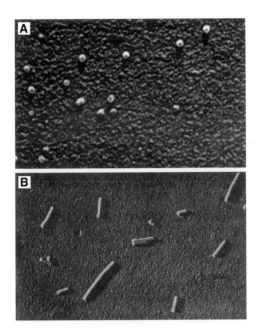

Figure 2-3. Reconstitution of TMV from protein and RNA. (*A*) Electron micrograph of TMV protein. (*B*) Isolated TMV RNA and isolated TMV protein can recombine. Occasional long particles are the same length as TMV. (Reprinted from Fraenkel-Conrat and Williams 1955.)

When pure RNA was applied to scarified (scratched) tobacco leaves, infections occurred, although with less efficiency than with the whole virus (Table 2-1). (Scratching of the leaf surface, which in nature is often accomplished through punctures introduced by insects, is necessary for the virus or the RNA to gain access to underlying cells.) Trace amounts of ribonuclease had no effect on the whole virus but completely killed the infectivity of the pure RNA.

Table 2-1. Phenol-extracted RNA from TMV proved to be infectious (indicated by numbers of lesions per 30 leaves)

	Ribonucleic acid	Tobacco mosaic virus
Normal	488	629
With normal serum	180	117
With antiserum	145	0
With ribonuclease	0	473
After ultracentrifugation	367	31
After 48 hr at 20°C	2	130

Antiserum blocked whole-virus infection but not RNA infection, whereas the opposite was true for ribonuclease. Total infectivity of RNA was obviously less: 10 mg/mL RNA and 1–27 mg/mL vines were used. (Reproduced, with permission, from Gierer and Schramm 1956 [©Macmillan].)

Moreover, the virus produced by infectious RNA was the normal, protein-coated, rod-shaped virus from which the RNA came (Gierer and Schramm 1956).

Fraenkel-Conrat and colleagues took the next and definitive step in proving the genetic role of RNA in TMV infection. They took the phenol-extracted RNA from strain HR (which gives only small local lesions and not a systemic infection) and the protein coat from regular TMV (which gives a systemic infection) and reconstituted fully infectious virus that produced progeny virus with exclusively HR characteristics. The reverse experiment with regular TMV RNA and HR protein produced regular TMV (Fraenkel-Conrat and Singer 1957; Fraenkel-Conrat et al. 1957a,b).

Clearly, the viral RNA carried the genetic information for the virus protein. No DNA was involved. However, it still remained unclear how the "genetic" and protein-coding functions of RNA could be realized in a cell.

Before leaving the plant viruses, an often overlooked 1948 paper should be mentioned that emphasized the importance of timing in research discoveries. Roy Markham and colleagues (Markham et al. 1948) at the Plant Virus Research Unit of The Molteno Institute at Cambridge University found two types of particles in plants infected with turnip yellow mosaic (TYM) virus. Both a "top" component and a "bottom" component from a centrifugal step of a preparation could crystallize, and both appeared to be identical in the electron microscope. However, only the "bottom" component contained RNA, and only the "bottom" component was infectious. However, TYM virus is a "spherical" virus that cannot be taken apart and reassociated like TMV, and the importance of Markham's finding simply lay fallow until the mid 1950s.

RNAs in Bacteriophage Infection

Elliot Volkin and Lazarus Astrachan reported experiments in 1956 that were ahead of their time in so far as their interpretation went, but they were also very important in the ultimate recognition of mRNA (Volkin and Astrachan 1956a,b). In the early and mid 1950s, the biochemistry of cells infected with bacteriophage was actively pursued in a number of laboratories. One early result was that when virulent ("T-even") bacteriophages T2, T4, or T6 infected *Escherichia coli* cells, the "total RNA synthesis" in the infected cells quickly came to a halt. The "synthesis of RNA" was measured as a continuing increase in total RNA in the uninfected growing cultures, and it ceased following infection. As would be appreciated only later, the increase represented mainly ribosomal RNA (rRNA).

Volkin and Astrachan found, however, that such bacteriophage-infected cells were still capable of taking up radioactive phosphate ($^{32}PO_4$) (Fig. 2-4)

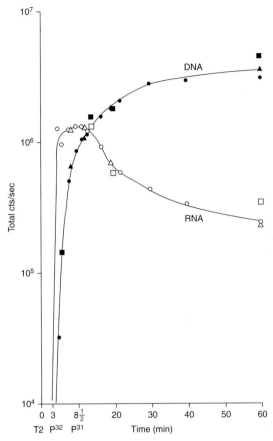

Figure 2-4. Unstable RNA in bacteriophage-infected cells. *E. coli* cells infected with T2 bacteriophage were exposed to $^{32}PO_4$ from 3 to 8.5 min after infection; samples were taken and RNA nucleotides were extracted by alkali treatment of perchloric acid precipitate, leaving DNA behind. Radioactivity accumulation in RNA was very prompt ($\sim$1–2 min after labeling), although no increase in total RNA occurred (data not shown). Phage DNA is formed mainly between 5 and 10 min after infection, causing uptake of $^{32}PO_4$ into DNA. The three different symbols represent three independent experiments. (Solid symbols) Radioactivity in DNA, (open symbols) radioactivity in RNA. (Redrawn, with permission, from Volkin and Astrachan 1957 [©The Johns Hopkins Press].)

and incorporating it into some type of unidentified RNA, although the total amount of RNA in the infected cells was constant. They concluded that there was an unstable RNA fraction that was synthesized rapidly and degraded rapidly so as not to change the total RNA content in the infected cell. At this time, no unstable macromolecule had ever been observed in cells. Most important, the percentage of labeled phosphate in the four RNA nucleotides from bacteriophage-infected cells (C17, A29, U29, and G25) was very different from the percentage in growing bacteria (C23, A23, U22, G32). And the distribution of the bases in the rapidly labeled RNA fraction from the infected cells was very similar to the distribution of C, A, U, and G in the bacteriophage DNA (substituting U for T and C for hydroxymethylcytosine, which replaces C in T-even phage DNA).

A technical note is appropriate here to explain how Volkin and Astrachan determined the "base composition" of the RNA from phage-infected cells. First of all, as was common at the time, they extracted the RNA from the total cell macromolecule precipitate with alkali, which degrades the RNA into its constituent mononucleotides. If such a $^{32}PO_4$-labeled RNA sample is examined for base distribution, the following occurs, although it was not yet recognized in the 1950s: A $^{32}PO_4$ is incorporated into a growing chain as a 5' nucleoside monophosphate derived from the α phosphate of a 5' nucleoside triphosphate. Alkaline cleavage transfers the incorporated α phosphate to either the 2' or 3' position of the ribose in the preceding nucleotide in the chain. However, regardless of which 5' triphosphate brought the ^{32}P into the RNA, the distribution of label in the four nucleotides of long molecules averages out to give an accurate picture of the "base composition" of the RNA, because there is no restriction of neighboring nucleotides. This type of analysis proved to be very important in many later experiments.

Volkin and Astrachan extended this experiment to T7 bacteriophage that had a distinguishably different base distribution and came to the same result: The RNA base distribution produced in the T7-infected cells resembled that of the T7 DNA (Volkin and Astrachan 1957). They were careful in their conclusions about what these findings meant but were very definite about the fact that they had described a metabolically active (made rapidly and destroyed rapidly) bacteriophage-related RNA.

PROTEIN SYNTHESIS: WHERE AND HOW DOES IT OCCUR?

Aside from the largest intracellular structures (mitochondria, the nucleus, and its chromosomes), functional divisions (organelles) within any eukaryotic cell type were not appreciated until after the end of World War II.

Two changes in the study of cell constituents led to the discovery of the different roles of RNA in protein synthesis. The key events were biochemically effective mammalian cell fractionation to remove cytoplasmic material from the nucleus, coupled with electron microscopy of subcellular structures in the cytoplasm that were found to contain RNA. Finally, biochemists found that certain of these subcellular structures could make protein in the test tube.

Brachet and Caspersson: Relating Cytoplasmic RNA to Protein Synthesis

Using only staining techniques, Jean Brachet, a Belgian embryologist, made the first suggestions of a relationship between cytoplasmic RNA and protein synthesis by as early as 1941 (Brachet 1941). Two of his observations are most often quoted. He noted the correlation of the very intense basophilic (RNA) staining of tissues such as liver and pancreas that by this time were thought by physiologists to make large amounts of protein for export. Torbjörn Caspersson had also called attention to the rich cytoplasmic basophilia of cells in onion root tips and embryonic cells, both of which contain rapidly growing cells and therefore synthesize proteins rapidly (for review, see Brachet 1955). Perhaps the most important point Brachet established was that in both enucleated amoebae and enucleated *Acetabularia* (a giant alga), protein synthesis could continue, in the latter case for days to weeks (Brachet and Chantrenne 1951). Virtually all subsequent investigators who uncovered the roles of RNA in protein synthesis referred back to the work of Brachet and Caspersson. But there was no hint in these early days of a mechanistic connection between RNA and protein formation. Cytologists were limited to examining cells after they were stained with dyes, in an effort to discern something about the internal biochemistry of cells. Nucleic acids were stained with basophilic dyes. One such, the Feulgen stain, introduced in 1914 by Robert Feulgen (1914), specifically stained nuclei. But it was not until 1930, when the Levene laboratory (Levene and Bass 1931) showed that deoxyribose was what the Feulgen stain recognized, that it became customary to think of a separation between DNA in the nucleus and the other prominent basophilic staining substance, RNA, which was mostly found in the cytoplasm. From this era of research, only a single ray of light emerged that was pertinent to solving how proteins were made.

Cell Fractionation and the Electron Microscope: Albert Claude, George Palade, and Keith Porter

In the late 1930s and early 1940s, Albert Claude (a Belgian native), at The Rockefeller Institute in New York, developed a technique of rupturing cells from mammalian tissues and separating nuclei and various cytoplasmic fractions by differential centrifugation (Claude 1941, 1948). Claude made his extracts by grinding tissues in a meat grinder or with fine sand, and nuclei largely survived. His earliest separations of cytoplasmic material, described in 1941, revealed that mitochondria (diagnosed by staining with the vital dye methyl green) were the largest organelles and that various other fractions could be collected at increasing times (forces) of centrifugation. He ultimately named the next-largest structures *microsomes*, not distinguishing clearly between the apparently different particle sizes. He found later that the microsomes were rich in phospholipids (membranes) and particulate material that contained RNA and proteins. The nucleic acid in the particles was entirely RNA (Feulgen negative).

In the early 1950s, George Palade and Keith Porter, also at The Rockefeller Institute, made dramatic improvements in methods of cell fixation and observation by electron microscopy of very thin tissue sections. This allowed visualization of cytoplasmic structures never before recognized, including very regularly sized, small, round granules both inside folded cell membranes (later named *endoplasmic reticulum*) in cell sections and in the microsomes of cell fractions. The particles ultimately proved to be rich in RNA.

Palade and Philip Siekevitz (1956) made an important contribution in characterizing these RNA-containing particles. First of all, with improved electron microscopic techniques, the small, dense, RNA-containing granules in the membrane-associated structures both *inside* cells and in microsome fractions were carefully measured to be very regular in size (100–150 Å in diameter) (Fig. 2-5). Moreover, the RNA-containing particles could be released intact in free form by treating microsomes with a detergent (deoxycholate) that dissolved membranes. Virtually all of the microsome-associated RNA occurred in these small particles.

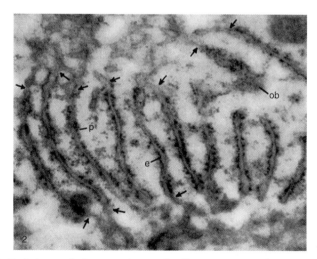

Figure 2-5. Early (∼1955) electron micrograph of hepatocyte cytoplasm showing stacks of "rough" endoplasmic reticulum. Dark-staining particles labeled "p" are ribosomes whose products would be secreted into the cisternae (spaces between the flattened membranous discs). (Reprinted, with permission, from Palade and Siekevitz 1956.)

Protein Synthesis in Cell-Free Extracts: The Zamecnik Influence

Radioactive amino acids became available soon after World War II, and a number of laboratories both in the United States and Great Britain found in 1950–1952 that microsomes prepared by Claude's techniques had a higher capacity than other cytoplasmic fractions to incorporate radioactive amino acid into proteins.

Paul Zamecnik and colleagues at Massachusetts General Hospital in Boston became the leading laboratory studying amino acid incorporation into microsomes, both in broken cell preparations and in the animal. The RNP (ribonucleoprotein) portion of the microsomes released by deoxycholate was more than five times as active (per milligram of protein) as any other cell fraction in incorporating amino acids into protein. This was true both in animal (Fig. 2-6) and cell-free preparations (Keller et al. 1954; Littlefield et al. 1955).

Siekevitz, who began in the Zamecnik group, made an important discovery early in the course of cell-free amino acid incorporation (Siekevitz and Zamecnik 1951). The original liver fractionation procedure of Claude did not clearly separate mitochondria and microsomes. If the two were separated by more careful differential centrifugation, Siekevitz found that

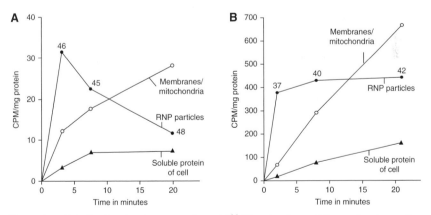

Figure 2-6. Relative in vivo incorporation of [^{14}C]leucine into different fractions of rat liver. (*A*) Intravenous injection of radioactive leucine as a pulse label (trace amount of [^{14}C]leucine). (*B*) Long label (80 times the total amount of [^{14}C]leucine; thus linear total incorporation continued for >1 h). In both cases, RNP particles in microsomes were clearly labeled before other cell fractions. Labeled protein then appeared in other parts of the cell. The numbers above each sample in the RNP particle fractions represent the percentage of RNA in the fraction. (CPM) Counts per minute. (Modified, with permission, from Littlefield et al. 1955 [©American Society for Biochemistry and Molecular Biology].)

both were required for maximum amino acid incorporation. By this time, mitochondria were suspected of being "energy factories." Therefore, a source of energy was required for cell-free protein synthesis. As improved fractionation methods removed mitochondria more efficiently, the Zamecnik group later used an ATP-regenerating system instead of mitochondria and discovered a requirement not only for adenosine triphosphate (ATP) but also for guanosine triphosphate (GTP) (Zamecnik and Keller 1954). The need for ATP for active amino acid incorporation led Siekevitz and Zamecnik to suggest that the idea expressed by Fritz Lipmann, the famous German biochemist who settled in America, might have merit. It was Lipmann who first established ATP as the chief internal energy source in cells and suggested that energy would be required to form a peptide bond (Lipmann 1949, 1953; Siekevitz 1952).

Ribosomes Get a Name

In the mid 1950s, broken cell preparations with bacteria were also shown to be capable of incorporating amino acids and to contain similar dense RNA granules. At a 1958 meeting, Richard Roberts of The Carnegie Institution of Washington suggested that the ubiquitous, small, dense, RNA-containing granules be called *ribosomes*, and the new name stuck (Roberts 1958). The Roberts group also conducted definitive experiments showing ribosomes to be the site of protein synthesis in growing bacterial cells (Fig. 2-7). They labeled bacteria (*E. coli*) for only a few seconds with ^{35}S sulfate, which was enzymatically converted into sulfur-containing amino acids. They then broke the cells and, using zonal sedimentation, a technique that separates particles on the basis of size, found that the great majority of the newly labeled growing protein chains was attached to structures that sedimented like ribosomes (i.e., contained most of the ultraviolet [UV]-absorbing capacity of the extract). In fact, the "ribosome-associated" radioactive amino acids were associated with UV-absorbing particles equal to and larger than (i.e., sedimenting faster than) single ribosomes. These were probably polyribosomes, a point not recognized to be important until several years later.

When unlabeled sulfate and unlabeled sulfur-containing amino acids were added, the labeling of protein chains was abruptly halted, and after a few minutes no radioactivity remained with the ribosomes (Fig. 2-7). These experiments produced strong supporting evidence that the ribosome was the site of protein synthesis and that new chains were completed in a few minutes. The newly named ribosome was therefore accorded the honor of being accepted as the site of protein synthesis in all types of cells.

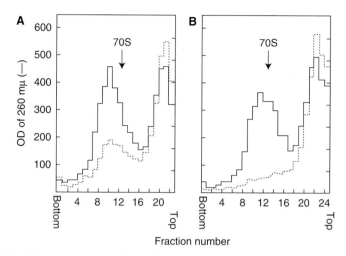

Figure 2-7. Sedimentation analysis of disrupted *E. coli* locates protein incorporation on ribosomes. (*A*) $^{35}SO_4^-$ was added to a growing culture of sulfur-depleted *E. coli* to label sulfur-containing amino acids. Fifteen seconds later, the cells were chilled, and cell extracts were prepared and centrifuged through a sucrose gradient. The samples were assayed for UV absorption ([unbroken line] OD_{260nm}, mainly due to total RNA in the cell, which is ~85% rRNA) and radioactivity ([broken line] cpm or counts per minute). Single ribosomes ($\rightarrow$) sediment at 70S. (*B*) As in *A*, but chased with unlabeled sulfate for 2 min before analysis. Note the transfer of radioactivity from the 70S–85S region to the slowly sedimenting region. (Modified, with permission, from McQuillen et al. 1959.)

The Template Idea

Despite this progress in locating the ribosomes as the structures in/on which protein synthesis occurs, what actually happened as amino acids were incorporated into proteins was still very much unknown by the late 1950s. These were the first several years after the Watson-Crick discovery focused attention on the fact that somehow DNA as the genetic material must control the order of amino acids in proteins. Much discussed was the idea of a "template." But this notion was ill-defined and meant different things to different people. Most hypotheses included nucleic acids (DNA, RNA, or both), but which one furnished the "template" and how it was furnished were subject to controversy. The earliest ideas (Caspersson 1947; Gamov 1954) envisioned protein synthesis on the surface of DNA in the nucleus. Immunologists conjectured that antibodies, proteins with exquisite specificity, had to be formed around the antigen in a "complementary" fashion (for review, see Frank Putnam in a memoir to the biochemist/immunologist Felix Haurowitz [Putnam 1994]). Physical chemists suggested that nucleic acids could, by unspecified affinities similar to molecules in a crystal, form a template that could directly attract specific amino acids. Aside from how any template was constructed, there was controversy about whether protein synthesis was stepwise or whether peptides were first formed and then aligned at the "template" and assembled en bloc, as if by a stamping machine.

The idea of peptide intermediates gained some acceptance because of one of its stout defenders, Christian Anfinsen, together with his colleague Daniel Steinberg at The National Institutes of Health. Anfinsen later won a Nobel Prize for describing the spontaneous refolding of the short enzyme ribonuclease into an active state from the denatured state. Steinberg went on to do important work on serum proteins that carry lipids and their relationship to cardiovascular disease (Steinberg 1993). The experiments that Anfinsen and Steinberg (and other collaborators) reported attempted to determine whether a radioactive amino acid (e.g., leucine) entered a particular secreted protein, ovalbumin, equivalently in all sites within a newly synthesized chain. Their results showed severalfold differences in use of the labeled amino acid in different sections of ovalbumin. They concluded that synthesis of the molecule occurred by linking blocks of peptide intermediates whose pool sizes differed (Steinberg and Anfinsen 1952; Steinberg et al. 1956). Other investigators using a similar approach (e.g., Sidney Velick at Washington University in St. Louis [Heimberg and Velick 1954; Simpson and Velick 1954]) came to the conclusion that the newly incorporated amino acids in new proteins were evenly distributed. Because stepwise synthesis of proteins was later proven to be correct, it is unclear how the extremely carefully obtained but misleading results from the Anfinsen laboratory occurred.

In the early literature on templates for protein synthesis that involved nucleic acids, the investigator who came closest to proposing a mechanism resembling the eventual truth was Alexander Dounce, professor of biochemistry at Rochester University. He argued in his initial brief article (Dounce 1952) that Brachet's idea of a connection between cytoplasmic RNA and protein synthesis was compelling. In a second short article (Dounce 1953), he states "If we accept for the purposes of argument . . . that genes are composed of deoxyribonucleic acid, then it could conceivably happen that the deoxyribonucleic acid gene molecules would act as templates for ribonucleic acid synthesis, and that the ribonucleic acids synthesized on the gene templates would then in turn become templates for protein synthesis. . . ." He closes this short note by saying "The template hypothesis constructed by me was published mainly in the hope of promoting thinking and experimentation. . . . It will be gratifying if this aim is achieved." That hope was surely realized (but not by Dounce), although almost none of the workers who solved the problem refer to Dounce in their original papers.

Discovery of tRNA and Amino Acid–Activating Enzymes

A big advance in the discovery of the specific ordered steps that occur during protein synthesis and in the clear identification of the first discrete species of RNA involved in protein synthesis was made by biochemists in the Zamecnik group in 1956–1957. By this time, all investigators using in vitro incorporation of labeled amino acids agreed that the ribosome did contain RNA and was the site of protein synthesis. Yet no specific RNA class that took part in protein synthesis had been identified. The break achieved by Zamecnik and collaborators concerned an event in protein synthesis that did *not* occur on the ribosome.

Ribosomes were first recognized by their property of being large enough to be sedimented into a pellet by high-speed ultracentrifugation (i.e., at 100,000g). For the pelleted ribosome-containing fraction to perform incorporation of amino acids, however, the nonsedimentable "soluble" cell fraction containing an ATP-generating system was also required. Elizabeth Keller and Mahlon Hoagland in the Zamecnik laboratory played a major part in determining the several roles of the soluble fraction (Hoagland et al. 1956). First, the aforementioned requirement for ATP could be satisfied with an ATP-generating system (usually creatine phosphate and creatine phosphokinase). In addition, when the pH of the soluble portion of the cell extracts was adjusted to a slightly acidic condition, pH 5.1, a precipitate formed. In this precipitate were enzymes that used ATP to "activate" amino acids (Hoagland 1955; Hoagland et al. 1956). The initial reaction to detect activation either used an amino acid–dependent ATP $\rightarrow$ PP (pyrophosphate) exchange or hydroxylamine (NH_2OH) to trap amino acid hydroxamates created at the expense of ATP. (For a detailed discussion of the discovery of amino acid activation, see Rheinberger 1997.) The Zamecnik group—with Hoagland leading the way—then found there to be an *acceptor* in the pH 5 supernatant (in place of the artificial acceptor hydroxylamine) (Hoagland et al. 1957, 1958). But this acceptor was not a protein. It was heat resistant and could be separated from protein in the "pH 5 supernatant" by the phenol procedure that Gierer and Schramm had described in 1956. (Robert Holley, the biochemist who eventually sequenced the first tRNA, showed that the reactions that Hoagland was studying were inhibited by ribonuclease [Holley 1957].) The acceptor of activated amino acids was clearly an RNA molecule present in the 100,000g supernatant. Therefore, it was called *soluble RNA*. There was no such acceptor capacity in RNA extracted from microsomes.

Most important, when this RNA fraction was first incubated with ("loaded with") a labeled amino acid and then added to the cell-free amino acid incorporation system, the labeled amino acid was promptly incorporated

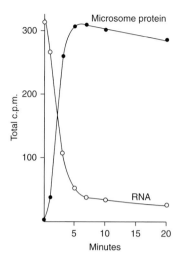

Figure 2-8. Incorporation of labeled amino acid from soluble RNA. A pH 5–soluble RNA fraction was loaded with radioactive leucine by "pH 5" enzyme activity. The labeled ^{14}C leucine/soluble RNA complex was then added to a cell-free microsomal protein synthesis system from rat liver. The label was promptly transferred to microsomal protein. (Redrawn, with permission, from Hoagland et al. 1958 [©American Society for Biochemistry and Molecular Biology].)

into protein (Fig. 2-8) (Hoagland et al. 1957, 1958). This pH 5 soluble RNA was the first specific functional class of cellular RNA to be identified, although the exact size or other characteristics were not identified at this point. It was only known to have a very small sedimentation constant. (Only in the early 1960s, after the discovery of messenger RNA, did the terms *transfer RNA* and *ribosomal RNA* come into common use.)

Paul Berg soon found that a purified valine-activating enzyme from *E. coli* was specific for attaching valine to the soluble RNA. The total cell extract contained ATP-dependent enzymes that could activate all of the other amino acids he tested. He further showed that, during purification, different valine- and methionine-activating activities could be separated and that each attached its specific activated amino acid to the 2′ or 3′ hydroxyl of the terminal adenosine 5′ phosphate (AMP) in the soluble RNA. It appeared likely that individual enzymes existed for activating each amino acid (Berg 1956; Berg and Ofengand 1958).

By the late 1950s, the stage was set to come to grips with how soluble RNA (tRNA) could, together with the ribosome, actually make proteins. A big question remained: What furnished the "information" to order the amino acids properly? Even with the lesson that virus RNA could somehow specify its own specific protein, confusion still reigned about the detailed mechanism of protein synthesis and the "informational" role of RNA. The universal working hypothesis at the time was that the RNA in the ribosomes did the job.

Straws in the Wind, but Still No Idea of mRNA

Even Francis Crick, in a very famous 1958 paper delivered at the Society for Experimental Biology in Britain in 1957, did not get everything right, although this paper brought together the threads of the argument—DNA to RNA to protein (Crick 1958). He cited Vernon Ingram's 1957 work on glutamic acid replacement by valine in sickle cell hemoglobin (Ingram 1957) (see Chapter 3, p. 104) as proof that, in general, the sequence of nucleotides in DNA would specify individual amino acids in protein. He summarized his earlier prediction that an "adaptor" molecule would be required to go from the four-base language of DNA (or RNA) to the 20-amino-acid language of protein.

This prediction had been made in privately circulated letters (papers) before the discovery of tRNA (Crick 1988, p. 95). Crick's theory of the "adaptor" is widely regarded as one of the few strictly theoretical ideas in molecular biology that was later proven by experiments to be true. He originally proposed that the adaptor might be a trinucleotide (of RNA) that carried amino acids. He suggested, as had many others, that the ribosome with its known RNA content somehow contained the genetic message copied from DNA. Base pairing between the adaptor and the "informational rRNA" would select the right amino acid to add to a growing polypeptide chain.

By the time of the 1957 conference, Crick accepted the possibility that the newly discovered *soluble RNA* could perform the adaptor function. But, like all others at the time, he continued to assume that the rRNA would have the instructive role in getting the order of the amino acids correct.

In the 1958 paper, Crick raised questions mainly on structural grounds about how tRNA and rRNA could perform the task of protein synthesis. He pondered the difficulties imposed if the RNA in ribosomes had to encode many different proteins and somehow still fold to create such a regular structure as the ribosome. Furthermore, he puzzled over how a molecule the size of tRNA ($\sim$100 nucleotides long) could deliver its amino acid cargo possibly deep into a ribosome at the site of protein synthesis. He continued to speculate that an amino acid bound, e.g., to a trinucleotide could more easily find its way into the interstices of the ribosome. Therefore, perhaps tRNAs were able to give up a trinucleotide during their participation in protein synthesis. He came close but did not suggest that a loop on the tRNA might do just what he wanted. Finally, he did not suggest that yet another RNA, an mRNA, would solve the problem of carrying information.

Other currents of thought were beginning to question not the role of the ribosome as a platform for protein synthesis but, rather, the role of rRNA in possessing the necessary instructional capacity to specify different proteins.

Erwin Chargaff had originally noticed that although the base-pairing principle $A = T$ and $G = C$ was true for a wide range of DNAs from different organisms, the $(A + T)/(G + C)$ content of different organisms varied widely (Chargaff 1951). Andrei Belozersky and Alexander Spirin in 1958 (Belozersky and Spirin 1958) reported that, just in different species of bacteria, the $G + C$ content of DNA changed greatly, from 30% to 74%. However, the distribution of bases in total bacterial cell RNA was fairly constant, with the $G + C$ content ranging only from 52% to 59% of the total base distribution. If most of the DNA encoded information to make protein, why was there no match in the base distribution between the DNA and total bacterial cell RNA? Furthermore, the rRNA, along with the much shorter tRNA, comprised the vast majority of the RNA of the cell and was quite uniform in size (Hall and Doty 1959; Kurland 1960).

Interest returned to the Volkin and Astrachan results (1956a,b) discussed earlier (p. 64–65): Total RNA synthesis (meaning mostly rRNA and tRNA) stopped in *E. coli* infected with T2 or T7 bacteriophages, but incorporation of $^{32}PO_4$ into RNA continued. Most important, the radioactively labeled RNA had a nucleotide distribution like that of T2 or T7 bacteriophage DNA, not like that of *E. coli*.

Finally, in experiments reported in 1960, Sol Spiegelman and colleagues (including Masayasu Nomura and Ben Hall) at the University of Illinois (Nomura et al. 1960) found that the briefly labeled RNA in T-even-infected bacteria (the Volkin-Astrachan-type RNA) was largely associated with the sedimentable ribosomal fraction of infected *E. coli* extracts. Moreover, preexisting *E. coli* RNA, presumably mostly rRNA, had uniform sizes (16S and 23S) in sucrose gradient sedimentation analysis, whereas the newly labeled T2 RNA averaged ~12S and had a broader sedimentation profile.

THE JACOB-MONOD HYPOTHESIS: OPERONS AND mRNA

At this point (~1958–1960) in the understanding of protein synthesis, biochemists had addressed only the *mechanics* of this complicated process. Assuredly, progress had been made. Proteins were linear arrays of amino acids (at least before they folded into three-dimensional [3D] structures) (Pauling and Corey 1951; Pauling et al. 1951). Each protein had a specific amino acid order (Sanger and Tuppy 1951a,b). Ribosomes contained RNA, and tRNA carried to ribosomes activated amino acids that were then added to a growing polypeptide chain. Furthermore, virus RNA somehow had a "specific" informational role in an infected cell (Gierer and Schramm 1956). As crucial and important as each of these experiments and conclusions was, most scientists in the 1953–1960 era had not concentrated on the ultimate question that lies beyond simply how to string amino acids together: How does a cell *choose* which proteins to make? In the late 1940s and early 1950s, bacterial genetics became concerned with this problem of choice. What controlled the action of different genes at different times?

It was mainly from two masters of bacterial genetics and bacterial physiology, François Jacob and Jacques Monod, that answers about gene control arrived: There are genes to make protein—structural genes—and genes to regulate other genes—regulatory genes. The thrust of this idea led Jacob and Monod to see a need for an intermediate between the genes, on one hand, and the tRNA and ribosomes, on the other hand. They correctly proposed the existence of a "messenger" between DNA and the protein-synthesizing machinery. By 1959, the experiments and ideas of Jacob and Monod stimulated a search for mRNA molecules, and soon thereafter, mRNA was identified in bacteria.

Genetics of Gene Control: First Step to mRNA

Jacques Monod had come into science just before World War II. For his thesis work at The Sorbonne, he studied the growth and use of various sugars by *E. coli*. This topic was the suggestion of André Lwoff (already at The Pasteur Institute). Lwoff, in his own inimitable style, twice described his early interactions with Monod (Lwoff 1966, 2003).

When placed in a medium containing both glucose and other sugars, the bacteria first grew and exhausted the glucose, paused, and then "adaptively" used the second sugar—producing the famous "diauxie" or double growth curve (Fig. 2-9; Monod 1941, 2003). Lwoff suggested to Monod that he was observing what had been first called in a Swedish Ph.D. thesis "enzyme adaptation" (Karstrom 1939). According to Lwoff, he gave his copy of the thesis to Monod, never to see it again (Lwoff 2003). After the war, Monod returned to again study how adaptive enzyme synthesis might work. Monod showed in 1946 that the ability to use lactose was a stable genetic characteristic of *E. coli* and chose to concentrate on this system (Monod and Audureau 1946).

Lactose is a disaccharide of galactose and glucose (Fig. 2-10). When given this complex sugar as a carbon source, lactose-positive *E. coli* respond by producing β-galactosidase (β-gal), which splits the disaccharide. This allows already-present enzymes to convert galactose to glucose, which is metabolized to provide energy to the cell. Humans have a similar enzymatic capacity to use

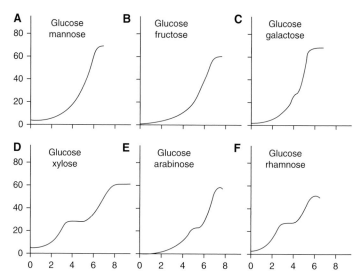

Figure 2-9. Growth of *E. coli* in the presence of different carbohydrate pairs serving as the only source of carbon in a synthetic medium. (*A*) Glucose and mannose, (*B*) glucose and fructose, (*C*) glucose and galactose, (*D*) glucose and xylose, (*E*) glucose and arabinose, and (*F*) glucose and rhamnose. Four of the six sugar pairs showed "diauxie," wherein glucose was exhausted before the second sugar was used. The *x* axis shows growth measured by an increase in light scattering (Klett readings); the *y* axis represents hours. (Redrawn, with permission, from Monod 2003 [©ASM Press].)

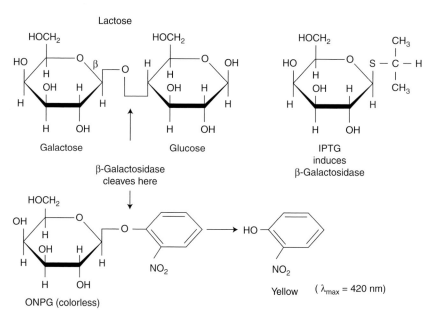

Figure 2-10. Structures of lactose, IPTG (isopropylthiogalactoside) and ONPG (orthoni-trophenyl-β-D-galactopyranoside), used to induce and assay β-galactosidase.

lactose, the major sugar in milk. Rare human mutations in any of the three enzymes necessary to convert galactose to glucose lead to galactosemia, which can result in severe neurological damage. Much more common in humans is the inability to efficiently split lactose, leading to the much milder condition of lactose intolerance, manifested as gastrointestinal discomfort after ingesting milk or milk products.

A very important characteristic of the manufacture of newly made β-gal enzyme by *E. coli* had early caught the attention of Monod. Within a few minutes of adding lactose by itself to a culture, the bacteria made β-gal at the maximum rate. Equally important, β-gal synthesis stopped within the same short few minutes when the inducer lactose was removed (Monod et al. 1952). Evidently, something crucial for new enzyme synthesis formed quickly and disappeared quickly when the enzyme was no longer needed.

A situation that originally seemed to be the inverse of β-gal induction involved enzymes required for the synthesis of the amino acids. For example, when tryptophan is in the medium, bacteria do not make the set of tryptophan-synthesizing enzymes. However, when no tryptophan is available, the bacteria quickly begin to make the complete set of enzymes that synthesize tryptophan. Other instances of quickly inducible enzymes to make amino acids in *E. coli* had come to light in the 1950s (Monod and Cohn 1952; Adelberg and Umbarger 1953; Cohen et al. 1953; Monod and Cohen-Bazire 1953; for review, see Cohen 1965). In all of these cases—as was the case with β-gal synthesis—there was a quick stopping and starting of enzyme synthesis. Thus, "on-off" switches for enzyme synthesis seemed to be common in *E. coli*. However, it was the intense study of the genetic regulatory mechanisms of β-gal formation that led the way in solving the gene control question.

Uncovering the *lac* Operon: Coordinating Gene Control

It is necessary in discussing how *E. coli* gene control works to start with what may seem to be a digression. After Joshua Lederberg's discovery of bacterial conjugation in 1947 (see Chapter 1, p. 40), particular strains that were specifically gifted at performing these transfers were selected. Such high-frequency transfer (Hfr) strains bore a "factor," an F factor or "sex factor," that integrated into the bacterial chromosome. Moreover, each of these special Hfr strains always started "injecting" genes of the male into the female from a particular point on the *E. coli* chromosome, continuing the transfer in an ordered fashion until the majority of the *E. coli* genes were transferred. Elie Wollman and François Jacob discovered that gene transfer by conjugation could be interrupted without harming the recipient cells by agitating the culture in a Waring blender. Jacob explains in *The Statue Within* (Jacob 1988, p. 279) that Elie had the idea of trying to break the mating cells apart *during* gene transfer. Jacob furnished the Waring blender, a kitchen adornment that he had brought for his wife, Lise, to use for pureeing food for their children. "Lise detested the machine," which then ended up stored in François's laboratory for use, as it turned out, in a most informative bacterial genetics experiment. Putting the mating culture in the blender at various times after beginning the experiment, they could then test which genes had been transferred to the genetically deficient female cells. The transfer was so precise (taking ∼100–110 min for the whole process) that the transfer of specific genes could be accurately timed. Using this technique, by the mid 1950s Wollman and Jacob had created a gene order for virtually the entire *E. coli* map, including the position of the *lac* locus (the genes involved in allowing lactose use) (for review, see Jacob and Wollman 1961).

A second technical advance became very important. "Gratuitous" inducers were discovered. The β-D-galactosyl linkage in lactose is mimicked in these compounds by a galactosyl linkage to a sulfur atom. These compounds easily enter the cell and, like lactose, induce the formation of β-gal, but they are not split by β-gal. One inducer in particular, IPTG (Fig. 2-10), became the favorite for inducing β-gal. The measurement of β-gal activity is performed with ONPG (Fig. 2-10), which is split by β-gal, giving a very stable yellow color for analysis.

In 1954, with knowledge of gratuitous inducers, a second gene was discovered for a "permease" that allows *E. coli* to make use of lactose. Using radioactive [^{14}C]thiomethyl-β-D-galactoside (TMG), also a gratuitous inducer in which an —S—CH$_3$ replaces the S-isopropyl group on IPTG (Fig. 2-10), Monod's group discovered that cells already induced by lactose transported the labeled TMG vastly faster than uninduced cells (Cohen and Monod 1957). Mutants were found that lacked this accelerated transport phenotype. However, with very high TMG (or IPTG) concentrations, these mutants could make β-gal. Thus, their β-gal–forming capability was intact, but they lacked the accumulation capacity. Monod reasoned that because this was a mutable function, it was a separate gene encoding a separate protein and it was *coordinately* induced by lactose. They named this function *galactoside permease* and suggested that it was a membrane protein. (Some years later, Eugene Kennedy identified and purified galactoside permease, which truly is a membrane protein [Fox and Kennedy 1965].)

Of great importance, in conjugation experiments, the gene for the permease was transferred very soon after the β-gal gene. The genes were therefore very close on the chromosome. These genes and their encoded proteins were not only induced together, but also their synthesis declined together after the removal of lactose. Finally, a third enzyme, thiogalactoside transacetylase, was also found to map to the *lac* locus and to be induced in parallel along with β-gal and the permease. The gene for this protein is listed as *a* in the group of lactose-induced genes: *z* for β-galactosidase (β-gal), *y* for permease, and *a* for transacetylase. The function of the a protein was not clear, although it may be a detoxification mechanism for the bacterium. Aside from being shown to be induced as part of the lactose group of genes, it has little further relevance to our discussion and most frequently was not assayed in experiments.

With this much known about enzyme induction, Jacob and Monod reviewed the types of mutants that affected β-gal, permease, and transacetylase formation (Jacob and Monod 1959). The collection included mutant cells that made no enzyme but could be induced to admit lactose. Other mutants could not admit lactose—permease mutants—but still could make the β-gal enzyme following induction by the gratuitous chemical inducers. Thus, in these two types of mutants, the genes encoding β-gal and permease had suffered independent mutations that did not affect each other.

The genes for β-gal and permease (and acetylase) were designated to be *structural genes* because they specified information for making a protein. In addition, other mutants made both β-gal and permease *without* an inducer. Such mutants suggested that there must be a "regulatory" gene that controlled induction. In extensive recombination analyses performed by conjugation, all of these regulatory mutants mapped to a single region very close to the structural genes at the *lac* locus. This gene for inducibility was named *i* (i^+ is the normal form of the gene). Cells that were mutant, i^-, were said to be *constitutive* (needed no inducer) for β-gal and permease synthesis. Thus, the effect of the wild-type *i* gene appeared to be that of a negative regulator. The i^+ gene product was normally present and functioned to prevent β-gal and permease synthesis. The inducer (lactose or IPTG) somehow overcame this negative effect.

The mutant collection contained many independent mutants in the *i* gene as well as in the genes for β-gal and permease. That is, all three genes occupied considerable "genetic space." The whole locus was diagrammed in order as *i*, *z*, *y*, and *a*. These facts were all clear by the mid 1950s, and although the *z*, *y*, and *a* genes clearly specified proteins, exactly how the *i* gene performed its negative regulatory function was a crucial point that was to be settled only after a few more years passed.

Prophage, Lysogeny, and λ Bacteriophage

A completely different line of experiments done by Jacob and Wollman (and later with other collaborators) uncovered another instance of apparently negative or repressive action for one gene in a completely different set of genes. The majority of lytic bacteriophages inject their DNA into the bacterial cell and in 10–15 min direct the production and release of ~100 new bacterio-phages. Temperate bacteriophages, however, can not only go through a repro-ductive cycle yielding a new burst of phage, but in a minority of infections, the phage DNA becomes a part of the bacterial chromosome *without* killing the cells. Such cells survive infection and continue to grow without any obvious difficulties. Moreover, they are immune to further infection by the bacterio-phage whose DNA they are now carrying.

This phenomenon had been well studied at The Pasteur Institute, having been observed in the 1930s. Elie Wollman's parents had studied what they called "hereditary contagion," a phenomenon of occasional spontaneous lysis that was an inherited property of certain bacterial cultures (Wollman and Wollman 1936, 1938).

Beginning in 1949, Andre Lwoff and his colleagues performed the experi-ments that explained this puzzle. First, they chose a culture of *Bacillus mega-terium* that exhibited this phenomenon of occasional lysis accompanied by production of bacteriophage. Using a micromanipulator and single cells in microdrops, they observed individual cells. Occasionally, one of these cells "disappeared." When this happened, the microdrop then contained ~100 new phage. These were newly produced phage particles because artificially breaking up even large numbers of the bacterial cells did not release any bacteriophage.

In an effort to get all of the cells in a culture to produce phage, a thin layer of a growing *B. megaterium* culture was irradiated with UV light. This resulted in virtually all of the cells lysing within 1 h, releasing ~100 new particles per cell. Lwoff wrote, "As far as I can remember this was the greatest thrill—molecular thrill—of my scientific career" (Lwoff 1966).

The "hidden" virus DNA was named *prophage*. Phages capable of this act of incorporation were called *temperate bacteriophages*. The introduction of the virus genome into the bacterial cell genome established a state called *lysogeny*, and the quiescent cell/virus state was called the *lysogenic state*.

One of the most widely studied temperate phages for *E. coli* is bacterio-phage λ, originally isolated in the Lederberg laboratory (actually by Esther Lederberg, Joshua Lederberg's wife) (Lederberg 1951). Two further facets of temperate phages and the lysogenic state are pertinent here. First, it is possible to study λ-phage genetics by recombination between mutant phage strains. Second, mutants of bacteriophage λ that can no longer lysogenize *E. coli* were isolated, and these mutations mapped to a single genetic locus on the bacteriophage DNA (Kaiser 1957; Kaiser and Jacob 1957). It seemed that one gene (ultimately referred to as the λ *repressor*) in the λ bacteriophage controlled the possibility of lysogeny. Thus, like the *i* gene in the *lac* region of the *E. coli* chromosome, the bacteriophage gene necessary for lysogeny acts negatively: There is a repressor gene whose expression is required for entry into the lysogenic state. When this gene is mutant, the phage is said to be *virulent*.

Jacob and Wollman (1956) performed a very illuminating experiment with *E. coli* cells that harbored λ bacteriophage. They mated a male (DNA donor) lysogenic cell carrying the λ genome with a female (DNA recipient) cell lacking the λ genome. When the segment of the male genome carrying λ DNA (previously mapped in the genome by recombination) entered the recipient female cell, the newly introduced λ virus genome was activated immediately. The production of ∼100 virus progeny resulted in killing the recipient cell. Jacob and Wollman called this "zygotic induction."

Because nonlysogenic "virulent" mutants of λ existed, they concluded that some product of the integrated DNA of the lysogenic bacteriophage λ had previously repressed the virus growth in the male cell. This product was presumably absent (or not made in sufficient quantity) during the initial stages of most infections by the virulent λ-phage genome. Thus, when the normal (wild-type) λ-phage DNA that resided in the male chromosome entered a new cell environment, it was freed of the negative repressing function, leading to new virus production.

The Jacob-Monod Alliance Breaks Open Negative Gene Regulation: The PaJaMa Experiment and Its Follow-Up

By 1956, the Jacob and Monod laboratories at The Pasteur Institute had clearly established that two regulatory genes existed in two different cases: (1) The *i* gene in the *E. coli* genome negatively controlled (repressed) β-gal, permease, and transacetylase synthesis, and this negative effect was overcome by inducers. (2) A repressor gene in λ bacteriophage was required for lysogeny, and some event such as UV irradiation, chemical treatment, or zygotic induction could release the genomically incorporated bacteriophage λ DNA to direct new bacteriophage synthesis. But a unifying idea for these observations was still lacking.

The genetic techniques perfected by Jacob and Wollman had been used to align the genes at the *lac* locus, but a more intensive use of genetic tools was required to get a better understanding of how gene function was controlled. As Jacob phrased it in his autobiographical memoir *The Statue Within*, "Jacques and I decided, therefore, to work together more closely on the study of his system: the metabolism of lactose, milk sugar, in the colon bacillus" (Jacob 1988, p. 290).

In the fall of 1957, the Paris group was joined by a visitor, Arthur Pardee, on a sabbatical from his teaching and research at the University of California at Berkeley. Pardee had been working on bacterial gene induction in parallel with the Paris labs for some years, quite often lagging a bit behind the flamboyant and charismatic Paris duo. Pardee was (and is, at this writing) an excellent experimentalist with—Jacob says—"an original mind" and "an air of being a good little boy that hid a remarkable experimenter" (Jacob 1988, p. 291).

Soon after Pardee arrived, an ingenious experimental attack on the question of "gene control" was designed. The first experiment that Pardee and Jacob performed together (Jacob 1988, p. 291) was a success. A male *E. coli* strain that was inducible for β-gal formation (i^+, z^+ genetically) was mated with a female strain that was both noninducible and unable to make β-gal (i^-, z^- genetically). What would happen? (The reader might keep in mind the zygotic induction experiment just explained on p. 83.) The injection of the "inducible" *lac* genes from the male required ∼20 min. In this first trial experiment, Pardee and Jacob struck pay dirt: Almost as soon as the male *lac* region entered, β-gal synthesis began at the *maximum* rate. But no inducer had been added. This initial experiment was very surprising. How was the enzyme made with no inducer from a genome with an i^+ gene? Obviously, no direct positive effect of a chemical inducer was required to release new enzyme synthesis directed by the male genome in its new environment.

This initial experiment was refined to include several additional facets leading to a famous summary experiment (Fig. 2-11). The basic genetic setup was as described above: A male that was inducible (i^+, z^+) for β-gal synthesis was used to transfer its DNA into female recipients that were *not* inducible and were unable to make β-gal (i^-, z^-). No inducer was added at the beginning of the gene transfer. The following genetic refinement was introduced: The male cells that were used were sensitive to a bacteriophage (T7) and were also sensitive to the antibiotic streptomycin. The female cells were resistant to both T7 bacteriophage and streptomycin. This genetic wrinkle allowed the male cells to be removed after injecting the *lac* locus by treating the mating pairs with bacteriophage T7 and streptomycin after 20 min.

As noted in the previous experiment, enzyme synthesis began at the maximal rate shortly after entry of the male i^+, z^+ genes and continued for up to 2 h, again *without any inducer*. At this point, one-half of the culture was allowed to continue; the other half was withdrawn, and inducer was added. In the culture with no inducer, new enzyme synthesis was soon halted, and the enzyme level was constant for several hours. However, in the culture to which inducer was added, enzyme synthesis continued for several hours.

The genes responsible for enzyme synthesis came from the male chromosome and, without inducer, functioned maximally for a time (up to 2 h). But *without an inducer*, the negative, repressive effect of the i^+ gene, also on the

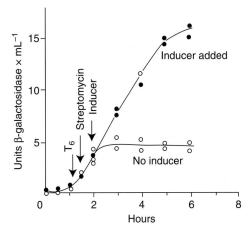

Figure 2-11. The PaJaMa (Pardee, Jacob, Monod) experiment. Synthesis of β-gal by merozygotes formed by conjugation between inducible, galactosidase-positive males and constitutive, galactosidase-negative females. Male ($Hfri^+z^+T6^sS^s$) and female ($F^-i^-z^-T6^rS^r$) bacteria grown in a synthetic medium containing glycerol as a carbon source were mixed in the same medium (time 0) in the absence of inducer. In such a cross, the first females receive the *lac* region (encoding galactosidase) from the males at minute 20. The rate of enzyme synthesis is determined from enzyme activity measurement on the whole population, to which streptomycin and phage T6 are added at times indicated by arrows to block further formation of recombinants and induction of the male parents. (At 2 h, the culture was divided, and inducer was added to one-half.) See the text for a full description of the "PaJaMa experiment." (Redrawn, with permission, from Pardee et al. 1959 [©Elsevier].)

male chromosome, took over and stopped further enzyme synthesis. However, when inducer was present, the repressive effect was blocked and β-gal enzyme synthesis continued at the maximal rate. Clearly, the inducer acted somehow on the *i* gene or on its product that was coming from the male chromosome. And this product of the *i* gene, whatever it was, had the capacity to act negatively on the z^+ gene, blocking its ability to cause β-gal synthesis. This powerful experiment became so famous it received a special name—the PaJaMa experiment (for Pardee, Jacob, Monod).

Jacob wrote that "On a late July Sunday in 1958," he put two and two together (Jacob 1988, p. 297). The i^+ product and the λ-repressor gene product had identical properties. Their products, whatever they were, acted at the level of blocking the synthesis of the products of structural genes: in the case of the lactose system, genes that encoded three proteins, and in the case of λ, a whole series of genes that led to virus production. Jacob became convinced that there must be a "cytoplasmic" product of these *regulatory* genes that somehow blocked either the action or the formation of a "cytoplasmic" product of the *structural* genes. Because this negative effect acted on more than one closely linked gene in each case, Jacob favored—but could not at this point argue strongly for—the idea that the repression might occur at the level of the DNA by blocking the function of a section of the DNA. It took Monod a while, but by the end of the summer, he had adopted Jacob's ideas.[1]

Heading toward the mRNA Hypothesis

The first short publication describing the separate function of structural and regulatory genes was sent off to press before Pardee left Paris in the fall of 1958 (Pardee et al. 1958).

Great progress had occurred, but big questions remained. What was the nature of the i gene product or, for that matter, the product of the z^+ gene that instructed the protein synthesis machinery to make β-gal? Did the i^+ gene product block the formation or the action of the z^+ gene product? Whatever the z^+ gene product was, its effect was short lived because enzyme synthesis ceased soon after the i^+ gene entered the female cell. To Monod in particular, this cessation of β-gal synthesis unless an inducer was added in the PaJaMa experiment recalled the cessation of synthesis of other induced enzymes. For example, this prompt switch was true for the set of enzymes required for tryptophan synthesis or the coordinately regulated enzymes required for histidine synthesis studied so successfully at about the same time by Bruce Ames at The National Institutes of Health in the United States (Ames and Garry 1959; Ames et al. 1960). Monod wanted a general explanation for all of these "on-off" switches.

Jacob and Monod decided first to consider the i^+ gene product and its site of action. Could it be proved or at least suggested that it worked at the level of DNA? Once again, genetics came to the rescue. With genetic sleight of hand, Jacob went looking for a genetic locus (presumably a site on the DNA) that when mutated would allow uninduced (i.e., "constitutive") β-gal synthesis *and* that was genetically separable from the i^+ gene. He found such mutants, although they were rare, and their map position lay between the i^+ gene and the z^+ gene. These mutants were constitutive for β-gal synthesis, but their i genes were intact because they could repress synthesis of β-gal from a *lac* locus (i^-, z^+) that was brought in during conjugation.

---

[1] The Hungarian physicist Leó Szilárd became famous for his introduction, along with Albert Einstein, of the idea of an atomic bomb to President Franklin Delano Roosevelt and later for his important efforts to control bomb making. In the 1950 postwar years, Szilárd, with his keenly analytical powers, became a regular visitor to important biology laboratories around the world, and his incisive intellect was greatly appreciated. He had argued strongly to Jacob and Monod that the i gene and enzyme repressors in general acted negatively (see Monod's Nobel lecture in *Origins of Molecular Biology* [Monod 2003]).

The new genetic locus that mapped within the *lac* region was christened "*o*" for *operator*, and the coordinated expression of the entire *lac* region was termed an *operon*. Whatever the product of the i^+ gene, it was envisioned as acting on the DNA site represented by operator mutants. Thus, the *lac* locus was now written in the order *i*, *o*, *z*, *y*, and *a* (Jacob et al. 1960).

After Pardee went home, Jacob redid the PaJaMa experiments with additional time points just to be sure. The improved precision simply strengthened the same conclusions: As soon as the male delivered the z^+ gene into an i^- environment, β-gal synthesis started at maximal rates.

After he returned to California, Pardee, together with his student Monica Riley, completed another experiment that had been conceived in Paris and that added a final twist to the PaJaMa experiment. They conducted the gene transfer (mating) part of the experiment in medium containing $^{32}PO_4$ so that any transferred DNA (which was known to be newly synthesized during the transfer) would have become highly labeled in the phosphate that links the DNA nucleotides together. They allowed enzyme synthesis to begin and then froze individual portions of the now ultimately lethally labeled *functioning* z^+ gene on the male chromosome (Fig. 2-12). Samples without $^{32}PO_4$ were also frozen. One unlabeled sample and one labeled sample were immediately thawed to begin the experiment. Synthesis was slightly less in the immediately thawed labeled sample right at the beginning, but the rate of enzyme increase in the two (labeled and unlabeled) during 50 min was almost the same. When they allowed the radioactive $^{32}PO_4$ to decay in the other frozen samples, a new fact was added to the picture. As decay proceeded for up to 10 d, the cells that were making enzyme before they were frozen, now after thawing, made less and less enzyme the longer the decay had occurred. The frozen sample without $^{32}PO_4$ behaved just as the other unlabeled sample had 10 d earlier (Fig. 2-12). (^{32}P decays to sulfur with a 14-d half-life, breaking DNA chains so that they can no longer function.) The DNA of the induced z^+ genes had to remain intact for the continued synthesis of β-gal. This was the best evidence yet that an *unstable* product that came from an *intact* z^+ gene directed synthesis of the enzyme (Riley et al. 1960; Riley and Pardee 1962).

By the spring of 1959, Jacob and Monod were ready to present their ideas more fully to the world (Jacob and Monod 1959). Weaving all of the experiments and interpretations together, they concluded that structural genes took their instructions from regulatory genes. These instructions were delivered to a site on the DNA, the operator, where regulatory gene products acted negatively. Relief from repression allowed production from the structural gene of a specific unstable molecule that, to be continuously produced, required that the DNA of the structural gene be intact. At the time they began writing,

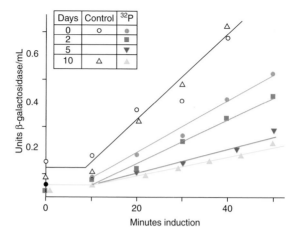

Figure 2-12. Kinetics of enzyme (β-gal) formation as influenced by decay of incorporated ^{32}P. A male strain of *E. coli* (i^+z^+) that had been grown for two to three generations on $^{32}PO_4$ was mixed with ("mated with") nonradioactive female (i^-z^-) bacteria. (A control culture without $^{32}PO_4$ was also mated with the same female strain.) Mating was arrested at 35 min, and zygotes were allowed to develop further for 25 min before storage in liquid nitrogen. Labeling was continued during this time in order to allow enzyme induction from the male DNA to begin in the presence of $^{32}PO_4$. The $^{32}PO_4$ was then removed and samples of both cultures were frozen. One sample of labeled and one of unlabeled was immediately thawed to measure rate of synthesis of β-gal. Samples were then tested for β-gal synthesis after 2, 5 and 10 d. No $^{32}PO_4$ times 0 (open circles) and 10 d (open triangles). For the $^{32}PO_4$ culture time 0 (orange circles), after 2 d (blue squares), after 5 d (inverted green triangles), after 10 d (yellow triangles). The control cells (no $^{32}PO_4$) made β-gal at the same rate before and after being frozen for 10 d (open circles and triangles). The ^{32}P-labeled cells lost enzyme-forming ability as a function of time. (Redrawn, with permission, from Riley et al. 1960 [© Elsevier].)

in the winter of 1959–1960, what was to become the masterful summary of all of the thoughts and evidence (Jacob and Monod 1961), they could easily envision a polynucleotide from the *i* gene binding to a DNA site, but it was as yet unclear that proteins could find discrete sites on DNA. Thus, they seriously considered that the *i* gene product might be an RNA. Whatever its nature, following binding to the operator, the repressor product *inhibited* the production of a "cytoplasmic" (i.e., a product free of DNA) unstable product of the structural gene that dictated to the protein synthesis machinery to make a specific protein.

In late 1959 and early 1960, both Jacob and Monod gave lectures in England and the United States. Jacob called the proposed unstable product of the structural gene X but was not definite, at least in public, that it was mRNA. In their magisterial full-length 1961 paper (Jacob and Monod 1961) that they had begun to assemble in the fall of 1959, they predicted a "messenger" that went from the gene to the protein-synthesizing apparatus. In the *E. coli* cell, this messenger had to have a short life span because, as mentioned repeatedly, induced enzyme synthesis occurred very quickly in the presence of inducer and stopped very quickly when inducer was removed. But as of April 1960, they knew of no biochemical evidence for a short-lived RNA molecule.

Proposing the Existence of mRNA

Although the idea of mRNA had great logic behind it, no useful idea for experimental proof had yet occurred. Jacob had presented the findings and interpretations already in September 1959 to a small gathering in Copenhagen that included not only Monod but Watson, Crick, Sydney Brenner, Seymour Benzer, and others. According to Jacob, "Jim Watson spent most of the session reading a newspaper" (Jacob 1988, p. 311). Jacob's ideas were not immediately embraced. Francis Crick wrote "Sydney and I knew there had to be a messenger to convey the genetic message of each gene from the DNA in the nucleus to the ribosomes in the cytoplasm, and we assumed that this had to be RNA. [By this time, Crick and Brenner shared an office and constantly discussed science.] In this we were right. Who would have been so bold as to say that the RNA we saw there [i.e., the rRNA in the ribosome] was *not* the messenger but that the messenger was another kind of RNA, as yet undetected, turning over rapidly and thus probably there in small amounts?" (Crick 1988, p. 140). Jim Watson wrote, "no one then had any compelling reason to take my hypothesis seriously but by November 1952, I liked it well enough to print DNA → RNA → protein on a small piece of paper which I taped on a wall above my writing table in my rooms at Clare College" (Watson 2006). But even 8 years later, Watson also had not tumbled to the idea of an unstable intermediate to explain regulated gene expression and its necessary connection to ribosomal protein synthesis in *E. coli*.

In the spring of 1960, Jacob was visiting Cambridge University. Brenner arranged another discussion in his rooms at King's College that included Francis Crick, Ole Maaløe (the Danish microbiologist who had arranged the September 1959 meeting), Alan Garen (a bacterial geneticist who would codiscover and characterize suppressor mutations in tRNA) and his wife Susan, and Leslie Orgel (an organic chemist and close friend of Crick). Once again, the purpose was to discuss all of the results of the Paris group and the Jacob and Monod conclusion about an unstable intermediate. In his lucid and engaging autobiography *The Statue Within*, Jacob sets the scene (Jacob 1988, p. 311). Both Crick (Crick 1988, p. 119) and Brenner (Brenner 2001, p. 74) also tell the same story. Jacob had just finished describing the ^{32}P experiment in which continued production of the induced β-gal was killed by ^{32}P decay in the DNA. "At this point, Francis and Sydney hopped to their feet. Began to gesticulate. To argue at top speed in great agitation" (Jacob 1988, p. 311).

Jacob (whose English is very good but perhaps not good enough for the speed of the Brenner/Crick conversation), after a minute, caught the drift—the two had gone back to the Volkin and Astrachan experiments on RNA synthesis in T2 bacteriophage-infected *E. coli*: *T2 RNA copied from T2 DNA was a labile intermediate*. It was what Jacob had referred to as X. It was "T2 mRNA," almost all of the people in the room shouted simultaneously. The idea was likened to a tape coming from a tape recorder; a base-paired RNA copy of DNA (thus preserving the genetic information) was grabbed onto by the *pre-existing E. coli* ribosomes to dictate the production of bacteriophage protein. Copies of any bacterial gene could be used in the same way to make the corresponding bacterial protein. This mechanism would account for the rapidly induced synthesis of enzymes such as β-gal and the quick cessation of enzyme synthesis once production of the unstable messenger had stopped. Further messenger synthesis would be inhibited by the repressor following removal of the inducer.

Suddenly, the Jacob and Monod ideas had a roomful of converts. That evening at Crick's house, The Golden Helix, Jacob and Brenner planned a definitive experiment. They decided that the most persuasive experiment would be to follow up the Volkin and Astrachan experiments.

The Caltech Experiment: Strong Evidence for mRNA

Brenner and Jacob were both scheduled to be in Pasadena, California, at The California Institute of Technology (Caltech) in a few weeks' time. There, the definitive experiment was to be done with the new equilibrium sedimentation technique that Matthew Meselson, Frank Stahl, and Jerome Vinograd had invented (Meselson et al. 1957) and used earlier for separating old and new DNA containing heavy or light isotopes (Meselson and Stahl 1958). (In fact, Rick Davern and Meselson had shown that, by labeling RNA and then growing cells for several generations, *E. coli* ribosomes were conserved, implying that their RNA was stable [Davern and Meselson 1960].) Ribosomes with the "normal" ^{12}C or ^{14}N in their proteins and nucleic acids were easily separable from ribosomes with ^{13}C and ^{15}N. The idea was that cells grown in heavy medium and then switched to regular medium would allow heavy, preexisting ribosomes to be distinguished from any new "bacteriophage-specific" ribosomes that hypothetically might be formed in the infected cells. Moreover, if ^{32}PO$_4$ or ^{14}C in RNA precursors, e.g., uracil, was added during bacteriophage infection, the association of new bacteriophage-directed RNA with old ribosomes could be discerned. (The labeled RNA precursor [$^{32}PO_4$ or ^{14}C uracil] was added for only 2 min at 5–7 min after T4 infection.) T4, a close cousin of T2, was to be used because Brenner's laboratory was accustomed to using it. All of the events connected with these groundbreaking experiments are described in considerable (and amusing) detail in Brenner's own words in *My Life in Science* (Brenner 2001, pp. 77–83) and by Judson in *The Eighth Day of Creation* (Judson 1996, pp. 423–426). Initial success was achieved on Jacob's last day or two at Caltech, and, as predicted, the new T4 RNA was associated with old ribosomes (Fig. 2-13).

Brenner returned to his laboratory in Cambridge to repeat the experiment and add some controls. The published experiments showed that (1) heavy and light ribosomes were clearly separated; the ribosome required a high Mg^{++} concentration to withstand disintegration in the high cesium chloride concentration of the equilibrium density gradient (and still some breakdown to ribosomal subunits occurred). (2) As was earlier thought to be the case, absolutely no new ribosomes were made in the infected cell; *all* of the ribosomes in the infected cell were heavy. (3) The newly made T4 phage RNA molecules labeled with [^{14}C]uracil were associated with the *old* ribosomes. Moreover, the radioactive T4 RNA appeared to be unstable, i.e., it was gone within 10 min if the label was removed. The conclusion followed that all protein synthesis in infected cells (which was known to be largely if not completely bacteriophage specific) occurred on the preexisting ribosomes. Brenner, Jacob, and Meselson concluded by the summer of 1960 that, indeed, the newly made T4 bacteriophage messenger RNA (for which the mRNA abbreviation became adopted) had programmed the preexisting ribosomes to make bacteriophage proteins.

Although most of the scientific world (and the world's science writers) cite the Brenner, Jacob, and Meselson paper as "proof" of the existence of mRNA, it was, of course, proof that Volkin and Astrachan's T-phage-specific RNA was an mRNA.

Members of Jim Watson's laboratory, particularly Alfred Tissières, an accomplished biochemist from Switzerland, had been studying ribosomes with an eye on solving the problem of how genes controlled protein synthesis. As had a number of other laboratories found by this time, they discovered that 70S ribosomes were very regular structures with two subunits, a 50S and a 30S. The larger contained a major RNA of 23S, and the smaller contained a 16S RNA (Tissières and Watson 1958). (Watson's long-standing interest in RNA is evident in his 1962 Nobel lecture printed in *Science* in 1963 [Watson 1963]).

François Gros from The Pasteur Institute, who knew of Jacob and Monod's conclusions about an unstable intermediate, joined the Watson

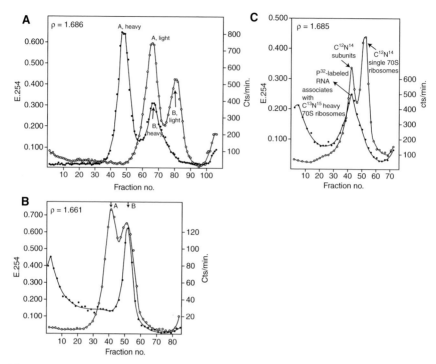

Figure 2-13. Demonstration that RNA made after bacteriophage T4 infection binds to preexisting *E. coli* ribosomes. The experiment required three steps (*A–C*). (*A*) Ribosomes from cells grown in "heavy" ([filled circles] $^{15}N^{13}C$, detected by ^{32}P counts per minute [right *y* axis]) or in "light" ([open circles] $^{14}N^{12}C$, detected by UV absorption at 254 nm [left *y* axis]) media were sedimented to equilibrium in CsCl gradients (density, $\rho = 1.686$, as indicated in figure). Ribosomes and their subunits formed two bands: Band A represents the 50S and 30S ribosomal subunits, and band B represents the 70S ribosomes. A clear separation was observed between heavy and light ribosomes and ribosomal subunits. However, note that the heavy B band sediments at about the same density as the light A band, which is purely fortuitous. (Separate experiments [data not shown] had proven that ribosomal subunits banded at the A position and, under certain salt conditions [high Mg^{++}], single 70S ribosomes banded at position B.) (*B*) T4-specific RNA binds to 70S ribosomes (band B). Cells infected with T4 bacteriophage were labeled with [^{14}C]ura-cil for 2 min (3–5 min after infection). (There was no density label in this experiment.) Ribosomes were purified and banded in CsCl. (The A band is ribosomal subunits; the B band is 70S ribosomes identified by OD_{254} [E 254]). (Filled circles) ^{14}C-labeled RNA (presumed to be all T4-specific based on the Volkin and Astrachan [1956a,b] results) was associated with the B band containing 70S ribosomes. When labeled uracil was replaced with unlabeled uracil, radioactivity decreased fivefold in 16 min (data not shown). (*C*) T4 RNA associates with preexisting ribosomes. For this experiment, heavy-labeled cells that had been infected with T4 and labeled for 2 min with $^{32}PO_4$ were mixed with a 50-fold excess of light cells grown in normal medium ($^{12}C^{14}N$) to furnish an OD marker (E 254) shown on the axis of *C*. The labeled (filled circles) T4 RNA associates with a band of the same density as the heavy B (70S) band but not the light B band (see *A* for comparison). (Modified, with permission, from Brenner et al. 1961 [©Macmillan].)

laboratory at Harvard University in May 1960 for a short sabbatical. Gros, assisted by several others, showed that in uninfected *E. coli*, rapidly labeled RNA species with an average size of 10S–12S existed and that at least some of this material was associated with ribosomes. It was obviously a different size from that of rRNA (Fig. 2-14). This experiment in uninfected cells (published in the same issue of *Nature* as the Brenner, Jacob, and Meselson paper; Gros et al. 1961) was parallel to the experiment that Nomura, Hall, and Spiegelman had reported in 1960 for T2 RNA (Nomura et al. 1960). The Harvard experiment showed that the most rapidly labeled RNA in a bacterial cell associated with ribosomes but was not the same size as rRNA. All of these experiments strongly supported the Jacob-Monod conclusion of an unstable mRNA intermediate between genes and ribosomes.

Molecular Hybridization Strengthens the Proof of mRNA

Despite this surge of experiments on RNA both in growing and infected *E. coli*, it was widely recognized that an assay for specific mRNAs was needed. This was soon supplied.

Paul Doty, a physical chemist who studied properties of DNA, and his younger colleague, Julius Marmur, who had studied transforming DNA with Rollin Hotchkiss at The Rockefeller Institute, made a striking observation in 1960 (Doty et al. 1960; Marmur and Lane 1960). They knew that if double-stranded DNA was heated, the two strands came apart—they "melted." This led to a characteristic increase in absorbance of UV light because when the bases of the DNA are in separated strands, they absorb more light. Amazingly, if the melted DNA from bacteria or bacteriophages was allowed to cool, the separated strands apparently reassociated: The UV absorbance returned to what it was before melting. This phenomenon was called *reannealing*. Marmur showed that in DNA taken from pneumococci, the pneumonia-causing organism that Avery, MacLeod, and McCarty had used—the separated "single strands"—had no transforming ability. But following reassociation, the reconstituted DNA regained its transforming capacity. Marmur concluded that the single strands of DNA must have found their *correct* partner to reconstitute this biological behavior.

Built on this idea of reannealing, Ben Hall, a student of Doty's, joined with Sol Spiegelman, who had been studying T2-specific RNA, to adapt the reassociation technique to detect and eventually measure specific RNA. They melted T2 DNA, and during the cooling and reannealing process, added RNA extracted from radioactively labeled, T2-infected bacteriophage cells. If the T2 RNA was indeed copied from the T2 DNA, it should find its correct spot on the T2 DNA during the reassociation period and bind specifically.

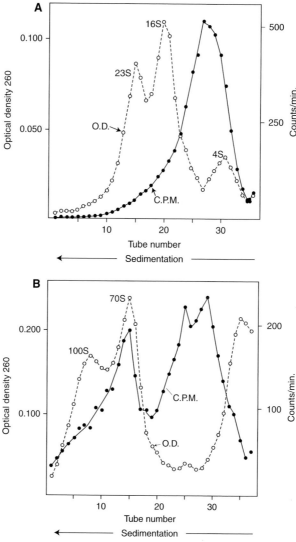

Figure 2-14. Pulse-labeled RNA is not the same size as rRNA but can associate with ribosomes. Growing *E. coli* cells were labeled with [^{14}C]uracil for 20 sec. (*A*) RNA was extracted with phenol from one part of the culture and subjected to sucrose zonal sedimentation (fraction 1 contains the largest particles). The labeled RNA (closed circles; [C.P.M.] counts per minute) is smaller than the ribosomal 16S and 23S RNA. (*B*) Another portion of the culture was subjected to cell breakage, and the whole extract was sedimented through a sucrose gradient to test for ribosome association. Some of the briefly labeled RNA (open circles; C.P.M.) was associated with the 70S and 100S ribosomes. (Modified, with permission, from Gros et al. 1961 [©Macmillan].)

The "annealed" RNA might be protected against RNase treatment, which destroys single-stranded RNA (Hall and Spiegelman 1961). This procedure worked and was the first experimental *molecular hybridization* of nucleic acid. (It was later shown that an RNA:DNA hybrid is more stable to melting than DNA:DNA. This is why the Hall and Spiegelman RNA/DNA hybridization worked, even though both DNA strands and the T2-specific RNA were present in one solution.) One of the early experiments, in which the Spiegelman laboratory used the technique, confirmed that RNA that hybridized to phage DNA carrying the *lac* operon was made after induction but not before. Therefore, mRNA was indeed copied from the β-gal gene during enzyme induction (Hayashi et al. 1963). This molecular hybridization technique has been adapted in countless ways in subsequent decades to show the relatedness of one nucleic acid sample (RNA or DNA) to another. Moreover, this is the fundamental first step that allows enzymes to use short DNA primers to make long DNA copies in unlimited amounts. Nucleic acid hybridization remains a mainstay in most modern nucleic acid experiments.

The 1961 Cold Spring Harbor Meeting Concludes an Era

In June 1961, the annual Cold Spring Harbor Symposia on Quantitative Biology was entitled "Cellular Regulatory Mechanisms." The meeting served as the climax of acceptance of the mRNA idea, at least among the cognoscenti who attended the meeting. Jacob and Monod essentially "held class," thoroughly describing their ideas about repressors blocking the synthesis of mRNA for induced enzymes. Brenner described the density experiments showing that preexisting ribosomes plus bacteriophage-specific mRNA made bacteriophage protein. (The paper of Brenner, Jacob, and Meselson describing these experiments had just been published in May 1961.) Many supportive papers were presented, and the feeling was that the time had arrived to move on to other recognized but as yet unsolved problems. The nature of the repressor had to be proved, and the biochemistry of regulated RNA copying in bacteria remained to be solved. And, of course, the generality of mRNA in nucleated cells, and especially its application to the changing protein expression that was already known to occur during embryogenesis and in specialized adult animal cells, had to be explored.

Perhaps the biggest *immediate* unsolved mystery was how the information in the tape, the mRNA, was interpreted by bacterial ribosomes and tRNAs to cause the addition of amino acids in the correct order during protein synthesis. Was there a "genetic code" that could be deciphered? The final ideas behind the mRNA prediction came from experiments between 1957 and 1960, and already by the summer of 1961, the curtain was lifted on a solution to the genetic code—the information spelled out in mRNA—that provided the direction for ribosomes and tRNA to get the amino acid sequence in protein correct. That story is told in Chapter 3.

REFERENCES

Adelberg EA, Umbarger HE. 1953. Isoleucine and valine metabolism in *Escherichia coli*. V. α-Ketoisovaleric acid accumulation. *J Biol Chem* **205:** 475–482.

Ames BN, Garry B. 1959. Coordinate repression of the synthesis of four histidine biosynthetic enzymes by histidine. *Proc Natl Acad Sci* **45:** 1453–1461.

Ames BN, Garry B, Herzenberg LA. 1960. The genetic control of the enzymes of histidine biosynthesis in *Salmonella typhimurium*. *J Gen Microbiol* **22:** 369–378.

Bawden FC, Pirie NW. 1937. The isolation and some properties of liquid crystalline substances from solanaceous plants infected with three strains of tobacco mosaic virus. *Proc R Soc Lond B Biol Sci* **123:** 274–320.

Belozersky A, Spirin A. 1958. A correlation between the composition of deoxyribonucleic acid and ribonucleic acids. *Nature* **182:** 111–112.

Berg P. 1956. Acyl adenylates: The interaction of adenosine triphosphate and L-methionine. *J Biol Chem* **222:** 1025–1034.

Berg P, Offengand EJ. 1958. An enzymatic mechanism for linking amino acids to RNA. *Proc Natl Acad Sci* **44:** 78–86.

Black FL, Knight CA. 1953. A comparison of some mutants of tobacco mosaic virus. *J Biol Chem* **202:** 51–57.

Brachet J. 1941. La détection histochimique et le microdosage des acides pentosenucléiques. *Enzymologia* **10:** 87–96.

Brachet J. 1955. The biological role of the pentose nucleic acids. In *The nucleic acids: Chemistry and biology* (ed. E Chargaff, JN Davidson), Vol. 2, pp. 475–519. Academic Press, New York.

Brachet J, Chantrenne H. 1951. Protein synthesis in nucleated and non-nucleated halves of *Acetabularia mediterranea* studied with carbon-14 dioxide. *Nature* **168:** 950. doi: 10.1038/168950a0.

Brenner S. 2001. *My life in science* (as told to Lewis Wolpert) (ed. EC Friedberg, E Lawrence). Biomed Central, London.

Brenner S, Jacob F, Meselson M. 1961. An unstable intermediate carrying information from genes to ribosomes for protein synthesis. *Nature* **190:** 576–581.

Brown DM, Todd AR. 1955. Nucleic acids. *Annu Rev Biochem* **24:** 311–338.

Caspersson T. 1947. The relation between nucleic acid and protein synthesis. *Symp Soc Exp Biol* **1:** 129–151.

Chargaff E. 1951. Structure and function of nucleic acids as cell constituents. *Fed Proc* **10:** 654–659.

Chargaff E, Davidson JN. 1955. The biological role of the pentose nucleic acids. In *The nucleic acids* (ed. E Chargaff, JN Davidson), Vol. 2, pp. 475–419. Academic Press, New York.

Claude A. 1941. Particulate components of cytoplasm. *Cold Spring Harbor Symp Quant Biol* **9:** 263–271.

Claude A. 1948. Studies on cells: Morphology, chemical constitution, and distribution of biochemical functions. *Harvey Lect* **48:** 121–164.

Cohen GN. 1965. Regulation of enzyme activity in microorganisms. *Annu Rev Microbiol* **19:** 105–126.

Cohen SS, Stanley WM. 1942. The molecular size and shape of the nucleic acid of tobacco mosaic virus. *J Biol Chem* **144:** 589–598.

Cohen GN, Monod J. 1957. Bacterial permeases. *Bact Rev* **21:** 169–194.

Cohen M, Cohen GN, Monod J. 1953. L'effet inhibiteur spécifique de la méthionine dans la formation de la méthionine-synthase chez *E. coli. C R Acad Sci* **236:** 746–748.

Crick FHC. 1958. On protein synthesis. *Symp Soc Exp Biol* **12:** 138–163.

Crick F. 1988. *What mad pursuit: A personal view of scientific discovery.* Basic Books, New York.

Davern CI, Meselson M. 1960. The molecular conservation of ribonucleic acid during bacterial growth. *J Mol Biol* **2:** 153–160.

Deichmann U. 1999. *Biologists under Hitler* (translated by Dunlap T). Harvard University Press, Cambridge, MA.

Doty P, Marmur J, Eigner J, Schildraut C. 1960. Strand separation and specific recombination in deoxyribonucleic acids: Physical chemical studies. *Proc Natl Acad Sci* **46:** 461–476.

Dounce AL. 1952. Duplicating mechanism for peptide chain and nucleic acid synthesis. *Enzymologia* **15:** 251–258.

Dounce AL. 1953. Nucleic acid template hypotheses. *Nature* **172:** 541. doi: 10.1038/172541a0.

Feulgen R. 1914. Über die "Kohlenhydratgruppe" in der echten Nuckleinsäure. *Z Physiol Chem* **92:** 154–158.

Fox CF, Kennedy EP. 1965. Specific labeling and partial purification of the M protein, a component of the β-galactoside transport system of *Escherichia coli. Proc Natl Acad Sci* **54:** 891–899.

Fraenkel-Conrat H, Singer B. 1957. Virus reconstitution. II. Combination of protein and nucleic acid from different strains. *Biochem Biophys Acta* **24:** 540–548.

Fraenkel-Conrat H, Williams RC. 1955. Reconstitution of active tobacco mosaic virus from its inactive protein and nucleic acid components. *Proc Natl Acad Sci* **41:** 690–698.

Fraenkel-Conrat H, Singer B, Williams RC. 1957a. Infectivity of viral nucleic acid. *Biochim Biophys Acta* **25:** 87–96.

Fraenkel-Conrat H, Singer BA, Williams RC. 1957b. The nature of the progeny of virus reconstituted from protein and nucleic acid of different strains of tobacco mosaic virus. In *The chemical basis of heredity* (ed. B Glass), pp. 501–517. Johns Hopkins Press, Baltimore.

Gamow G. 1954. Possible relationship between deoxyribonucleic acid and protein structures. *Nature* **173:** 318. doi: 10.1038/173318a0.

Gierer A, Schramm G. 1956. Infectivity of ribonucleic acid from tobacco mosaic virus. *Nature* **177:** 702–703.

Gros F, Hiatt H, Gilbert W, Kurland CG, Risebrough RW, Watson JD. 1961. Unstable ribonucleic acid revealed by pulse labeling of *Escherichia coli. Nature* **190:** 581–584.

Hall BD, Doty P. 1959. The preparation and physical chemical properties of ribonucleic acid from microsomal particles. *J Mol Biol* **1:** 111–126.

Hall BD, Spiegelman J. 1961. Sequence complementarity between T2 DNA and T2 specific RNA. *Proc Natl Acad Sci* **47:** 137–146.

Hayashi M, Spiegelman S, Franklin NC, Luria SE. 1963. Separation of the RNA message transcribed in response to a specific inducer. *Proc Natl Acad Sci* **49:** 727–736.

Heimberg M, Velick SF. 1954. The synthesis of aldolase and phosphorylase in rabbits. *J Biol Chem* **208:** 725–730.

Heppel LA, Rabinowitz JC. 1958. Enzymology of nucleic acids, purines and pyrimidines. *Annu Rev Biochem* **27:** 613–647.

Hoagland MB. 1955. An enzymic mechanism for amino acid activation in animal tissues. *Biochem Biophys Acta* **16**: 288–289.

Hoagland MB, Keller EB, Zamecnik PC. 1956. Enzymatic carboxyl activation of amino acids. *J Biol Chem* **218**: 345–358.

Hoagland MB, Zamecnik PC, Stephenson ML. 1957. Intermediate reactions in protein biosynthesis. *Biochem Biophys Acta* **24**: 215–216.

Hoagland MB, Stephenson ML, Scott JF, Hecht LI, Zamecnik PC. 1958. A soluble ribonucleic acid intermediate in protein synthesis. *J Biol Chem* **231**: 241–257.

Holley RW. 1957. An alanine-dependent, ribonuclease-inhibited conversion of AMP to ATP, and its possible relationship to protein synthesis. *J Am Chem Soc* **79**: 658–662.

Ingram VM. 1957. Gene mutations in human haemoglobin: The chemical difference between normal and sickle cell haemoglobin. *Nature* **180**: 326–328.

Jacob F. 1988. *The statue within: An autobiography* (translated by Philip F). Basic Books, New York.

Jacob F, Monod J. 1959. Genes of structure and genes of regulation in the biosynthesis of proteins. *C R Hebd Seances Acad Sci* **249**: 1252–1284.

Jacob F, Monod J. 1961. Genetic regulatory mechanisms in the synthesis of proteins. *J Mol Biol* **3**: 318–356.

Jacob F, Wollman E. 1956. Processes of conjugation and recombination in *Escherichia coli.* I. Induction by conjugation or zygotic induction. *Ann Inst Pasteur* **91**: 486–510.

Jacob F, Wollman E. 1961. *Sexuality and genetics of bacteria.* Academic Press, New York.

Jacob F, Perrin D, Sanchez C, Monod J. 1960. Operon: A group of genes with the expression coordinated by an operator. *C R Hebd Seances Acad Sci* **250**: 1727–1729.

Judson HF. 1996. *The eighth day of creation: Makers of the revolution in biology, expanded edition.* Cold Spring Harbor Laboratory Press, Cold Spring Harbor, NY.

Kaiser AD. 1957. Mutations in a temperate bacteriophage affecting its ability to lysogenize *Escherichia coli. Virology* **3**: 42–61.

Kaiser AD, Jacob F. 1957. Recombination between related temperate bacteriophages and the genetic control of immunity and prophage localization. *Virology* **4**: 509–521.

Karstrom H. 1939. Enzymatische adaptation bei mikroorganismen. *Ergeb Enzymforsch* **7**: 350–376.

Keller EB, Zamecnik PC, Loftfield RB. 1954. The role of microsomes in the incorporation of amino acids into proteins. *J Histochem Cytochem* **2**: 378–386.

Knight CA. 1952. The nucleic acid of some strains of tobacco mosaic virus. *J Biol Chem* **197**: 241–248.

Kurland CG. 1960. Molecular characterization of ribonucleic acid from *Escherichia coli* ribosomes: I. Isolation and molecular weights. *J Mol Biol* **2**: 83–91.

Lederberg EM. 1951. Lysogenicity in *E. coli* K12. *Genetics* **36**: 560

Levene P, Bass LW. 1932. *Nucleic acids.* Chemical Catalogue, New York.

Lipmann F. 1949. Mechanism of peptide bond formation. *Fed Proc* **8**: 597–602.

Lipmann F. 1953. On chemistry and function of coenzyme A. *Bacteriol Rev* **17**: 1–16.

Littlefield JW, Keller EB, Gross J, Zamecnik PC. 1955. Studies on cytoplasmic ribonucleoprotein particles from the liver of the rat. *J Biol Chem* **217**: 111–123.

Lwoff A. 1966. The prophage and I. In *Phage and the origins of molecular biology* (ed. J Cairns, et al.), pp. 88–99. Cold Spring Harbor Laboratory, Cold Spring Harbor, NY.

Lwoff A. 2003. Jacques Lucien Monod 1910–1976. In *Origins of molecular biology* (ed. A Ullman), pp. 1–46. ASM Press, Washington, DC.

Markham R, Matthews REF, Smith KM. 1948. Specific crystalline protein and nucleoprotein from a plant virus having intact vectors. *Nature* **162**: 88–90.

Marmur H, Lane D. 1960. Strand separation and specific recombination in deoxyribonucleic acids: Biological studies. *Proc Natl Acad Sci* **46:** 453–461.

McQuillen K, Roberts RB, Britten RJ. 1959. Synthesis of nascent protein by ribosomes in *Escherichia coli*. *Proc Natl Acad Sci* **45:** 1437–1447.

Meselson M, Stahl FW. 1958. The replication of DNA in *Escherichia coli*. *Proc Natl Acad Sci* **44:** 671–682.

Meselson M, Stahl FW, Vinograd J. 1957. Equilibrium sedimentation of macromolecules in density gradients. *Proc Natl Acad Sci* **43:** 581–588.

Monod J. 1941. *Recherches sur la croissance de cultures bactériennes.* Hermann, Paris.

Monod J. 2003. Nobel lecture (reprinted). In *Origins of molecular biology* (ed. A Ullman), pp. 275–317. ASM Press, Washington, DC.

Monod J, Audureau A. 1946. Mutation et adaptation enzymatique chez *Escherichia coli-mutabile*. *Ann Inst Past* **72:** 868–879.

Monod J, Cohn M. 1952. Biosynthesis induced by enzymes; enzymatic adaptation. *Adv Enzymol* **13:** 67–119.

Monod J, Cohen-Bazire G. 1953. The effect of specific inhibition in biosynthesis of tryptophan-desmase by *Aerobacter aerogenes*. *C R Hebd Seances Acad Sci* **236:** 530–532.

Monod J, Pappenheimer AM Jr, Cohen-Bazire G. 1952. The kinetics of the biosynthesis of β-galactosidase in *Escherichia coli* as a function of growth. *Biochem Biophys Acta* **9:** 648–660.

Nomura M, Hall BD, Spiegelman S. 1960. Characteristic of RNA synthesized in *Escherichia coli* after bacteriophage T2 infection. *J Mol Biol* **2:** 306–326.

Northrop JH. 1930. Crystalline pepsin. *J Gen Physiol* **13:** 739–766.

Palade GE, Siekevitz P. 1956. Liver microsomes: An integrated morphological and biochemical study. *J Biophys Biochem Cytol* **7:** 171–200.

Pardee AB, Jacob F, Monod J. 1958. Sur l'expression et le role des alleles "inductible" et constitutif dans la syntheside la B galactosidase chez des zygotes d'*Escherichia coli* [The role of the inducible alleles and the constitutive alleles in the synthesis of β-galactosidase in zygotes of *Escherichia coli*]. *C R Hebd Seances Acad Sci* **246:** 3125–3128.

Pardee AB, Jacob F, Monod J. 1959. The genetic control and cytoplasmic expression of "inducibility" in the synthesis of β-galactosidase by *E. coli*. *J Mol Biol* **1:** 165–178.

Pauling L, Corey RB. 1951. The pleated sheet, a layer configuration of polypeptide chains. *Proc Natl Acad Sci* **37:** 251–256.

Pauling L, Corey RB, Branson HR. 1951. The structure of proteins: Two hydrogen bonded helical configurations of the polypeptide chain. *Proc Natl Acad Sci* **37:** 205–216.

Putnam FW. 1994. Felix Haurowitz—March 1, 1896–December 2, 1987. *Biog Mem Nat Acad Sci* **64:** 135–163.

Rheinberger H-J. 1997. *Toward a history of epistemic things: Synthesizing proteins in the test tube,* pp. 114–142. Stanford University Press, Stanford, CA.

Riley M, Pardee A. 1962. β-Galactosidase formation following decay of ^{32}P in *Escherichia coli* zygotes. *J Mol Biol* **5:** 63–75.

Riley M, Pardee AB, Jacob F, Monod J. 1960. On the expression of a structural gene. *J Mol Biol* **2:** 216–225.

Roberts RB, ed. 1958. Introduction. In *Microsomal particles and protein synthesis.* Pergamon Press, New York.

Sanger F, Thompson EOP. 1953. The amino acid sequence in the glycyl chain of insulin. *Biochem J* **53:** 353–356; 366–374.

Sanger F, Tuppy H. 1951a. The amino-acid sequence in the phenylalanyl chain of insulin. I. The identification of lower peptides from partial hydrolysates. *Biochem J* **49:** 463–481.

Sanger F, Tuppy H. 1951b. The amino-acid sequence in the phenylalanyl chain of insulin. II. The investigation of peptides from enzymic hydrolysates. *Biochem J* **49:** 481–490.

Siekevitz P. 1952. Uptake of radioactive alanine in vitro into the proteins of rat liver fractions. *J Biol Chem* **195:** 549–565.

Siekevitz P, Zamecnik PC. 1951. In vitro incorporation of 1-^{14}C-DL-alanine into proteins of rat liver granular fractions. *Fed Proc* **10:** 246–247.

Simpson MV, Velick SF. 1954. The synthesis of aldolase and glyceraldehyde-3-phosphate dehydrogenase in the rabbit. *J Biol Chem* **208:** 61–71.

Stanley SM. 1935. Isolation of a crystalline protein possessing the properties of tobacco mosaic virus. *Science* **81:** 644–645.

Steinberg D. 1993. Modified forms of low-density lipoprotein and atherosclerosis. *J Internal Med* **233:** 227–232.

Steinberg D, Anfinson CB. 1952. Evidence for intermediates in ovalbumin synthesis. *J Biol Chem* **199:** 25–42.

Steinberg D, Vaughan M, Anfinson CB. 1956. Kinetic aspects of assembly and degradation of proteins. *Science* **124:** 389–395.

Sumner JB. 1926. The isolation and crystalization of the enzyme urease. *J Biol Chem* **69:** 435–441.

Tissières A, Watson JD. 1958. Ribonucleoprotein particle from *E. coli*. *Nature* **182:** 778–780.

Volkin E, Astrachan L. 1956a. Phosphorous incorporation in *Escherichia coli* ribonucleic acid after infection with bacteriophage T2. *Virology* **2:** 149–161.

Volkin E, Astrachan L. 1956b. Intracellular distribution of labelled ribonucleic acid after phage infection of *Escherichia coli*. *Virology* **2:** 433–437.

Volkin E, Astrachan L. 1957. RNA metabolism in T2-infected *Escherichia coli*. In *The chemical basis of heredity* (ed. WD McElroy, B Glass), pp. 686–694. Johns Hopkins Press, Baltimore.

Watson JD. 1963. Involvement of RNA in the synthesis of proteins. *Science* **140:** 17–26.

Watson JD. 2006. Prologue to first edition (reprinted). In *The RNA world*, 3rd ed. (ed. RF Gesteland, et al.), pp. xv–xxiii. Cold Spring Harbor Laboratory Press, Cold Spring Harbor, NY.

Watson JD, Crick FH. 1953a. Genetical implications of the structure of deoxyribonucleic acid. *Nature* **171:** 964–967.

Watson JD, Crick FH. 1953b. Molecular structure of nucleic acids; a structure for deoxyribose nucleic acid. *Nature* **171:** 737–738.

Wollman E, Wollman E. 1936. Recherches sur le phenomene Twort d'Herelle (bacteriophagie ou autolyso- hereditocontagieusea). *Ann Inst Pasteur (Paris)* **56:** 137–170.

Wollman E, Wollman E. 1938. Recherches sur le phenomene Twort d'Herelle (bacteriophagie ou autolyso-hereditocontagieuse). *Ann Inst Pasteur* **60:** 13–57.

Zamecnik PC, Keller EB. 1954. Relation between phosphate energy donors and incorporation of labeled amino acids into proteins. *J Biol Chem* **209:** 337–354.

Zimmern D, Hunter T. 1983. Point mutation in the 30-K open reading frame of TMV temperature-sensitive assembly and local lesion spreading mutant Ni 2519. *EMBO J* **2:** 1893–1900.

3

After mRNA

The Genetic Code, Translation, and the Biochemistry of Controlled RNA Synthesis in Bacteria

KNOWING THE STRUCTURE of DNA not only immediately provided a plausible scheme for the replication of the genetic materials but also quickly stimulated speculation regarding other critical issues (Watson and Crick 1953). It was immediately assumed that the *genetic information* that specified protein sequence was written in the sequence of the bases in DNA. But how? Scientists interested in mathematical representations of information began thinking and talking about a "genetic code" in which the four nucleic acid bases in DNA would contain information that could be translated into the 20 amino acids in protein. It was the *nature* of the code—not the operational mechanics of how the code was translated—that mainly attracted this group.

A BRIEF HISTORY OF IDEAS REGARDING THE GENETIC CODE

One of the first individuals to prod Francis Crick and James Watson to think about the code problem was George Gamow. Gamow, a highly imaginative nonconformist Ukrainian physicist who had escaped from Stalin's antiscience Soviet Union in 1933, was a professor of physics at George Washington University in Washington, DC. With his student Ralph Alpher, he proposed the still-accepted theory of the original synthesis of hydrogen and subsequent fusion of hydrogen nuclei to form helium in the early universe. (For an excellent nontechnical explication and amusing historical accounting of the famous

Alpher, Bethe, and Gamow paper [Alpher et al. 1948], see pp. 306–322 in *Big Bang* [Singh 2004]. Reading the authors as α, β, and γ was irresistible to Gamow, although Hans Bethe had nothing to do with Alpher's work as a graduate student with Gamow.) Soon after the DNA structure papers were published, Gamow wrote a long letter to Watson and Crick in July 1953 regarding how information might be encoded in a four-letter code. Both Crick and Watson met with and had extensive correspondence with Gamow. (See pp. 91–96 in Crick's *What Mad Pursuit* [Crick 1988] and J.D. Watson's *Genes, Girls, and Gamow* [Watson 2001], in which more than a dozen letters from Gamow were published.)

Gamow proposed several coding schemes. One involved a physical fit of amino acids into cavities formed by bases in DNA (Gamow 1954). Another featured four overlapping bases encoding amino acids, which, however, would restrict neighboring amino acids. Gamow's persistent intellectual pursuit of the problem resulted in the formation of a notable meeting of the minds of scientists interested in a code of life. He proposed to Watson that 20 individuals (equal to the number of amino acids used in proteins) join in an "RNA Tie Club." This was before messenger RNA (mRNA) was established as actually specifying the amino acids in proteins. Members of the RNA Tie Club apparently never all met together, but a few papers by its members were circulated—one of which was very important: Crick's paper on the *adaptor hypothesis*, which was circulated in the summer of 1955 (Watson 2001, p. 165). In this paper, Crick proposed that translation from the language of nucleic acid to the language of protein required an "adaptor," most likely a short oligonucleotide. Such an adaptor would allow the machinery of the cell to use the information in DNA to specify the alignment of amino aids in proteins.

Crick and Sydney Brenner provided functional leadership to this group. Brenner published a paper completely disproving a code of completely overlapping triplets of nucleotides (Brenner 1957). By 1957–1958, Crick favored a code with three nucleotides in a continuous string specifying each amino acid. Such a "commaless" code that read from a fixed starting point would specify a specific protein (Crick et al. 1957; for review, see Crick 1966b).

Although these early discussions of the code stirred the intellectual pot, no definitive progress toward an actual solution of the code was realized. Pessimism reigned, and it appeared that the difficult problem of "breaking the code" might require decades. This pessimism was codified in an oft-quoted article by Crick (1959). It appeared that the theorists would have to wait for the biochemists. At this point, the only fact pertinent to a specific change in the code, surprisingly enough, came from a human protein.

SICKLE CELL HEMOGLOBIN: MENDELIAN MUTATION CAN AFFECT A SINGLE AMINO ACID

A landmark achievement in biochemistry that began in 1949 and concluded in 1957 gave a solid foundation to the idea of a *code*. Sickle cell anemia, a disease indigenous in sub-Saharan Africa, is due to a single recessive gene mutation that is inherited in a strict Mendelian fashion. Heterozygotes are affected but not nearly so severely as homozygotes. Red blood cells with the sickle defect exhibit a shape change with only a slight decrease in oxygen tension. This causes capillary clogging and more rapid than normal red blood cell destruction, with a resulting severe anemia. In 1949, Harvey Itano and Linus Pauling discovered a lesser charge on sickle cell hemoglobin than on normal hemoglobin (Itano and Pauling 1949; Pauling et al. 1949). Indeed, the homozygous normal hemoglobin and the sickle cell disease hemoglobin were completely separable based on charge, and the heterozygote (sickle cell trait) had an equal mixture of both (Fig. 3-1). Pauling pronounced sickle cell anemia a "genetic disease," the first case in which a simple change in a particular protein had been linked to a heritable disease.

Vernon Ingram, a chemist working at the Medical Research Council for the Study of Molecular Structure of Biological Systems in Cambridge, England (where so much had already happened to inaugurate molecular biology), took on the task of examining hemoglobins, after discussions with Crick (Olby 2009) on the importance of understanding the code for finding amino acid change(s) in a genetically modified protein. He developed a reproducible two-dimensional (2D) (chromatographic and electrophoretic) technique for

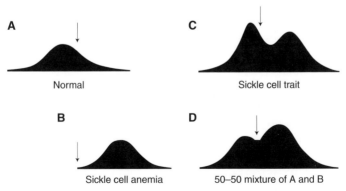

Figure 3-1. Charge difference between normal and sickle cell hemoglobin. Electrophoretic separation (electrofocusing) of normal and sickle cell hemoglobin and demonstration that heterozygotes (sickle cell traits) have about one-half of each electrophoretic type. (Redrawn, with permission, from Pauling et al. 1949 [©AAAS].)

separating and recovering tryptic peptides from the whole hemoglobin molecule. When comparing normal and sickle cell hemoglobin, he discovered that only one peptide had a charge difference. A glutamic acid in the normal version of this peptide was replaced by a valine in the sickle cell globin (Fig. 3-2). This very clear result in 1957 convincingly illustrated that a mutation in a *single* gene could result in a change in a *single* amino acid in the sequence of a protein—a human protein no less (Ingram 1956, 1957). Ingram was emboldened to state that "this is the smallest alteration (of a protein) possible—only one amino acid is affected...it may well be that this involves a replacement of no more than a single base pair in the chain of DNA of the gene" (Ingram 1957, p. 327). Thus, searching for mutational changes that changed amino acids was at least possible.

Knowing Ingram's results, Brenner and Cyrus Levinthal, a physicist turned molecular biologist, decided to try extensive mutagenesis in bacteria or bacteriophage by chemicals whose nucleic acid base targets were known. Brenner began to study mutant T4 phage coat protein, and Levinthal studied alkaline phosphatase. Such directed mutagenesis should cause predictable, specific base changes that could be subsequently correlated with specific amino acid changes. They hoped in this way to try to break the code. Such approaches would be expected at least to gradually shed light on the code, however long it might take.

Crick provided an illuminating review of the thoughts regarding the code in this early period on the occasion of the 1966 Cold Spring Harbor Symposium entitled *The Genetic Code* (Crick 1966b). The first break in this monumentally important problem that had attracted mathematicians, physicists, and geneticists was made quite unexpectedly by a young biochemist in the summer of 1961.

MARSHALL NIRENBERG, THE INITIAL CODE BREAKER

Marshall Nirenberg, after obtaining his Ph.D. in 1957, came to the National Institutes of Health (NIH) in Bethesda, Maryland and worked for 3 years on problems in enzymology, demonstrating his biochemical talents along the way. In his own words, Nirenberg was interested in "exploring an important problem" (Nirenberg 2004). He was thinking of leaving NIH to join Bob Krooth at the University of Michigan to work on human genetics, or perhaps staying at the NIH. If he remained at the NIH, he would try to work on mechanisms of protein synthesis, a problem he recognized as perhaps the most important unsolved fundamental question in cellular biochemistry. When he was offered a position to remain at the NIH in the summer of 1960 by

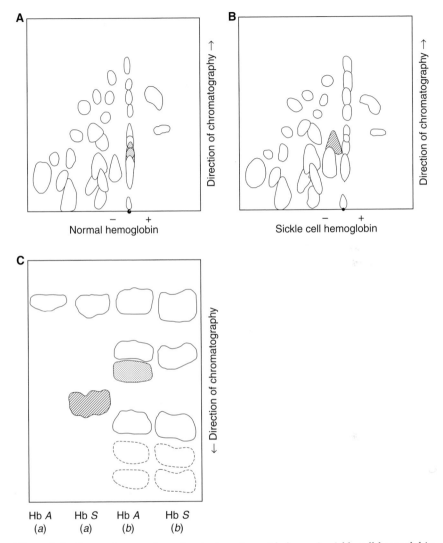

Figure 3-2. Vernon Ingram showed a single amino acid change in sickle cell hemoglobin using the fingerprint technique. Two-dimensional chromatograms of peptides in tryptic digests of normal (*A*) and sickle cell (*B*) hemoglobin revealed a single peptide with a greater negative charge. The peptides were separated first by electrophoresis, the paper was dried and turned 90°, and separation was performed by ascending chromatography. (*C*) Regions from the first chromatograms (shaded) containing the peptides with different charges were further separated ([Hb *A*] normal; [Hb *S*] sickle cell). The stippled and striped peptides were sequenced, revealing the substitution of valine for glutamic acid as the only change in Hb *S*. (Redrawn, with permission, from Ingram 1956 [©Macmillan].)

Gordon Tomkins, himself a very gifted scientist, Nirenberg gladly accepted
and remained (Martin 1984).[1]

During 1959, Nirenberg had realized that bacterial extracts were much
simpler than the animal tissue extracts made famous in the Paul Zamecnik
laboratory for performing cell-free protein synthesis. He therefore undertook
the task of making a workable, cell-free, bacterial protein synthesis system.
Nirenberg had first optimized conditions for protein synthesis using *Bacillus
cereus*, the bacterium in which the induction of penicillinase had been well
studied. He hoped that he might both induce and produce penicillinase in
the test tube. After Heinrich Matthaei joined him, they switched to *Escherichia
coli* (Martin 1984), possibly because of what Nirenberg had learned at a
summer conference of work using *E. coli* extracts by Alfred Tissières and colleagues at the Harvard Biological Laboratory (Tissières et al. 1960). And, of
course, *E. coli* was the most commonly used bacterium. In the modest
recounting of his success, Nirenberg outlines several simple modifications
that he and Matthaei used to fully develop the *E. coli* cell-free protein synthesis system (Nirenberg 2004), as follows:

1. The task of making extracts capable of protein synthesis was arduous.
   They tested various conditions to determine whether extracts could be frozen in small amounts and remain active. If so, they could easily start an
   experiment every day. When β-mercaptoethanol (used by enzymologists
   to prevent unwanted, damaging –SH interactions) was added before freez-
   ing, the extracts retained their protein-synthesizing ability after thawing.

2. The PaJaMa (Pardee, Jacob, Monod) experiment from the Pasteur group
   had been published in 1959 (Pardee et al. 1959), and rumors of the
   mRNA idea had already reached the NIH by the fall of 1960. If ribosomes
   were already engaged in protein synthesis on some preexisting RNA when
   cells were disrupted, the best way to get ribosomes ready to accept and
   translate newly added RNA was to let them finish their already started
   task. Thus, Nirenberg and Matthaei simply incubated the *E. coli* extracts
   first under conditions that allowed protein synthesis, to allow the already
   started reaction to finish. Like others at the time, they assumed that a
   high-speed pelleted fraction of a bacterial extract would contain an RNA

[1] Like Marshall Nirenberg, I was doing postdoctoral training at the NIH during this time. I sat down at
a cafeteria table with Marshall in late July 1960 just before I left to join François Jacob in Paris for the
continuation of my postdoctoral training. Marshall had already started to work on cell-free protein
synthesis, and he fervently pointed out to me how lucky I was. On that day, he understood much
more clearly than did I the power of the recent work in the Paris laboratories. Soon after my lunch
with Marshall, I left for my year in Paris, and by the time I returned in June 1961, Nirenberg was
about to become famous.

fraction that stimulates protein synthesis. Therefore, they extracted RNA with phenol from sedimented ribosomes from growing cells and then added this to the preincubated protein synthesis system. They could actually show a small but reproducible increase in incorporation of amino acids into protein by the now presumably "empty" ribosomes (Fig. 3-3A; Matthaei and Nirenberg 1961a,b). Thus, they had a system that would respond to free RNA by increasing incorporation. In the first of these papers (Matthaei and Nirenberg 1961a), they also describe a simple filtration technique to trap trichloroacetic acid–precipitated protein, providing an assay for radioactively labeled, in vitro synthesized protein. This technique was much simpler than the clumsier radioactive protein assay methods generally used at the time.

3. Radioactively labeled samples of each of the 20 amino acids were prepared so as not to rely on the incorporation of a single or even a few amino acids in measuring protein synthesis. Thus, they could independently test the incorporation of each labeled amino acid in the presence of the 19 other unlabeled amino acids. With this system, they were ready to study the stimulation of protein synthesis by added RNA in a cell-free system.

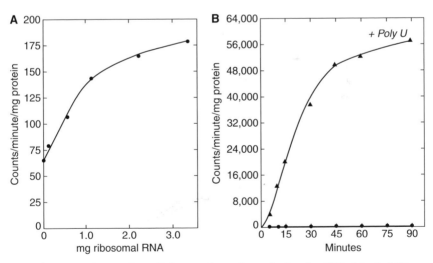

Figure 3-3. Stimulation of *E. coli* ribosomal protein synthesis using RNA. Marshall Nirenberg and Heinrich Matthaei developed a soluble system using *E. coli* components to produce protein in vitro. Two RNA samples were added to the *E. coli* extracts: (*A*) the total RNA from pelleted ribosomes (labeled Mg ribosomal RNA on abscissa), which presumably included a small amount of mRNA, and (*B*) 10 μg of poly(U). The poly(U) increased incorporation 10,000 times better than the cell RNA used in *A*. (Redrawn, with permission, from Nirenberg and Matthaei 1961.)

Because tobacco mosaic virus (TMV) RNA had genetic capacity, Nirenberg obtained some TMV RNA and added it to the *E. coli* protein synthesis
system. A large ($\sim$20-fold) stimulation in incorporation was obtained
(Nirenberg 2004). Nirenberg told Heinz Fraenkel-Conrat, the TMV expert
(see Chapter 2), of this result and went off to his lab in Berkeley, California
in May 1961 to characterize TMV coat protein that might be made with the
new system. Before Nirenberg left Bethesda, he had obtained, from Leon Heppel and Maxine Singer, a sample of the synthetic polyribonucleotide poly(U),
just a string of Us made with polynucleotide phosphorylase (see p. 111).
Nirenberg had not had a chance to use the poly(U) in the protein synthesis
system when he went to Berkeley.

Soon after he arrived in Berkeley, Nirenberg received an excited phone
call from Matthaei. Nirenberg (perhaps) had originally intended to use the
poly(U) as a negative control for "real" protein synthesis directed by TMV
RNA (Martin 1984). But the news Matthaei conveyed was that the poly(U)
stimulated far better than anything else they had ever used (Fig. 3-3B). *However, it only stimulated the incorporation of one amino acid—phenylalanine.* It
was clear to both Nirenberg and Matthaei that this was the first step in solving
the genetic code. UUU (or some larger combination of Us) encoded phenylalanine. Nirenberg promptly returned to Bethesda, and soon they discovered
that poly(C) also stimulated incorporation of only one amino acid, this time
proline. Poly(G) did not work, but physical chemists had already shown that
the bases in poly(G) formed interacting ("stacking") structures; thus, that was
no surprise.[2]

Nirenberg had made no public presentation of these remarkable findings.
In August 1961, he went to the International Congress of Biochemistry in
Moscow, one of the earliest openings for American scientists in large numbers
to go to the Soviet Union of those days. There, Nirenberg presented a brief
talk that was attended by several scientists from Harvard, where, as noted
above, cell-free protein synthesis in *E. coli* extracts was also being studied in
the Watson laboratory (Tissières et al. 1960). Nirenberg only described the
poly(U) stimulation of polyphenylalanine incorporation into protein, which
by this time had been thoroughly checked and was now safely headed for

---

[2]Nirenberg (2004) relates an interesting story illustrating how luck plays a part in so many discoveries.
Poly(A), which was successfully used a bit later, stimulates the incorporation of lysine. Both Mirko
Beljanski, as a visitor in Severo Ochoa's lab, and Alfred Tissières, in the Watson laboratory at Harvard,
had tried in vitro stimulation of protein synthesis by poly(A) in 1959–1960. But they were unsuccess-
ful because the polylysine that is encoded by poly(A) is soluble in 10% trichloroacetic acid, the agent
all had used to precipitate and therefore collect for measurement the newly formed protein after the
in vitro synthesis reactions. Thus, Nirenberg's choice of poly(U) and then poly(C) was indeed
fortuitous.

publication. It was submitted by a former associate director of the National Institutes of Health, Joseph Smadel, to the *Proceedings of the National Academy of Sciences* before Nirenberg left for the Soviet Union. In that paper, Nirenberg and Matthaei state, "One or more uridylic acid residues therefore appear to be the code for phenylalanine" (Nirenberg and Matthaei 1961).

The message in Nirenberg's brief talk was relayed to Francis Crick, who was also at the Moscow meeting. Crick immediately recognized the momentous nature of this discovery. Nirenberg was invited to present his findings to a larger audience (a "plenary" session of which Crick was the chair). To at least a select set of the biochemists at this meeting—those interested in genetics, gene control, and protein synthesis—Nirenberg's talk was the hit of the meeting. Many questions remained unanswered. Aside from the large remaining task of obtaining a complete set of nucleotides encoding all of the amino acids, it was not yet clear that the code was really a triplet commaless code.

BACTERIOPHAGE GENETICS PROVES THE TRIPLET CODE

In early 1961, Crick himself had actually gone into the laboratory, temporarily abandoning his formally singular role as the reigning molecular biology theoretician, to try to solve the proposed triplet nature of the code. The logic of these genetic experiments used dyes that insert (intercalate) into DNA. The first dye used was proflavin, which was shown to be mutagenic by Bob DeMars during his thesis work with Salvador E. Luria (DeMars 1953). When DNA with an intercalated dye is copied, errors result that were thought to be (Brenner et al. 1961)—and most likely were, on the basis of Leonard Lerman's experiments (Lerman 1961)—insertion of a nucleotide or, after the cell attempts a correction, deletion of a nucleotide.

Following this reasoning, Sydney Brenner, Seymour Benzer, and Leslie Barnett (at the time, a technician) had started to collect proflavin-induced mutants in the late 1950s (Brenner et al. 1958). Crick became interested in using such mutants himself to test his and Brenner's ideas regarding the genetic code (Crick 1988, pp. 122–136; Brenner 2001, pp. 90–99). As we noted above, overlapping triplet codes had been ruled out (Brenner 1957), and a commaless triplet code seemed most likely (Crick et al. 1957). They knew of one other important characteristic of large collections of bacteriophage mutants: Revertants very often mapped closely to the site of an original mutant (Crick 1959), suggesting a single base change that could be reversed.

Benzer had defined the *r*II genetic locus in T4 bacteriophage by growth on particular strains of *E. coli*. Wild-type phage grew on either *E. coli* B or *E. coli* K. *r*II mutants grew on B but not on K. Crick and his colleagues made many additional mutants induced by proflavin in the *r*II locus and selected only those that mapped in a limited region. By examining nearly 100 of these very closely spaced mutants by pairwise crosses, they found some pairs could recombine and grow on *E. coli* B or K, but other pairs could not. The mutants were divided into two groups: pluses (+) or minuses (−). A + could recombine to give growth on *E. coli* B or K with a −, but two pluses or two minuses failed to yield growth. Moreover, cells infected with three pluses or three minuses yielded recombinants (less frequently, because two recombinations were required) that grew both on strains B and K. They reasoned that single-base additions or single-base deletions could cause the *r*II mutation, disrupting the reading of the code that was read in groups of three nucleotides ("inframe"). The reading of the nucleotide code after a deletion (nominally, a minus) of a base could be put back inframe by an addition of a base (nominally, a plus; they, of course, did not know which of their mutants actually represented an addition or deletion). Likewise, it would be possible for three deletions to correct the reading of the code just as would three additions of a base.

These experiments clinched the idea of a commaless triplet code that must be read from a fixed starting point three bases at a time. Crick had personally started to perform the recombination experiments himself in May and was halfway through in the summer of 1961 when he first learned of Nirenberg and Matthaei's initial results. He and the remainder of the genetics community could see the detailed solution of the coding problem coming from biochemists using synthetic polyribonucleotides of varying composition. But the great theorist was intensely proud of his personal experimental contribution that the code was a triplet code and very likely commaless (Crick 1988, Chapter 12). This very important step forward in understanding the nature of the code was published in the fall of 1961 (Crick et al. 1961).

USING POLYRIBONUCLEOTIDES TO DESIGNATE CODONS: THE OCHOA AND NIRENBERG DUEL

After returning home from Moscow and his cordial reception by Crick, Nirenberg (only 34 at the time) soon learned that Severo Ochoa, a Nobelist in 1959 and head of a large laboratory at New York University Medical School, was either already working on the code or was just about to start. This was a situation that stimulated discussion at the time among biologists of the possibility of predatory behavior of a great senior biochemist toward a young competitor.

Ochoa had won his Nobel Prize along with Arthur Kornberg with the following citation: "for their discoveries of the mechanisms in the biological synthesis of ribonucleic acid and deoxyribonucleic acid." In fact, what Ochoa and his colleagues had done was to purify an enzyme whose real function in the bacterial cell is to *degrade* RNA. The enzyme was polynucleotide phosphorylase, which conducts phosphorolysis of RNA, yielding nucleoside diphosphates (Fig. 3-4A). By running the degradation process backward with high concentrations of nucleoside diphosphates, the enzyme will synthesize synthetic polyribonucleotides. But the reaction requires no template and was probably oversold as a possible mechanism for making RNA in cells. For example, Ochoa's Nobel Lecture of 1959 (reprinted in Ochoa 1964) continued to hold out hope that polynucleotide phosphorylase was the "real" synthetic enzyme for RNA in cells (Fig. 3-4B).

Nevertheless, the polynucleotide phosphorylase reaction was indispensable for making the synthetic polyribonucleotides that Nirenberg and Matthaei had used. Moreover, if the nucleoside diphosphate used in the reaction is uridine diphosphate (UDP), the polyribonucleotide produced is poly(U); if adenosine diphosphate (ADP) is used, then poly(A) is produced; and if both ADP and UDP are used, poly(AU) results. Thus, by using mixed polyribonucleotides, other triplets of the code could be determined.

Within 2 years after receiving his Nobel Prize, Ochoa and his colleagues had started to work on in vitro protein synthesis, and they clearly recognized the important possibility that mixed polyribonucleotides made by polynucleotide

---

A "The reaction, which requires magnesium ions and is reversible, can be formulated by the equation

$$n\,X - R - P - P \overset{Mg^{++}}{\rightleftharpoons} \left(X - R - P\right)_n + n\,P$$

where R stands for ribose, P–P for pyrophosphate, P for orthophosphate, and X for one or more bases including, among others, adenine, hypoxanthine, guanine, uracil, or cytosine. In the reverse direction, the enzyme brings about a cleavage of polyribonucleotides by phosphate, i.e., a phosphorolysis, to yield ribonucleoside diphosphates."

B "The occurrence of polynucleotide phosphorylase in nature appears to be widespread enough to warrant the assumption that this enzyme may be generally involved in the biosynthesis of RNA."

Figure 3-4. Quotes from Severo Ochoa's 1959 Nobel Lecture. (*A*) Ochoa's diagram and description of the polynucleotide phosphorylase reaction. (*B*) Although the majority of the lecture describes the practical usefulness in producing polyribonucleotides for study, Ochoa's concluding remarks included this quote. (Excerpt reproduced, with permission, from The Nobel Foundation, http://nobelprize.org/nobel_prizes/medicine/laureates/1959/ochoa-lecture.pdf.)

phosphorylase offered in solving the details of the genetic code. Although they had begun in vitro protein synthesis studies earlier, they energetically launched into decoding studies after the Nirenberg and Matthaei papers were published.

This situation greatly depressed Nirenberg because he believed that his breakthrough might be taken over by the large Ochoa laboratory. But the administration at the NIH came to Nirenberg's rescue and provided additional resources. Almost immediately, talented younger investigators joined his laboratory (Martin 1984; Nirenberg 2004). Both the Nirenberg and Ochoa laboratories began to use long synthetic polyribonucleotides in the *E. coli* protein-synthesizing system and quickly obtained provisional evidence for an additional 10–15 codon assignments (e.g., see Nirenberg et al. 1963; Speyer et al. 1963).

But the exact sequence of the polyribonucleotides produced by polynucleotide phosphorylase was not known. For example, a polyribonucleotide containing both U and C stimulated the production of a polypeptide product containing both labeled phenylalanine and proline. This was expected because UUU encoded phenylalanine and CCC encoded proline. However, a long polyribonucleotide containing U and C also caused incorporation of labeled serine and leucine. Thus, one or more of the other six possible combinations of Us and Cs—UUC, UCC, UCU or CCU, CUU, CUC—specified these other two amino acids, but it was not certain which triplet encoded which amino acid.

Two advances occurred to allow further progress on assigning individual codons to particular amino acids. One of these came when Nirenberg and Philip Leder, a postdoctoral fellow, discovered that RNA trinucleotides (each specifically synthesized in a known order) would bind a specific transfer RNA (tRNA) molecule to a ribosome. (This result recalls Crick's initial adaptor hypothesis; see p. 102.) This complex—ribosome/trinucleotide/tRNA^X, where X is a specific amino acid—could be retained by a nitrocellulose filter, whereas tRNA^X by itself would flow through (Leder and Nirenberg 1964). This allowed the assignment of several dozen triplets to specific amino acids, but not all triplets led to successful complex formation, and others were weak in this test.

KHORANA'S SYNTHESIS OF ORDERED RNA TEMPLATES

The final decisive approach to the biochemical solution of the code came from H. Gobind Khorana's laboratory. No technique to chemically synthesize RNA of known sequence had yet been developed. As of 1962, therefore, no protein synthesis stimulated by a mixed polyribonucleotide of known sequence had been achieved. Khorana painstakingly developed a technique that produced polyribonucleotides of known sequence but *not* using polynucleotide phosphorylase (Morgan et al. 1966; Khorana 1972).

Through a chemical synthesis strategy he had developed for DNA, Khorana made short complementary DNA strands of known sequence and then linked the short double-stranded units together (with the enzyme DNA ligase) to make DNA of known sequence. This synthetic DNA was then transcribed by RNA polymerase to make RNAs of known sequence. (The discovery in 1960 of RNA polymerase, the enzyme that transcribes DNA, is discussed in a following section.) For example, he made a polyribonucleotide with the known sequence of ACACACAC---. When translated, this polymer produced a polypeptide of alternating threonine and histidine, but it was impossible to tell which of the two triplets, ACA or CAC, coded for which amino acid. Apparently, a ribosome could start on this synthetic polyribonucleotide at any site (e.g., ACACACAC--- or ACACACAC---) and then continue reading the ACA or CAC codons alternately.

A polymer of AACAACAAC--- was then prepared. This polymer directed the synthesis of three kinds of chains: all arginine, all glutamine, or all threonine. Again, depending on where a ribosome started, it would read one of three repeating triplets (AACAACAA---, AACAACAA---, or AACAACAA---). Because ACA is the only triplet in common between the two polymers ACACACAC--- and AACAACAAC--- and both polymers led to threonine incorporation, ACA could be assigned to threonine. In addition, Khorana, the master organic chemist, also made all of the possible 64 triplets with known sequences that could be used in the Nirenberg filter- binding assays. Using both of these approaches, all of the amino acid coding triplets had been assigned by 1966. But three of the possible 64 triplets did not seem to code amino acids (Table 3-1).

THE CODE INCLUDES PUNCTUATION: START AND STOP SIGNALS

Because ribosomes can read synthetic polyribonucleotides randomly starting in any reading frame, something must guide natural mRNAs to start in frame at a "real" start site in order to make a protein correctly—and, of course, stop at a specific stop site to finish the protein correctly.

Termination Signals and Nonsense Codons

The code for stop signals was solved first. Genetic experiments in *E. coli*, especially in the laboratories of Sydney Brenner (Stretton et al. 1966) and

Table 3-1. The genetic code: RNA codons for amino acids

First position (5' end)	Second position				Third position (3' end)
	U	C	A	G	
U	Phe	Ser	Tyr	Cys	U
	Phe	Ser	Tyr	Cys	C
	Leu	Ser	Stop (och)	Stop (opal)	A
	Leu	Ser	Stop (amb)	Trp	G
C	Leu	Pro	His	Arg	U
	Leu	Pro	His	Arg	C
	Leu	Pro	Gln	Arg	A
	Leu	Pro	Gln	Arg	G
A	Ile	Thr	Asn	Ser	U
	Ile	Thr	Asn	Ser	C
	Ile	Thr	Lys	Arg	A
	Met (start)	Thr	Lys	Arg	G
G	Val	Ala	Asp	Gly	U
	Val	Ala	Asp	Gly	C
	Val	Ala	Glu	Gly	A
	Val (Met)	Ala	Glu	Gly	G

Although the genetic code is universal for most proteins in most cells, it has some variations—both in bacteria, where UGA = Trp (in mycoplasma), and in single-celled eukaryotes, where CUG = Thr (in yeast mitochondria) and UAA, UAG = Gln and UGA = Cys (in a variety of protists).

Alan Garen (Weigert et al. 1966), showed that the three triplets that failed to bind a labeled amino acid—UGA, UAG, and UAA—signaled *stop* in protein synthesis.

In certain defective bacteriophage or bacterial mutants, the normal strain of *E. coli* was incapable of making complete phage coat proteins (studied by Brenner) or enzyme (alkaline phosphatase) (studied by Garen). Instead, these mutants produced truncated, prematurely terminated proteins. (Later, these were called *nonsense* mutants because they did not cause the insertion of a wrong amino acid as did *missense* mutants, which incorporate errant amino acids.)

In selected *E. coli* strains, however, these nonsense mutants led to the production of enough complete protein to make some phage or detectable enzyme. This phenomenon is called *suppression* (the mutation is "suppressed"), and the *E. coli* strains are called *suppressor strains*. These strains have a rare mutant tRNA that recognizes the chain-terminating codon in the mRNA as a normal codon and therefore adds an amino acid to complete the protein. For example, the codon UAG, normally only a chain-terminating codon, is not recognized by a normal tRNA anticodon (see the next section for discussion of codon:anticoden recognition). A mutation from UGG (a tyrosine codon) to UAG would normally result in chain termination. In one suppressor strain, a minor tyrosine tRNA with an anticodon 3'-AUC-5' recognizes the 5'-UAG-3', thus inserting a tyrosine. (Discovery of anticodons in tRNA is described on pp. 106 and 107.) Thus, instead of chain termination, the whole protein is produced, albeit at a reduced level. The suppressor tRNA is made at a low level; otherwise, it might be lethal to the bacterium.

Analysis of different suppressor strains (and ultimately their tRNAs) proved the three stop codons—UAG, UAA, and UAG. The understanding of the suppressor tRNAs greatly aided future experiments because any mutation that could be corrected by a suppressor tRNA was presumed to represent a stop codon, thus identifying the gene as a protein-coding gene.

The Start Codon AUG and Initiator tRNA

Although it was recognized immediately after experiments involving in vitro transcription of mixed synthetic polyribonucleotides began that a precise location of protein start sites was necessary, it took until the late 1960s before this important problem was completely solved. The start codon is AUG in all types of cells. The final solution to this crucial problem rested on the isolation of a specific methionyl tRNA that is only used at a translation start site, where it is recognized by a special protein, an "initiation factor," for protein synthesis. This tRNA is referred to as $tRNA_i^{Met}$, signifying that it is the initiator tRNA (Clark and Marcker 1966). Methionine, of course, like all of the other amino acids, is also used internally in protein chains. The insertion of methionine at internal sites in a protein is handled by other tRNAs designated methionyl-tRNA, $tRNA^{Met}$, that are not recognized by the initiation factor.

With the establishment of three stop codons and AUG as the universal start codon, the genetic code was complete (Table 3-1). The code showed many amino acids to be specified by more than one codon; i.e., the code is "degenerate." This was not a surprise because there are 64 (4^3) possible codons and only 20 amino acids in proteins. Degeneracy (described below) was only cleared up by understanding exactly how tRNAs recognized codons. As we learn in the next section, translation of the code is firmly in the hands of the tRNA molecules—starting at a specific site, choosing the correct amino acid, and terminating after a specific amino acid.

The Code Is Universal

The original experiments of Nirenberg, Ochoa, and Khorana all used extracts of *E. coli*. Was the code revealed by these experiments only for bacteria or was it more general? This was first tested in extracts of rabbit reticulocytes, which were known to make mostly hemoglobin. Gunter von Ehrenstein in Fritz Lipmann's laboratory used activated *E. coli* tRNA as a source of amino acids and reticulocyte extracts to make hemoglobin. The hemoglobin sequence that these mammalian ribosomes and mRNA produced using *E. coli* tRNAs was correct as tested by Vernon Ingram's tryptic peptide analysis technique (von Ehrenstein and Lipmann 1961). Thus, it seemed very likely that the code was universal, a conclusion that has held true (Nirenberg et al. 1966). (As noted in the footnote to Table 3-1, there are some rare code variants in nature, but the vast majority of all mRNAs are read according to the "universal" code.)

HOW tRNA READS THE CODE IN AN mRNA MOLECULE

During the final stages of "code breaking," another landmark achievement was made in 1965 by Robert Holley and coworkers. They finished the complete sequence of the nucleotides in a tRNA molecule (Holley et al. 1965), that of an alanine tRNA from yeast. (Their laborious sequencing technique used a limited repertoire of ribonucleases and brief alkali treatment. The method was only useful for short RNAs like tRNA.) Using Watson-Crick base-pair rules, Holley and colleagues suggested that the single-stranded tRNA sequence could fold back on itself (using G:C and A:U base pairs) and form a 2D "cloverleaf" structure (Fig. 3-5A). The real structure, of course, is three-dimensional (3D), and it was a number of years before a 3D L-shaped tRNA structure was solved (Kim et al. 1973; Robertus et al. 1974). The L-shaped atomic structure (Fig. 3-5B) confirmed the base-pairing arrangement suggested originally by Holley and colleagues.

A non-base-paired loop projected at one end of Holley's "cloverleaf" structure. One of the codons for alanine was known by that time to be 5'-GCC-3', and in the projecting, non-base-paired loop of the alanine tRNA were three bases: 5'-IGC-3'. I represents inosine, a close relative of guanine (guanine minus a methyl group) that can form a base pair with C just as G does (see the next section on the "wobble" hypothesis). Therefore, the 5'-IGC-3' in the tRNA is the Watson-Crick complement, the "anticodon" of the alanine codon.

5'- I G C-3' anticodon
3'-C C G-5' codon

Many biochemical and biophysical studies have since proven that all tRNAs have a similar structure with an anticodon loop as proposed by Holley. Crystallographic structures of tRNA have reemphasized the similarity in overall shape of all of the tRNAs (Fig. 3-5B).

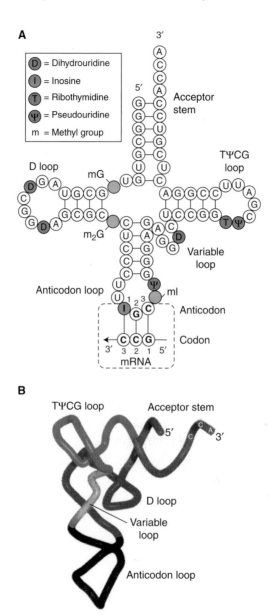

Figure 3-5. tRNA structure. (*A*) The stem–loop structure was proposed in 1965 by Holley et al. (1965) when the sequence of yeast phenylalanine tRNA was completed. (*B*) The crystal structure of tRNA (the backbone of which is shown here) confirmed the base pairing proposed by Holley and colleagues and showed the L shape that is characteristic of all tRNAs. (*A*, Modified from Lodish et al. 2008; *B*, reprinted, with permission, from Arnez and Moras 1997 [©Elsevier].)

"Wobble" Hypothesis and Degeneracy

The presence of inosine in the alanine tRNA anticodon led to the explanation of degeneracy in the code. Again, it was Francis Crick who correctly hypothesized how variations in the third coding position (such as that for inosine in the anticodon of alanine tRNA) would explain degeneracy (Crick 1966a).

Crick proposed that strict C:G and A:U base pairing was not required in the third codon:anticodon position but that a "wobble" would be allowed. Inosine can form reasonably strong pairing with three bases: U, C, or A (Fig. 3-6). Moreover, it was recognized earlier that G and U could pair almost as well as G and C. (In fact, in Holley's suggested cloverleaf structure, there were two G:U base pairs.) Thus, although the first two base pairs of the codon:anticodon are always "canonical" (i.e., G:C or A:U), the third interaction can be G:U, I:U, I:C, or I:A. This "wobble" allowance explains why 61 codons can exist for the 20 amino acids.

Direct Look at Protein Synthesis in Action

Thus, the broad picture for information transfer from DNA to RNA to protein was finally complete by 1966–1967. In fact, electron micrographs (Miller et al. 1970) of gently lysed *E. coli* show the whole process (Fig. 3-7). In bacteria, DNA is copied into mRNA, ribosomes bind quickly to the mRNA as it comes off the DNA, and each codon is read successively by the correct tRNA, which finds its place through its anticodon. The growing peptide is linked to one tRNA and is then transferred to the incoming tRNA. The process repeats itself until the stop codon is reached. All of this occurs while the tRNAs are associated with a surface on the large ribosomal subunit. (In Chapter 6, we refer to the current version of the operation of this RNA/protein machine at atomic resolution.) The primacy of RNA in information transfer and use seemed to be complete at this point.

HOWARD DINTZIS AND DIRECTION OF PROTEIN SYNTHESIS

We have left unattended a fundamental issue in protein synthesis that was not answered by bacterial genetics or *E. coli* biochemistry but was actually already solved during 1960. Because a protein has both an amino end (—NH$_2$) and a carboxyl end (—COOH) and proteins are synthesized stepwise, one amino acid at a time, as the ribosome successively reads each codon, the direction of synthesis could theoretically proceed either —NH$_2$ to

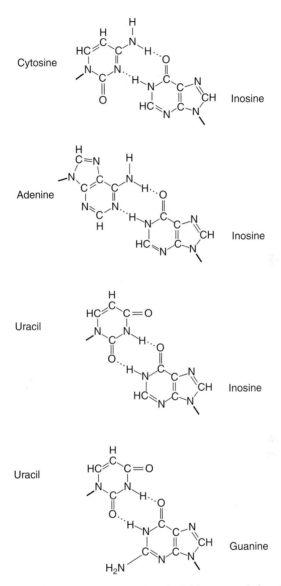

Figure 3-6. "Wobble" base pairs. In 1966, Francis Crick proposed that the structure of inosine would allow for pairing with three of the bases (cytosine, adenine, and uracil) in the third position of codons; this explained the degeneracy of the genetic code. The G:U base pair had already been recognized. (Modified, with permission, from Crick 1966a [©Elsevier].)

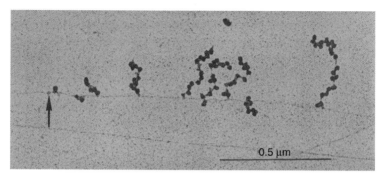

Figure 3-7. Coupled transcription/translation in bacteria is visualized. Oscar Miller and colleagues lysed *E. coli* cells and immediately collected the cell contents on electron microscope grids. They saw threads of mRNA still associated with DNA (thin lines), and ribosomes—several at a time—were already translating protein along the mRNA. Thus, in bacterial cells, the picture of information recovery and use, at least in broad outline, was complete: mRNA was made on demand; ribosomes recognized the 5′ end of the mRNA, bound, and began protein synthesis even before the mRNA had been completely synthesized. (In this photo, the arrow indicates a presumptive RNA polymerase [the faint disk to the left of the first ribosome]. The DNA thread at the top is being copied into mRNA, but the one at the bottom is not. Both are presumably double stranded.) (Reprinted, with permission, from Miller et al. 1970 [©AAAS].)

—COOH or vice versa. Howard Dintzis at The Massachusetts Institute of Technology (MIT), who had used rabbit reticulocytes in his earlier studies at the Medical Research Council (MRC) in Cambridge, solved this question in elegant experiments performed in 1959–1960 and reported in early 1961 (Dintzis 1961). (Note that the enzymology of tRNA activation was being done at about the same time as the Dintzis experiments and the transfer of an activated amino acid from its aminoacylated tRNA to the —COOH group of the most recently added amino acid on the peptidyl tRNA was not firmly established. In fact, as Dintzis pointed out, stepwise synthesis versus simultaneous joining of blocks of peptides was still being discussed.) Dintzis knew the details of Ingram's tryptic peptide analysis of globin (Ingram 1956) and could easily separate the tryptic digests of each of the α and β chains of hemoglobin, which represent 90% of the protein made in reticulocytes.

Dintzis introduced radioactive amino acids to the reticulocytes for short, increasing intervals (Fig. 3-8; actually, for convenience, he performed incorporation for intervals between 4 and 60 min at 15°C, at which temperature synthesis was one-fourth the rate at 37°C). At various intervals, he then collected the *finished* hemoglobin chains that had been released from ribosomes after labeling began (Fig. 3-8A), expecting that the peptides in the hemoglobin

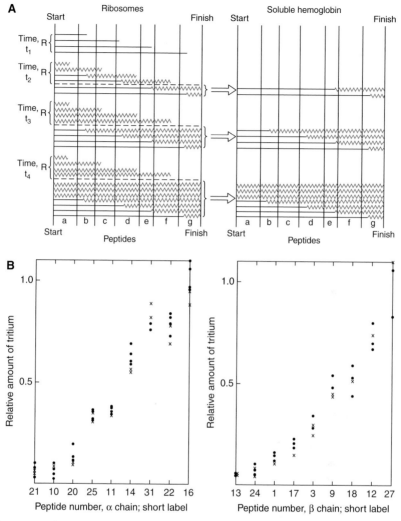

Figure 3-8. The Dintzis experiments proved that proteins are formed from the —NH₂ end to the —COOH end. (*A*) Model of protein synthesis showing nascent protein chains unlabeled (time t_1); times t_2, t_3, and t_4 show progression of nascent chains in the presence of labeled amino acids (wavy red lines). The black dashes separate nascent from finished chains. The letters at the bottom (a–g) represent ordered peptides. The diagram makes it obvious that radioactivity in *finished* protein chains would first appear in peptides g and in sequence back to peptide a. (*B*) There are two different globin chains in hemoglobin (α and β; it is actually a tetramer, $\alpha_2\beta_2$). Dintzis analyzed the appearance of label in finished chains of all of the tryptic peptides of α-globin or β-globin at several different times. The diagram shows an incubation (label time) of 7 min in two experiments (filled circles and ×s). The 7-min incorporation time approximates the distribution of label similar to time t_3 in *A*. Experiments of this sort showed unambiguously that protein synthesis proceeds in one direction. Dintzis then used an enzyme, carboxypeptidase, to remove the —COOH terminus from finished label chains (such as those labeled for times such as t_2 and t_3) and showed that the first labeled peptides were —COOH terminal. Clearly, synthesis proceeds from —NH₂ to —COOH. (Redrawn, with permission, from Dintzis 1961.)

chains would be labeled in a definite order if stepwise synthesis occurred. The newly finished α- and β-hemoglobin chains were separated and cleaved into peptides with trypsin. The peptides were separated so that he could determine the content of radioactivity in each peptide. With longer and longer label times, he discovered a definite order of increasing radioactivity in the peptides from each of the α and β chains (Fig. 3-8B). Accordingly, a stepwise synthesis from one end to the other was proved.

To complete the argument, Dintzis discovered that the peptide that was the first to be labeled in the shorter labeling interval was quickly removed by a carboxypeptidase that attacks from the —COOH end of a protein. The results clearly showed that the first labeled peptide to appear in finished chains was the —COOH terminal peptide. As the label time increased, the finished chains acquired label in a —COOH to —NH$_2$ direction. This was true for both the α- and β-hemoglobin chains. Clearly, globin synthesis starts at the amino terminus and proceeds toward and finishes with the carboxyl terminus. This very important fact has held true for every protein synthesized by a ribosome.

BIOCHEMISTRY OF REGULATED BACTERIAL mRNA SYNTHESIS

Despite much progress, important questions regarding the biochemistry of gene regulation were not yet answered. But the doors had been flung open, and thanks to additional bacterial genetics coupled with improving biochemistry of both proteins and nucleic acids, the study of regulated transcription both inside bacteria and in cell-free extracts advanced steadily.

RNA Polymerase: An Enzyme That Makes RNA

At virtually the same time that ideas about mRNA were being formulated, the discovery of an enzyme that can make RNA by copying a DNA sequence was made. Ochoa's polynucleotide phosphorylase, whose normal function is digesting RNA, was not the RNA polymerase. What begged for discovery was an enzyme that depended on DNA and copied the DNA faithfully into mRNA so that the protein-synthesizing machinery could be correctly programmed. Just at the time of discovery of mRNA, there came the first ray of light that such enzymes existed—and it was not from bacteria, but from broken cell preparations of rat liver.

Experiments in the 1950s using *autoradiography* showed that the nucleus of animal and plant cells incorporated ribonucleotides more actively than did the cytoplasm (see Chapters 1 and 4). When animal cells or tissue are ground up appropriately, most *nuclei* of cells survive and can be separated from the rest of the cell debris. Treating nuclei with solutions of ~0.4 M NaCl or more releases chromatin—chromosomes and associated proteins. Sam Weiss (Weiss and Gladstone 1959; Weiss 1960) discovered that such rat liver nuclear extracts would incorporate labeled ribonucleotides into RNA. Moreover, all four ribonucleoside *tri*phosphates (UTP, ATP, etc.) were required for incorporation—*not* ribonucleoside *di*phosphates (UDP, ADP, etc.), as were used with the Ochoa enzyme polynucleotide phosphorylase. This finding was very significant, even though Weiss was not able to solubilize the enzyme at the time.

Arthur Kornberg and colleagues had shown in the mid 1950s that an enzyme, DNA polymerase I, can elongate one DNA chain of a partial double-stranded DNA by copying the accompanying chain, and the DNA polymerase required triphosphates of *deoxy*ribonucleosides (Kornberg 1957). It was widely suspected that a true RNA polymerase that copied DNA would also use the energy stored in ribonucleoside triphosphates to accomplish linkage of the ribonucleoside monophosphate units into an RNA chain. Thus, when the reaction that Weiss reported was discovered to require all *four* triphosphates, it was widely believed that real RNA synthesis had been observed.

Soon after, Audrey Stevens and Jerard (Jerry) Hurwitz reported—in the same issue of the same journal—a cell-free reaction in extracts of *E. coli* that produced RNA, also from all four triphosphates (Hurwitz et al. 1960; Stevens 1960). Purification of this bacterial enzyme soon followed. It strictly required a DNA template in order to produce RNA. A second very important characteristic was noticed immediately: The *E. coli* RNA polymerase could start a new chain by itself. The first ribonucleotide was the 5′ end of the new chain (e.g., 5′pppApXp---). The enzyme could make very long RNA chains moving in a 5′-to-3′ direction while reading the DNA template from its 3′ end to its 5′ end. This ability to start a new chain was significant. DNA polymerase only added deoxyribonucleotides to already existing chains (DNA or RNA). Being able to start a new chain was needed if the RNA polymerase was to skip around the bacterial chromosome producing specific mRNA that the cell needed at any given moment.

Given that an RNA polymerase could copy DNA, a new goal became to prove in a test tube that regulatory proteins control RNA polymerase in the copying of specific genes. Before discussing the fulfillment of this goal, which did occur in the late 1960s and early 1970s, it is necessary to backtrack to a bit more bacterial genetics to explain the discovery of additional gene regulatory proteins that occurred in the late 1960s. Two very important general principles were uncovered during this time: (1) Positive transcription factors, as well as repressors, exist and (2) regulated transcription, even in the relatively simple bacterial systems, requires *several* proteins in addition to bare DNA and RNA polymerase.

UNCOVERING THE REQUIREMENTS FOR REGULATED RNA POLYMERASE ACTION

Jacob and Monod's logic was developed from bacterial systems in which repressors were the regulatory molecules—the *lac* operon, maintenance of lysogeny by repressor, and amino acid repression of amino-acid-synthesizing enzymes in rich medium (see Chapter 2). But it was just as logical that there should be positive activators that would help to direct the RNA copying machinery, including the now identified RNA polymerase, to start at a particular site on the chromosome.

Positive-Acting Transcriptional Proteins

In the early 1960s, Ellis Englesberg and coworkers had mapped a cluster of *E. coli* genes (an operon) required for the metabolism of the five-carbon sugar arabinose. Three of these genes encoded enzymes required to convert the arabinose into a readily usable metabolite (pyruvate) for the preexisting enzymatic machinery of the cells. However, one of these genes did not encode any apparent necessary enzyme. Products of this *AraC* gene were inactivated at 42°C, and ultimately, suppressible nonsense mutations in the gene were discovered (Englesberg et al. 1965; Irr and Englesberg 1970). Englesberg concluded that the *AraC* gene product was, in fact, a *positive-acting* protein, a gene regulator that was required to *activate* transcription of the *Ara* operon. Monod resisted the Englesberg interpretations largely on the basis, it seems, that negative control was so elegant. Once nature had solved such an important problem as gene regulation, would not that one logical solution be universal? Jacob thought differently; to him, "such uniformity seemed to me improbable" (Jacob 1988, pp. 319–320).

As has long since been proven, the *AraC* gene does encode a protein that, in the presence of arabinose, is positive acting. The positive action of AraC during transcription was clearly demonstrated, as we see below, in a cell-free system, showing that *E. coli* RNA polymerase needed AraC to copy mRNA from the *Ara* operon (Lee et al. 1974). (The complete function of the AraC protein and its role in regulating the *Ara* operon, which encodes the enzyme that metabolizes arabinose, was not worked out until the mid 1980s by Robert Schleif and colleagues [Dunn et al. 1984].) Other investigators from the late 1960s on showed genetically that numerous other operons in *E. coli* (e.g., those for the disaccharide maltose [Schwartz 1967] and the five-carbon sugar rhamnose [Power 1967]) also require their own positive-acting factors.

lac and λ Repressor Proteins Are Purified

An important initial goal of regulated transcription in a cell-free system was to prove without a doubt that the *lac* operon could be regulated by a repressor protein. By 1965, suppressor mutations had shown that the repressor was almost surely a protein (Horiuchi and Novick 1961; Sadler and Novick 1965). The existence of operator mutations (o^s) that were constitutive for β-gal synthesis, but still made the repressor, the *i* gene product, suggested that the lac repressor was a protein that bound to DNA, preventing *lac* operon mRNA transcription. But only with pure protein could proof of these ideas be achieved.

Walter Gilbert and Benno Müller-Hill knew that the lac repressor protein probably interacted with lactose or with a gratuitous inducer such as isopropylthiogalactoside (IPTG; see Chapter 2). Such compounds presumably caused the repressor to leave the DNA where it blocked synthesis of the *lac* operon mRNA. Gilbert and Müller-Hill, therefore, worked out a very sensitive chemical assay for the lac repressor protein using a radioactive inducer. It was expected that the number of repressor molecules in a cell might be very low. They found only ~10 molecules per cell based on the amount of radioactively labeled galactoside (IPTG) that bound (Gilbert and Müller-Hill 1966; for review, see Gilbert and Müller-Hill 1970). It was clear that the task of purifying the protein for study would be difficult. Gilbert and Müller-Hill then proceeded, patiently and ingeniously, to engineer *E. coli* genetically so that the engineered cells produced thousands of lac repressor molecules instead of the ~10 that the normal *E. coli* cells produce (Müller-Hill et al. 1968). At this stage, they purified the repressor protein.

Mark Ptashne, working as a graduate student in the same department at the Harvard Biological Laboratories, purified the λ repressor, the other much-studied negative-acting protein at the time (Ptashne 1967), which could be produced more easily in higher amounts. By the late 1960s, both of these proteins were available in pure form for testing in biochemical systems.

MORE AND DIFFERENT POSITIVE-ACTING
TRANSCRIPTIONAL PROTEINS

The next problem, solved through genetic manipulations, was to obtain DNA molecules in which the *lac* operon had been greatly enriched to furnish the DNA to transcribe in vitro. This was done by including the *lac* operon within a bacteriophage DNA. The purified phage DNA then became the template from which to transcribe β-gal mRNA. The object was to add *E. coli* RNA polymerase to the DNA containing the *lac* operon and make β-gal mRNA. To measure β-gal mRNA, the intent was to use a cell-free protein-synthesizing system to make β-gal protein. However, no such success was achieved initially. The RNA polymerase by itself produced no β-gal mRNA from the template DNA.

"Catabolite Repression" and the CAP Protein

The problem was that undiscovered positive-acting proteins were required. It was necessary to return to bacterial physiology and genetics for the answer to this puzzle. *E. coli* cells grown on glucose do not induce β-gal in response to lactose (for that matter, in the presence of glucose, the arabinose enzymes are not produced in response to arabinose either). This phenomenon had been noted in the 1940s by Monod (Chapter 2, Fig. 2-9; see also Fig. 1 in Monod 2003) and studied in detail as "catabolite repression" by Boris Magasanik (1968).

The missing link was discovered after the following observations were made. When *E. coli* cells run out of glucose, they produce an "alert signal," cyclic AMP (cAMP). This discovery was made in the laboratory of Earl Sutherland, who had discovered cAMP in animal cells. It is now known, of course, that cAMP is used as a signal, although somewhat differently, in all cells (for review, see Sutherland 1970). *E. coli* cells with a high level of cAMP respond to a wide variety of different sugars and digest and use them as a carbon source (for review, see deCrombrugghe and Pastan 1978).

The explanation for this change in "taste habits" of the bacterium was discovered in Ira Pastan's laboratory in the late 1960s and early 1970s. A cAMP-binding protein (first discovered by mutations in its gene) is required for *E. coli* to be induced to split lactose. This single protein is required in order for *E. coli* to use many different sugars, including arabinose. The protein is called CAP (cAMP, or catabolite, activator protein). It is now known that the CAP binds cAMP, changing its conformation to an activator that then binds and bends DNA (Schultz et al. 1991). The difference in conformation between the CAP protein with and without cAMP was demonstrated at the atomic level only recently (Popovych et al. 2009).

Cell-Free Gene Regulation

The proof of the role of CAP came from cell-free protein synthesis systems. First, broken cell preparations of *E. coli* could make β-galactosidase from bacteriophage DNA template *if* cAMP was added. Extracts from cells mutant in the CAP gene could not, but purified CAP protein plus cAMP added to the extracts allowed β-gal synthesis. Finally, Pastan's lab showed in a more purified system that RNA polymerase, cAMP, and CAP protein were all necessary. In this system, the lac repressor stopped synthesis, and the inducer, IPTG, caused de-repression (Zubay et al. 1970; deCrombrugge et al. 1971). Thus, the goal of proving that repressor protein serves to block RNA polymerase from making β-gal mRNA was finally reached in 1971.

The magic of the CAP protein was not confined to the *lac* operon. When the positive-acting AraC protein plus DNA containing the *Ara* operon and RNA polymerase were mixed, no *AraC*-specific RNA was formed. When the CAP and AraC proteins plus cAMP and arabinose were added, the polymerase now formed mRNA from the *Ara* operon (Lee et al. 1974). In this case, *two* positive-acting proteins—CAP and AraC, each bound to their small molecular ligands—were required to get the RNA polymerase to do its job of specific, regulated transcription. By the time of these experiments in the 1970s, the idea of positive-acting bacterial transcription factors was very widely acknowledged.

RNA POLYMERASE COMPOSITION AND DEFINITION OF PROMOTER SEQUENCES IN DNA

E. coli RNA polymerase, in early purified samples, had four protein chains $(\alpha_2\beta\beta')$ in its "core," but a fifth chain was also present in the most carefully purified polymerase, the so-called *holoenzyme*. The holoenzyme with the fifth subunit did the best job of copying DNA (Burgess et al. 1969). This extra subunit, named *sigma* (σ), is yet another required positive-acting factor. It is now known that bacteria have several σ factors, some of which serve specialized sets of genes such as those to survive nutritional stress or to facilitate spore formation.

By the late 1970s, when the first DNA sequencing techniques had been worked out by Allan Maxam and Walter Gilbert (Maxam and Gilbert 1977), the roles of the various positive-acting factors were more precisely described. Naturally, among the earliest regulatory sites to be fully explored were those in the *lac* operon regulatory region (Fig. 3-9). Techniques had been developed to detect the physical presence of proteins interacting with DNA (Siebenlist et al. 1980). Sites in DNA that were protected against chemical reactions by site-specifically bound protein were determined. Because techniques for site-specific mutagenesis of DNA had been developed (Hutchison et al. 1978), mutations of the contact sites themselves could be investigated to define regions necessary for specific DNA:protein associations.

The original, genetically identified *operator* site in the *lac* operon was right where RNA polymerase started transcription of the *lac* operon. The repressor

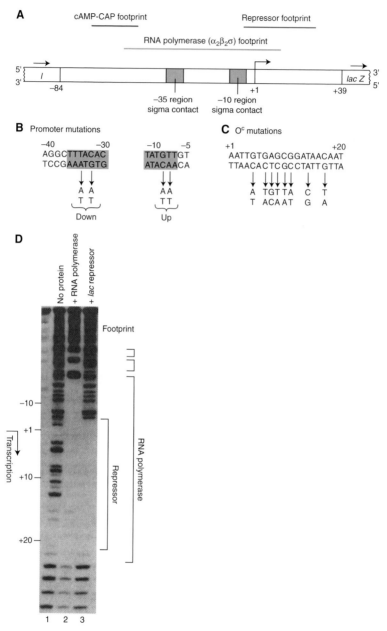

Figure 3-9. The *lac* operon and protein-binding sites circa 1978. (*A*) By the late 1970s, the DNA sequence in the immediate vicinity of the transcriptional start site (+1) of the *lac* operon was known, and DNase footprinting for determining DNA-binding sites of specific proteins had been achieved. The results were in accord with RNA polymerase (plus σ^{70}) requiring unobstructed access to the promoter region. (*See facing page for legend.*)

protein covers this site and is removed after binding with an inducer. Two important σ-binding sites are upstream (in the 5' direction) of the transcription start site (Fig. 3-9). With the repressor not present, the σ factor associated with RNA polymerase contacts nucleotides 10 and 35 base pairs from the beginning of transcription. These regions are referred to as the −10 and −35 regions, and they are similar in many other *E. coli* operons. But as in vitro synthesis showed, the binding of RNA polymerase, even with σ factors, is weak. The CAP protein binds to a region at −50 to −70 from the RNA start site; furthermore, the CAP protein binds directly to the RNA polymerase. Together, the complex of CAP plus polymerase with σ binds strongly, and this is what causes a high rate of *lac* mRNA production from the *lac* operon. Thus, *lac* operon transcription is a two-step process: Inducer (lactose) causes the repressor to come off the DNA, and then the RNA polymerase plus σ and CAP protein act to transcribe the operon rapidly.

Likewise, the CAP protein is required for arabinose metabolism, and there is an interaction between CAP and the AraC protein to promote transcription. (All of these more recent experiments are described in Busby and Ebright 1999; Bell and Lewis 2001; Darst 2001.)

These studies on gene regulation and the protein machinery that interprets DNA-binding sequences in bacteria represent an imposing edifice on which to consider the regulatory arrangements that are discussed in Chapters 4 and 5 in connection with eukaryotic transcription. However, the transcription of genes in cells of eukaryotes, including human cells, requires an even more elaborate set of positive-acting transcription factors to make mRNA, and negative restraint on transcription does not rest simply on negative-binding proteins blocking RNA polymerase access to start sites on DNA.

Figure 3-9. (*Continued*) With repressor bound to the *lac* operator, this access is blocked. (*B*) Nucleotide sequence in upstream regions and in positions of regions protected from DNase. Mutations in protein-binding sites show bases that increase (Up) or decrease (Down) transcription. The positive action of CAP occurs on the upstream side (for review, see Reznikoff and Abelson 1978). (*C*) The most advanced understanding of *lac* operon control involved biophysics studies, finally including crystallography of the dimeric *lac* repressor, which can form a tetramer. The binding of a dimer at the major operator shown here is abetted by a second dimer bound either at an upstream or downstream accessory site that keeps the local concentration of repressor high near the major physiologically important repressor site. (See Müller-Hill 1998 for discussion.) The O^C mutations that cause constitutive synthesis of β-gal mRNA are located just downstream from the transcription start. (*D*) Electrophoresis of DNase digestion products of the *lac* repressor shows cleavage with no protein (dark bands) and sites protected by the *lac* repressor or by RNA polymerase and σ^{70} (spaces between bands). (*A*−*C*, Modified from Lodish et al. 1995; *D*, reprinted, in part, from Schmitz and Galas 1979 [©Oxford University Press].)

COLINEARITY

In the 1960–1967 era, before DNA sequencing when biochemical genetics, largely with *E. coli* and its phages, was the reigning mode of research, one further idea and resulting type of experiment claimed a great deal of attention. If genes encoded the amino acid sequence of proteins, then multiple genetic mutations, closely mapped in a definite order, should correspond to changes in the amino acid sequence of the encoded protein. Seymour Benzer's elaborate recombinational analysis of the *r*II locus in bacteriophage T4 had originally stimulated this idea. By the late 1950s, Benzer's recombination map was so detailed that it was thought to represent the *r*II gene at the level of individual amino acids. Unfortunately, the rII protein remained an unknown entity (Benzer 1960).

Charles Yanofsky and colleagues tackled this problem of colinearity of gene and protein with the A and B subunits of tryptophan synthetase in *E. coli*. They collected many hundreds of carefully mapped mutants in this enzyme. They then purified normal and mutant proteins encoded by many of the mutants and located by protein sequence analysis the position of each amino acid change. They even obtained reverted mutants (mutants that regained enzyme activity) and mapped the amino acid changes in the reverted proteins (Yanofsky et al. 1964).

A similar set of experiments was performed with bacteriophage T4 coat proteins by Brenner and colleagues, particularly Anand Sarabhai and Tony Stretton (Sarabhai et al. 1964). All of this elegant and exhaustive work proved that the mutations in both of these proteins lined up perfectly with the genetic map. *Colinearity* between the gene and the protein sequence—the idea that was born with the Watson-Crick structure of DNA—was proven for bacteria and for phage. The "collinearity" doctrine seemed likely, like the code itself, to be universal. Adherence to this idea impeded the understanding of mammalian mRNA formation, as we see in Chapter 4.

WAS THE PARTY OVER?

Toward the close of the 1960s, at least some scientists felt a sense that the major fundamental questions of molecular biology had *all* been answered. Perhaps this was understandable. In 1950, many still doubted that DNA was the genetic material. Proteins were only recognized in the early 1950s to be long, ordered strings of amino acids, and it was accepted after George Beadle and Edward L. Tatum that somehow genes controlled protein function. But there was no hint in 1950 that in less than 20 years the mystery of heredity would be lifted with the solution of the structure of DNA, that the genetic code and the fundamentals of protein synthesis would be solved, or, finally, that a first understanding of genetic control would be uncovered. The major advances after 1953 vaulted RNA and the control of RNA synthesis to a very prominent, if not the most prominent, position aboard the steamroller that molecular biology had become. Monod and Jacob (1961) were waxing eloquent regarding how the insight into gene control would be soon extended to unlock the secrets of development in eukaryotes. They conceded that operons might not exist in eukaryotes. It was already known that in single-cell eukaryotes such as fungi (e.g., *Aspergillus*), genes for the synthesis of some amino acids were not clustered (Pontecorvo 1963). Thus, it was suggested by Jacob that each transcription unit might operate as an "operon" (Monod and Jacob 1961).

Little mentioned during the time of the solution of the code (1961–1966) was why RNA might be so intimately involved in translation. By 1968, however, three investigators (Woese 1967; Crick 1968; Orgel 1968) had all proposed a possibility that has assumed great currency in today's speculation: Because of RNA's central role not only in coding for amino acids but in the act of translation, perhaps RNA was the original informational molecule (see Chapter 6).

Much of the mystery that lingered through the 1940s and early 1950s had been dispelled. A sense of *triumphalism* prevailed. A prominent bacteriophage investigator, Gunther Stent, published a paper in *Science* in 1968 entitled "That Was the Molecular Biology That Was" (Stent 1968). Stent, a physicist convert to biology and acolyte of Max Delbrück, had been captivated, as had been Delbrück and even Niels Bohr, by the possibility of some "new physics" that might be required to explain biology. One gets the distinct impression of a letdown among the physicists turned biologists following the rapid advances of the 1953–1966 era, which left no need for a rescue of biology by new physics. The physicists themselves who had been attracted to biology (and there were many) either had become geneticists or had joined the cadre of dogged biochemists who showed that molecular biology could be explained by "simple" chemistry.

Stent wrote "By that time [mid-60s] many of the details of the genetic code were known, the *colinearity* [sic; my emphasis] of nucleotide sequence in DNA and amino acid sequence in protein had been finally proved, the structural details of the tRNA (Crick's postulated adaptor) and the mechanism by which it combines with its cognate amino acids had been worked out, and the *general enzymatic* and *informational* [my emphasis] mechanisms connected with the synthesis of DNA, RNA, and protein had been elucidated.

All hope that (physico-chemical) paradoxes would still turn up in the study of heredity had been abandoned long ago, and what remained now was the need *to iron out the details* [my emphasis]" (Stent 1968).

Unquestionably, there was breathtakingly rapid progress from 1953 to the late 1960s. What was often stated but often glossed over was that the great majority of all of this progress came from experiments with bacteria and their viruses. Assumptions were rampant that information storage, retrieval, and gene regulation (i.e., the details of RNA function) would be similar in all cells. Major surprises were still in store.

REFERENCES

Alpher RA, Bethe H, Gamow G. 1948. The origin of chemical elements. *Phys Rev* **73:** 803–804.

Arnez JG, Moras D. 1997. Structural and functional considerations of the aminoacylation reaction. *Trends Biochem Sci* **22:** 211–216.

Bell CE, Lewis M. 2001. The *lac* repressor: A second generation of structural and functional studies. *Curr Opin Struct Biol* **11:** 19–25.

Benzer S. 1960. Genetic fine structure. *Harvey Lect* **56:** 1–21.

Brenner S. 1957. On the impossibility of all overlapping triplet codons in information transfer from nucleic acid proteins. *Proc Natl Acad Sci* **43:** 687–694.

Brenner S. 2001. *My life in science* (as told to Lewis Wolpert) (ed. EC Friedberg, E Lawrence). Biomed Central, London.

Brenner S, Benzer S, Barnett L. 1958. Distribution of proflavin-induced mutations in the genetic fine structure. *Nature* **182:** 983–985.

Brenner S, Barnett L, Crick FHC, Orgel A. 1961. The theory of mutagenesis. *J Mol Biol* **3:** 121–124.

Burgess RR, Travers AA, Dunn JJ, Bautz EKF. 1969. Stimulating transcription by RNA polymerase. *Nature* **221:** 43–46.

Busby S, Ebright RH. 1999. Transcription activation by catabolite activator protein (CAP). *J Mol Biol* **293:** 199–213.

Clark BF, Marcker KA. 1966. The role of *N*-formyl-methionyl tRNA in protein biosynthesis. *J Mol Biol* **17:** 394–406.

Crick FHC. 1959. Biochemical activities of nucleic acids. The present position of the coding problem. *Brookhaven Symp Biol* **12:** 35–39.

Crick F. 1966a. Codon–anticodon pairing: The wobble hypothesis. *J Mol Biol* **19:** 548–555.

Crick FHC. 1966b. The genetic code—yesterday, today and tomorrow. *Cold Spring Harbor Symp Quant Biol* **31:** 3–9.

Crick FHC. 1968. The origin of the genetic code. *J Mol Biol* **38:** 367–379.

Crick F. 1988. *What mad pursuit: A personal view of scientific discovery*. Basic Books, New York.

Crick FH, Griffith JS, Orgel LE. 1957. Codes without commas. *Proc Natl Acad Sci* **43:** 416–421.

Crick FH, Barnett L, Brenner S, Watts-Tobin RJ. 1961. General nature of the genetic code for proteins. *Nature* **192:** 1227–1232.

Darst SA. 2001. Bacterial RNA polymerase. *Curr Opin Struct Biol* **1:** 155–162.

de Crombrugghe B, Pastan I. 1978. Cyclic AMP, the cyclic AMP receptor protein and their dual control of the galactose operon. In *The operon* (ed. JH Miller, WS Reznikoff), pp. 303–324. Cold Spring Harbor Laboratory, Cold Spring Harbor, NY.

de Crombrugghe B, Chen B, Anderson W, Nissley P, Gottesman M, Pastan I, Perlman R. 1971. *Lac* DNA, RNA polymerase and cyclic AMP receptor protein, cyclic AMP, *lac* repressor and inducer are the essential elements for controlled *lac* transcription. *Nat New Biol* **231**: 139–142.

DeMars RI. 1953. Chemical mutagenesis in bacteriophage T2. *Nature* **172**: 964. doi: 10.1038/172964a0.

Dintzis HM. 1961. Assembly of the peptide chains of hemoglobin. *Proc Natl Acad Sci* **47**: 247–261.

Dunn TM, Hahn S, Ogden S, Schleif RF. 1984. An operator at −280 base pairs that is required for repression of araBAD operon promoter: Addition of DNA helical turns between the operator and promoter cyclically hinders repression. *Proc Natl Acad Sci* **81**: 5017–5020.

Englesberg E, Irr J, Power J, Lee N. 1965. Positive control of enzyme synthesis by Gene C in the L-arabinose system. *J Bacteriol* **90**: 946–957.

Gamow G. 1954. Possible relationship between deoxyribonucleic acid and protein structures. *Nature* **173**: 318. doi: 10.1038/173318a0.

Gilbert W, Müller-Hill B. 1966. Isolation of the lac repressor. *Proc Natl Acad Sci* **56**: 1891–1898.

Gilbert W, Müller-Hill B. 1970. The lactose repressor. In *The lactose operon* (ed. JR Beckwith, D Zipser), pp. 93–109. Cold Spring Harbor Laboratory, Cold Spring Harbor, NY.

Holley RW, Apgar J, Everett G, Madison JT, Marquisee M, Merrill S, Penswick JR, Zamir A. 1965. Structure of ribonucleic acid. *Science* **147**: 1462–1465.

Horiuchi T, Novick A. 1961. A thermolabile repression system. *Cold Spring Harabor Symp Quant Biol* **26**: 247–248.

Hurwitz J, Bresler A, Diringer R. 1960. The enzymatic incorporation of ribonucleotides into polyribonucleotides and the effect of DNA. *Biochem Biophys Res Commun* **3**: 15–19.

Hutchison CA, Phillips S, Edgell MH, Gillam S, Jahnke P, Smith M. 1978. Mutagenesis at a specific position in a DNA sequence. *J Biol Chem* **253**: 6551–6560.

Ingram VM. 1956. A specific chemical difference between the globins of normal human and sickle-cell anaemia haemoglobin. *Nature* **178**: 792–794.

Ingram VM. 1957. Gene mutations in human haemoglobin: The chemical difference between normal and sickle cell haemoglobin. *Nature* **180**: 326–328.

Irr J, Englesberg E. 1970. Nonsense mutants in the regulator gene *araC* of the L-arabinose system of *Escherichia coli* B-r. *Genetics* **65**: 27–39.

Itano HA, Pauling L. 1949. Difference in electrophoretic behavior of sickle cell hemoglobin and normal human hemoglobin. *Fed Proc* **8**: 209. (http://oregondigital.org/cdm4/document.php?CISOROOT=/sickle&CISOPTR=62&REC=6.)

Jacob F. 1988. *The statue within: An autobiography* (translated by F Philip). Basic Books, New York.

Khorana HG. 1972. Nucleic acid synthesis in the study of the genetic code. In *Nobel Lectures, Physiology or Medicine 1963–1970* (http://nobelprize.org/nobel_prizes/medicine/laureates/1968/khorana-lecture.html). Elsevier, Amsterdam.

Kim SH, Quigley GJ, Suddath FL, McPherson A, Sneden D, Kim JJ, Weinzierl J, Rich A. 1973. Three dimensional structure of yeast phenylalanine transfer RNA; folding of the polynucleotide chain. *Science* **179**: 285–288.

Kornberg A. 1957. Enzymatic synthesis of DNA. *Harvey Lect* **53**: 83–112.

Leder P, Nirenberg M. 1964. RNA code words and protein synthesis. II. Nucleotide sequence of a valine code word. *Proc Natl Acad Sci* **52**: 420–427.

Lee N, Wilcox G, Grelow W, Arnold J, Cleary P, Englesberg E. 1974. In vitro activation of the transcription of araBAD operon by araC protein. *Proc Natl Acad Sci* **71**: 634–638.

Lerman LS. 1961. Structural considerations in the interactions of DNA and acridines. *J Mol Biol* **3:** 18–30.

Lodish H, Berk A, Kaiser CA, Krieger M, Scott MP, Bretscher A, Ploegh H, Matsudaira P. 2008. *Molecular cell biology,* 6th ed. Freeman, New York.

Magasanik B. 1968. Glucose effects inducer exclusion and repressions. In *The lactose operon* (ed. JR Beckwith, D Zipser), pp. 189–220. Cold Spring Harbor Laboratory, Cold Spring Harbor, NY.

Martin RG. 1984. A revisionist view of the breaking of the genetic code. In *NIH: An account of research in its laboratories and clinics* (ed. D Stetten Jr), pp. 281–296. Academic Press, New York.

Matthaei H, Nirenberg M. 1961a. Characteristics and stabilization of DNase-sensitive protein synthesis in *E. coli* extracts. *Proc Natl Acad Sci* **47:** 1589–1602.

Matthaei H, Nirenberg M. 1961b. The dependence of cell-free protein synthesis in *E. coli* upon RNA prepared from ribosomes. *Biochem Biophys Res Commun* **4:** 404–408.

Maxam AM, Gilbert W. 1977. A new method for sequencing DNA. *Proc Natl Acad Sci* **74:** 560–564.

Miller OC, Hankalo BA, Thomas CA. 1970. Visualization of bacterial genes in action. *Science* **169:** 392–395.

Monod J. 2003. Nobel lecture (reprinted). In *Origins of molecular biology* (ed. A Ullman), pp. 275–317). ASM Press, Washington, D.C.

Monod J, Jacob F. 1961. Teleonomic mechanisms in cellular metabolism, growth, and differentiation. *Cold Spring Harbor Symp Quant Biol* **26:** 389–401.

Morgan AR, Wells RD, Khorana HG. 1966. Studies on polynucleotides, LIX. Further codon assignments from amino acid incorporations directed by ribopolynucleotides containing repeating trinucleotide sequences. *Proc Natl Acad Sci* **56:** 1899–1906.

Müller-Hill B. 1998. Some repressors of bacterial transcription. *Curr Opin Microbiol* **1:** 145–151.

Müller-Hill B, Crapol L, Gilbert W. 1968. Mutants that make more lac repressor. *Proc Natl Acad Sci* **59:** 1259–1264.

Nirenberg M. 2004. Historical review: Deciphering the genetic code—a personal account. *Trends Biochem Sci* **29:** 46–54.

Nirenberg MW, Matthaei JH. 1961. The dependence of cell free protein synthesis in *E. coli* upon naturally occurring or synthetic polyribonucleotides. *Proc Natl Acad Sci* **47:** 1588–1602.

Nirenberg M, Jones OW, Leder P, Clark BFC, Sly WS, Pestka S. 1963. On the coding of genetic information. *Cold Spring Harbor Symp Quant Biol* **28:** 549–558.

Nirenberg M, Caskey T, Marshall R, Brimacombe R, Kellogg D, Doctor B, Hatfield D, Levin J, Rottman F, Pestka S, et al. 1966. The RNA code and protein synthesis. *Cold Spring Harbor Symp Quant Biol* **31:** 11–24.

Ochoa S. 1964. Nobel lecture (1959): Enzymatic synthesis of ribonucleic acid. In *Nobel lectures in physiology or medicine 1942–1962.* Elsevier, New York. (http://nobelprize.org/nobel_organizations/publications/lectures/index.html.)

Olby R. 2009. The genetic code. In *Francis Crick: Hunter of life's secrets,* pp. 221–238. Cold Spring Harbor Laboratory Press, Cold Spring Harbor, NY.

Orgel LE. 1968. Evolution of the genetic apparatus. *J Mol Biol* **38:** 381–393.

Osawa S, Jukes TH, Watanabe K, Muto A. 1992. Recent evidence for evolution of the genetic code. *Microbiol Rev* **56:** 229–264.

Pardee AB, Jacob F, Monod J. 1959. The genetic control and cytoplasmic expression of "inducibility" in the synthesis of β-galactosidase by *E. coli. J Mol Biol* **1:** 165–178.

Pauling L, Itano HA, Singer SJ, Wells AC. 1949. Sickle cell anemia, a molecular disease. *Science* **110:** 543–548.

Pontecorvo G. 1963. The Leeuwenhoek lecture: Microbial genetics: Retrospect and prospect. *Proc R Soc Lond B Biol Sci* **158:** 1–23.

Popovych N, Tzeng S-R, Tonelli M, Ebright RH, Kalodimos CG. 2009. Structural basis for cAMP-mediated allosteric control of the catabolite activator protein. *Proc Natl Acad Sci* **106:** 6927–6932.

Power J. 1967. The L-rhamnose genetic system in *Escherichia coli* K-12. *Genetics* **55:** 557–568.

Ptashne M. 1967. Isolation of the λ phage repressor. *Proc Natl Acad Sci* **57:** 306–313.

Reznikoff WS, Abelson JN. 1978. The *lac* promoter. In *The operon* (ed. JH Miller, WS Reznikoff), pp. 221–243. Cold Spring Harbor Laboratory, Cold Spring Harbor, NY.

Robertus JD, Ladner JE, Finch JT, Rhodes D, Brown RS, Clark BF, Klug A. 1974. Structure of yeast phenylalanine tRNA at 3Å resolution. *Nature* **250:** 546–551.

Sadler J, Novick A. 1965. The properties of repressor and the kinetics of its action. *J Mol Biol* **12:** 305–327.

Sarabhai AS, Stretton AO, Brenner S, Bolle A. 1964. Co-linearity of the gene with the polypeptide chain. *Nature* **201:** 13–17.

Schmitz A, Galas DJ. 1979. The interaction of RNA polymerase and lac repressor with the lac control region. *Nucleic Acids Res* **6:** 111–137.

Schultz SC, Shields GC, Steitz TA. 1991. Crystal structure of CAP–DNA complex: The DNA is bent by 90°. *Science* **253:** 1001–1007.

Schwartz M. 1967. Phenotypic expression and genetic localization of mutations affecting maltose metabolism in *Escherichia coli* K 12. *Ann Inst Pasteur* **112:** 673–698.

Siebenlist U, Simpson RB, Gilbert W. 1980. E. coli RNA polymerase interacts homologously with two different promoters. *Cell* **20:** 269–281.

Singh S. 2004. *Big Bang: The origin of the universe.* HarperCollins, New York.

Speyer JF, Lengyel P, Basitio C, Wahba AJ, Gardenr RS, Ochoa S. 1963. Synthetic polynucleotides and the amino acid code. *Cold Spring Harbor Symp Quant Biol* **28:** 559–568.

Stent G. 1968. That was the molecular biology that was. *Science* **160:** 390–395.

Stevens A. 1960. Incorporation of the adenine ribonucleotide into RNA by cell fractions from *E. coli*. *Biochem Biophys Res Commun* **3:** 92–96.

Stretton AOW, Kaplan S, Brenner S. 1966. Nonsense codons. *Cold Spring Harbor Symp Quant Biol* **31:** 173–180.

Sutherland EW. 1970. On the biological role of cyclic AMP. *J Am Med Assoc* **214:** 1281–1288.

Tissières A, Schlessinger D, Gros F. 1960. Amino acid incorporation into proteins by *Escherichia coli* ribosomes. *Proc Natl Acad Sci* **46:** 1450–1462.

von Ehrenstein G, Lipmann F. 1961. Experiments on hemoglobin biosynthesis. *Proc Natl Acad Sci* **47:** 941–950.

Watson JD. 2001. *Genes, girls and Gamow: After the double helix.* Alfred A Knopf, New York.

Watson JD, Crick FH. 1953. Molecular structure of nucleic acids; a structure for deoxyribose nucleic acid. *Nature* **171:** 737–738.

Weigert M, Gallucci E, Lanka E, Garen A. 1966. Characteristics of the genetic code in vivo. *Cold Spring Harbor Symp Quant Biol* **31:** 145–150.

Weiss SB. 1960. Enzymatic incorporation of ribonucleoside triphosphates into the interpolynucleotide linkages of ribonucleic acid. *Proc Natl Acad Sci* **46:** 1020–1030.

Weiss SB, Gladstone L. 1959. A mammalian system for the incorporation of cytidine triphosphate into ribonucleic acid. *J Am Chem Soc* **81:** 4118–4119.

Woese C. 1967. *The genetic code: The molecular basis for genetic expression.* Harper & Row, New York.

Yanofsky C, Carlton BL, Guest JR, Helinski DR, Henning U. 1964. On the colinearity of the gene structure and protein structure. *Proc Natl Acad Sci* **51:** 266–272.

Zubay G, Schwartz D, Beckwith J. 1970. Mechanism of activation of catabolite-sensitive genes: A positive control system. *Proc Natl Acad Sci* **66:** 104–110.

4

Gene Expression in Mammalian Cells

Discovery of RNA Processing, Genes in Pieces, and New RNA Chemistry

THE CENTRAL DOGMA EMERGING FROM THE WILDLY successful bacterial genetics and biochemistry of the 1950s and 1960s was the oft-quoted phrase "DNA makes RNA makes protein." The following three profound questions were answered, at least in bacteria, that established this epigram: (1) What did it mean that DNA (deoxyribonucleic acid) contained "information"? The answer was that the genetic code for protein sequences was enshrined in the base sequence of DNA. (2) How could the information be retrieved, especially in an orderly fashion? Gene-specific transcription by RNA polymerase into messenger RNA (mRNA) retrieved the information, and retrieval was regulated by site-specific DNA-binding proteins. (3) Once retrieved, how could the information be used to make protein? A combination of ribosomal RNA (rRNA) and transfer RNA (tRNA) interactions translated the mRNA into proteins. The answers to each one of these questions depended on discovering that there are different types of RNA molecules, with roles both of informational interlocutor and functional executor in using the information in DNA.

How much of this catechism was transferable to eukaryotic cells, particularly those of plants and animals? Certainly, it seemed possible in the early 1960s that the three major types of RNA molecules—rRNA, mRNA, and tRNA—and their rules of mechanical operation might be the same in all cell types. After all, the first discovered mutation resulting in a single amino acid change was in human β globin. Furthermore, it was established that *Escherichia coli* tRNA could assist the ribosomes and mRNA of rabbit reticulocytes (red blood cell precursors) to make correct globin protein. The first tRNA to be sequenced was a yeast tRNA, and it used the same code as the *E. coli* tRNA. Yeasts, of course, are eukaryotes. However, whether chromosome structure, mRNA formation, and gene regulation would turn out to be similar in molecular detail in bacteria and eukaryotes, no one knew.

EUKARYOTIC RNA AND CHROMOSOMAL INFORMATION IN 1960

What was known in 1960–1961—the years when mRNA took center stage—regarding RNA and protein synthesis in animal or plant cells? The only certainty was that rat tRNA and ribosomes (with their RNA content) cooperated to make protein in cell-free extracts. In the early 1960s, the existence of mRNA and the formulation of the first ideas regarding gene regulation were only known for bacteria.

By 1956–1958, infectious RNA from both plant and animal viruses had demonstrated that RNA could serve a genetic function in the appropriate eukaryotic host cell. Therefore, it was entirely reasonable that an mRNA function would exist in eukaryotic cells. All of this information made it seem likely to some that plant and animal cells would quickly be shown to be just "larger" bacteria operating their genes by the same biochemical/molecular rules.

But other information raised red flags. First, plant and animal cells are compartmentalized—the cytoplasm and the nucleus are separate. The DNA is in the nucleus, and the ribosomes were already known to be mostly (or possibly all) in the cytoplasm. Thus, unlike bacterial ribosomes, eukaryotic ribosomes were likely not allowed immediate access to mRNA as the mRNA was copied from DNA. Second, there is roughly 1000 times as much DNA in a mammalian cell as in a bacterium: What did this huge difference mean? In *E. coli*, virtually the entire genetic map was known in the mid 1960s, and most of the bacterial DNA was thought to code for protein. With $\sim3 \times 10^6$ to 4×10^6 base pairs (bp) of DNA per cell and three bases per amino acid, a bacterium could encode 2000–3000 different kinds of protein chains of ~500 amino acids each. This seemed a reasonable estimate for the number of proteins that an *E. coli* cell might make. Did our cells have 1000 times as much information? Did our chromosomes have information for 1000 times as many proteins?

The C-Value Paradox

There were strong hints that perhaps some new principles of information storage might be lying in wait. For some but not all eukaryotic organisms, a gradual increase in DNA content per cell—the so-called C value—matched the apparent complexity of the organism. Whereas bacteria had $\sim$3–4 million base pairs (megabases, Mbp) of DNA per cell, *Saccharomyces cerevisiae* (baker's yeast) had $\sim$12 million, *Drosophila melanogaster* had $\sim$180 million, and humans had 3 billion. These distributions seemed to be in general accord with an obvious increase in complexity.

But the DNA content per cell of some animals that are seemingly equally complex, such as different amphibians and reptiles (frogs, salamanders, and turtles), varies from 10-fold to 100-fold (Fig. 4-1). Moreover, some amphibian cells have 10–20 times as much DNA as human cells (for review, see Moore 1984). Plants are truly absurd in this regard: The total DNA content per cell in different plants varies 10-fold to 50-fold. Tulips, for example, have eight times as much DNA per cell as humans. It did not seem conceivable that tulips had information for eight times as many proteins as humans or that some amphibian chromosomes had 100-fold more protein-coding information than the chromosomes of other amphibians. At the very least, it seemed that not all DNA in eukaryotic organisms coded for proteins. What did these unexplained variations in DNA content portend for eukaryotic RNA synthesis and for the transfer of *useful* information from DNA to mRNA that was used to code for protein?

Unstable Nuclear RNA: Related to Cytoplasmic RNA?

By 1960, there was already a strong suggestion that RNAs synthesized in the nucleus were not simply dumped as unchanged molecules into the cytoplasm. After World War II, cell biologists used newly available radioactive molecules (most informatively, [14]C- and ^{3}H-labeled nucleosides and amino acids, as well as radioactive orthophosphate [32]PO_4) to label newly formed macromolecules (Smellie 1955). Early work with labeled precursors used the technique of autoradiography (where, under the microscope, 0.1-μm silver grains reported the location of atomic decay in a cell) (for review, see Rogers 1979). Incorporation of specific labels (e.g., thymidine) for brief periods showed very clearly that chromosomal DNA was made in the nucleus as expected. The fact that all of the cells in a growing culture of animal cells were not labeled immediately with ^{3}H thymidine led to the discovery of the phases of the eukaryotic cell cycle. The S phase (DNA synthesis) was separated from mitosis (M phase) by two gaps of no synthesis, G_1 after mitosis and G_2 after DNA synthesis.

But the most active cellular site of RNA synthesis (ribonucleoside incorporation) was also in the nucleus, even though histological staining techniques indicated that the majority of the cell's RNA was in the cytoplasm. The cytoplasm was the most active site of amino acid incorporation. Several investigators commented on the lengthy time that it took for the cytoplasm

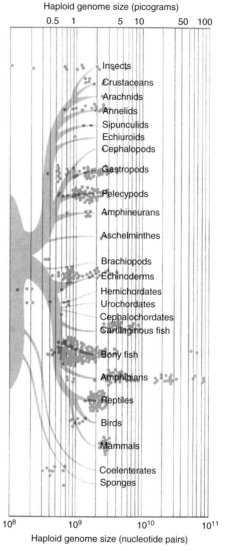

Haploid genome size (picograms)

Haploid genome size (nucleotide pairs)

Figure 4-1. DNA per haploid genome in a variety of animals. (Modified, with permission, from Britten and Davidson 1971 [©University of Chicago Press].)

to catch up with the nucleus in terms of the amount of labeled RNA—i.e., for the accumulation of labeled nucleosides in cytoplasmic RNA to match the distribution of total RNA indicated either by basophilic staining or cell fractionation and chemical measurement (for review, see Smellie 1955).

Henry Harris and coworkers, in particular, questioned the whole idea of the transfer of nuclear RNA to the cytoplasm (Harris 1959, 1964). In the late

1950s and early 1960s, the Harris group published several papers, the most convincing experiment in which they showed that if, after a brief exposure of cells to a labeled nucleoside (e.g., adenosine or uridine), the labeled compound was replaced with unlabeled nucleoside, there was a decrease in nuclear grains (in autoradiograms) without a commensurate increase in cytoplasmic grains (Fig. 4-2). This seemed to be clear evidence that some or most of the rapidly labeled nuclear RNA was unstable and certainly allowed the reasonable conclusion that the majority of the briefly labeled nuclear RNA was not transferred en bloc to the cytoplasm. Harris persisted throughout the 1960s in arguing against *any* transfer of nuclear RNA to the cytoplasm, even raising the possibility at one point that plant cells might make most of their RNA in the cytoplasm (Harris and LaCour 1963).

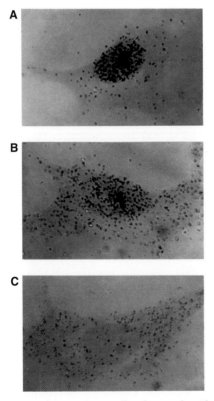

Figure 4-2. Autoradiograms of rat embryo cells after nucleoside label and chase. (*A*) Cultured rat embryo cells were labeled for 20 min with ^{3}H adenosine, showing the majority of label in the nucleus. (*B,C*) As in *A*, but the medium was then replaced with unlabeled adenosine and guanosine for 3 h (*B*) or 12 h (*C*). (Reprinted, with permission, from Harris 1959 [©The Biochemical Society].)

But autoradiographic experiments and even the crude early biochemical extractions of total cellular or nuclear RNA told nothing regarding potentially different kinds of RNA that might exist in the nucleus or the cytoplasm of the eukaryotic cell. For example, there might be a rapidly synthesized and degraded nuclear RNA that swamped out the ability to observe some RNA transfer to the cytoplasm. After all, the rapid synthesis and degradation of bacterial mRNA had just been proven. Clearly, a technique was needed that could reliably remove, in an undamaged fashion, all of the RNA from a eukaryotic cell (or from the nucleus and cytoplasm separately) to allow the careful characterization of the composition, the size, and, if possible, the site of synthesis of whatever RNA molecules that the cell contained.

As had been repeatedly emphasized by successful biologists in the past, the choice of experimental cell or organism was often the key to solving biological problems. Think of garden peas and *Drosophila* for genetics, *Neurospora* for uniting genetics and enzyme synthesis, and, of course, *E. coli* and its bacteriophages for the theory of the operon in gene control, the discovery of mRNA, and for breaking the genetic code.

The first consideration in studying the molecular steps in information transfer in animal cells was the tissue or cell type that was the most satisfactory for such studies. The original success with liver cell extracts in discovering tRNA and ribosomal protein synthesis necessitated starting each experiment with live animals, and the liver tissue was a mixture of cell types that did not grow or even survive for long outside of the body. Reticulocytes had obviously been put to good use as cells making mainly one protein, but mammalian reticulocytes have already extruded their nucleus. The overwhelming choice of the small group of researchers who made progress in early studies of mammalian cell molecular biology, especially RNA metabolism, was cultured mammalian cells and mammalian cells infected with viruses. The cells were easy to grow and furnished a source of homogeneous cells.

At the time (~1961), the only specific RNA molecules that served an informational or genetic function that could be identified in eukaryotic cells were virus-specific RNA molecules. This remained true until the late 1970s. A possible parallel with the thoroughgoing success with *E. coli* and its bacteriophages in understanding gene function was not lost on animal cell and animal virus investigators. So, what was known in the late 1950s regarding culturing animal cells and growing animal viruses?

A brief detour into the early history of cell culture and virus propagation, although admittedly not strictly germane to our main purpose, describes a very important era of biology and sets the stage for the original use, continued to the present day, of cultured human and mouse cells, both growing and infected with viruses, to study eukaryotic RNA.

EARLY CELL CULTURE: HARRISON AND CARREL

At the turn of the 20th century, most biologists believed that cells from an animal could not survive outside of the body. By 1905–1910, Ross Harrison (Fig. 4-3) at Johns Hopkins University (later at Yale) proved that this was untrue. He transplanted chick embryonic fragments into Petri dishes in the presence of tissue extracts. He was particularly interested in the survival and outgrowth of nerve cells from embryonic fragments. Eventually, Harrison established beyond a doubt that sections of an early embryo that were destined to give rise to nerve cells, but had not yet done so, would differentiate into nerve cells in culture (Harrison 1909, 1910). On the occasion of the publication of his 1909 article, an editorial in *Science* enthusiastically observed, "He has actually seen the fibers growing outwards in embryonic structures."

In addition to his many papers on the outgrowth of nerve fibers in culture, Harrison also described placing presumptive embryonic heart tissue into culture and later observing "beating" (i.e., regularly and spontaneously contracting) cells. Harrison proved that cell survival, even of specialized cells, could occur outside of the body of an animal. But he did not attempt to show or claim that any of the embryonic cells actually divided.

The next chapter in "cell culture," as such experiments became known, has an interesting if somewhat nefarious history. Alexis Carrel, a French surgeon, received the 1912 Nobel Prize in Medicine or Physiology for his surgical work on suturing blood vessels in human patients. These techniques were life-saving in the aftermath of serious traumatic injuries.

Figure 4-3. Ross Granville Harrison (1870–1959), the Yale and Johns Hopkins embryologist who observed nerve cell outgrowth in tissue culture. (Reprinted from Nicholas 1960 [©*Yale Journal of Biology and Medicine*].)

Carrel believed in the capacity of cells, even organs, to survive indefinitely out of the body. By 1908, he had relocated to The Rockefeller Institute for Medical Research in New York. There, he heard Harrison's 1908 Harvey Lecture on "tissue culture." By 1910, Carrel's laboratory group had maintained beating heart cell explants in culture for weeks and months. Carrel's autocratic behavior and his staff's fears (those of his technicians as well as his professional associates) were perhaps responsible for the claims that he kept these beating chicken heart cultures alive for *years*. In all likelihood, the cultures were replenished with new cells during the feeding of the explants with embryo extract that likely contained surviving cells. But Carrel's indefatigable role as publicist made popular the possibilities of "tissue culture."

Carrel's widespread fame and reputation came crashing down because of his dalliance with Fascism and eugenics in the 1930s. His likely collaborations with the Vichy government after the fall of France in 1940 were also well known. The problems in his life, in both the validity of his supposed long-term cultures and his unsavory politics, is chronicled by J.A. Witkowski in "Dr. Carrel's Immortal Cells" (Witkowski 1980). Through all of this difficulty, however, it remained widely accepted that animal cells could live in culture dishes, but regular, reproducible growth had not been demonstrated.

CONTINUOUS CELL CULTURES

After the conclusion of World War II, renewed efforts to culture cells from multicellular organisms finally met with success. The most useful of these techniques were with mouse cells and human cancer cells. Later, plant cells were also shown to be capable of growth, and, in fact, Frank Steward and colleagues (1964) showed regeneration of carrots from single cells.

Earle's Mouse L Cells

The first continuous cultured animal cell line was established at the National Cancer Institute in Bethesda, Maryland, in the laboratory of Wilton R. Earle in the late 1940s. Adult and mouse fibroblasts from skin explants gave rise to continuously growing cells that came to be known as L929 or simply mouse L cells. Earle and his colleagues, Katherine Sanford and Gwendolyn Likely, wished to prove that *single* cells from these cultures would divide and continue to grow, not just survive, outside of the body, thus furnishing a source of presumably genetically identical cells. An idea current at the time was that cultured cells produced products that helped to sustain one another. To provide the most favorable situation for isolated single cells, they placed individual cells in very narrow capillary glass tubes and observed the cells microscopically. Individual cells grew with a high efficiency in a medium devised by Earle consisting of horse serum, chick embryo extract, and a balanced salt solution (Fig. 4-4). This landmark result in 1948 settled the issue: Single mammalian cells were capable of growth and division outside of the body (Sanford et al. 1948).

Henrietta Lacks and HeLa Cells

The first continuously growing human cell culture was established by Dr. George Gey, a Baltimore physician, who with his wife Margaret Gey had for years attempted unsuccessfully to grow human cancer cells continuously in culture. In 1951, a young woman, Henrietta Lacks, was diagnosed with advanced carcinoma of the cervix of the uterus (Landecker 2007, pp. 127–128). At the time of operation to remove the tumor, a "quarter-sized" piece of the tumor was given to the Geys, who minced the tissue and placed it in culture on February 8, 1951. The tumor cells in this biopsy not only migrated out of the explanted tissue and attached to the walls of the culture tube but also grew to completely cover the surface of the tube. The initial report of this success is contained in an 18-line abstract published

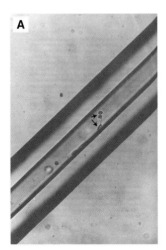

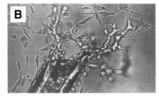

Figure 4-4. Origin of Earle's L cells. After a permanent mouse fibroblast cell culture was established, the Earle laboratory placed single cells in glass capillary tubes filled with culture medium (*A*) and then placed the tubes in a culture dish. They observed single cells and a round daughter cell (arrows), which ultimately overgrew the tip of the capillary tube (*B*). (Reprinted from Landecker 2007; photographs originally from Sanford et al. 1948; photographer: Wilton Earle, National Institutes of Health [NIH].)

in *Cancer Research* in 1952 (Gey et al. 1952). The abstract states (in part), "Thus far only one strain of epidermoid carcinoma has been established and grown in continuous roller tube cultures for almost a year. It grows well in a composite medium of chicken plasma, bovine embryo extract and human placental cord serum. The autologous normal prototype [i.e., the normal cervical epithelium that came along with the tumor cells] is most difficult to maintain under comparable conditions."

By now, more experiments may had been done on HeLa cells and with extracts from HeLa cells than with perhaps any other eukaryotic cell. Unfortunately, Henrietta Lacks did not live to see the importance that her continuously growing tumor cells have made to research. In today's medical culture, she would have to have signed a permission slip for the use of the original surgically removed tissue. Would she have done so? One can hope the answer would have been yes and also that somehow she and her survivors might have been properly compensated at a later date (Skloot 2010). We return to HeLa and L cells later.

VIRUS PRODUCTION IN PRIMARY CULTURED ANIMAL CELLS

For several decades before World War II, poliovirus epidemics resulted in thousands of deaths each year and many thousands of cases of permanent paralysis in the United States. Virologists were desperate to find a host other than live monkeys in which to produce poliovirus to possibly develop a vaccine. John F. Enders and his colleagues Thomas Weller and Frederick Robbins reported in 1949 that human embryonic tissue (from spontaneously aborted fetuses) would produce an outgrowth of undifferentiated cells in culture (a primary cell culture) in which poliovirus multiplied quite well (Enders et al. 1949). This was a crucial step to making a vaccine because it was previously believed that the poliovirus would only grow in and damage differentiated nerve cells, thus leading to paralysis.

In developing the poliovirus vaccine, Jonas Salk used primary cultures of monkey tissue to produce poliovirus (Youngner et al. 1952). Cells prepared from minced monkey testicles and later kidneys were the most successful. Contrary to conventional wisdom, Salk believed that sufficient virus material could be obtained in this way so that he could kill the virus (using formaldehyde treatment) and still stimulate an adequate immune response that would protect individuals. This proved to be the case. By 1952, Salk showed that he could protect monkeys from polio infection by immunization with a killed virus vaccine. The three major types of polio were included. (Type 1 caused the majority of paralytic disease; types 2 and 3, however, also caused several thousand cases each year.) Moreover, human volunteers who were inoculated with the killed virus developed antibodies that could block virus growth in cell culture and in animals (Salk 1953).

What other cultured cells might allow the production of poliovirus? Jerome Syverton and William Scherer, two microbiologists at the University of Minnesota, obtained HeLa cells from George Gey and showed in 1953 that these cells produced all three types of poliovirus very well (Fig. 4-5) (Scherer et al. 1953). There was some fear, however, that human cancer cells were not a safe, uncontaminated source for vaccine production, but HeLa cells were widely used in laboratory testing during the production of the first successful poliovirus vaccine.

PRECISE REPRODUCIBLE GROWTH OF MAMMALIAN CELLS AND VIRUSES

The success in animal cell culture techniques that made possible the exploration of basic molecular events including RNA biogenesis was based on the ability of reproducible cell growth and the uniformity of response to simultaneous infection by a chosen virus. Three scientists—Harry Eagle, Theodore

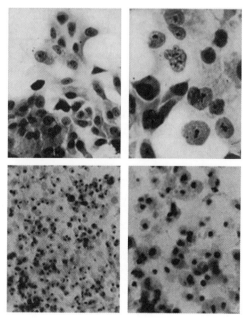

Figure 4-5. Poliovirus destruction of HeLa cells. Two magnifications of growing HeLa cells (*top*; *left*, 160×; *right*, 250×) and cells infected 48 h earlier with type 1 poliovirus (*bottom*; *left*, 65×; *right*, 160×). (Reprinted, with permission, from Scherer et al. 1953 [©Rockefeller University Press].)

Figure 4-6. Three architects of the use of modern culture of mammalian cells and viruses: (*Left to right*) Harry Eagle, Renato Dulbecco, and Ted Puck. The photo was taken in 1973 on the occasion of the three receiving the Louisa Gross Horwitz Prize at Columbia University. (Photo courtesy of Harry Eagle.)

Puck, and Renato Dulbecco (Fig. 4-6)—had more to do with launching this new era than any other investigators.[1]

Harry Eagle and the Development of "Eagle's Medium"

Throughout the years 1945–1953, Harry Eagle had been working on the effectiveness of penicillin on treponemal diseases including syphilis (caused by a spirochete, *Treponema pallidum*). He wished to determine how penicillin killed microorganisms but was not toxic to animals and humans. Some proteins in bacteria bound penicillin, and to test whether animal cells also had penicillin-binding proteins, he obtained strain L929 mouse fibroblasts from their "inventor" Wilton Earle (Eagle 1954).

Earle and his colleagues practiced tissue culture in the mode of Alexis Carrel, i.e., the laboratory was spotless and investigators wore caps, masks, gowns, and booties on their feet as if they were on the way to surgery (Landecker 2007, p. 88). *No* antibiotics were used in the culture medium in Earle's laboratory.

Harry Eagle was a vigorous scientist whose patience (or impatience) did not match well the more deliberate style of the Earle laboratory. Eagle wanted

---

[1] Because I worked in Eagle's laboratory in 1956–1960, in fact, studying poliovirus replication and knew the other investigators quite well, this section contains personal recollections interwoven with the scientific narrative.

cells when he wanted them. He therefore took some of the L929 cells and added penicillin and streptomycin to the medium so that the cultures would better withstand accidental contamination with bacteria (a practice still used in most cell culture laboratories around the world). He found that the mouse cells still grew perfectly well in Earle's complex medium (containing antibiotics plus serum and embryo extract that Earle added to a balanced salt solution) (Earle 1943).

By 1954, Eagle decided to define the medium in order to have controlled and reproducible growth. He could then study the metabolism and nutritional requirements of the cells under defined conditions. With the same idea in mind, he also obtained HeLa cells from George Gey, at the time the only continuously cultured human cell line. Eagle had earlier in his career developed a defined medium for a spirochete (a treponeme similar to *T. pallidum*, which causes syphilis). These organisms are a very fastidious and demanding class of bacteria. His group of technicians was quite up to the task of making dozens of complex media lacking individual components—amino acids, vitamins, salts, etc., to test what a mammalian cell needed to grow.

I had visited the Eagle laboratory in December 1954 to apply for a postdoctoral position because in medical school, I had worked with the killing of streptococci by penicillin in Robert Glaser's laboratory at Washington University in St. Louis, Missouri. Eagle did not accept me for the summer of 1955, but after I finished a year of internship, he did take me in July 1956. By this time, Eagle had stopped work on penicillin and switched entirely to work with cultured cells. He had defined the 13 amino acids, the eight vitamins, and the correct balanced salt solution sufficient to grow both L cells and HeLa cells. Although humans need only eight amino acids (see Chapter 1, p. 13), cells in culture, which are bathed in a large volume of medium per cell, require 13 amino acids. This paradox is due to the relatively slow rate of synthesis of the extra five amino acids and the rapid exchange of these amino acids with the medium. Most media in use today contain some of the "nonessential" amino acids such as glycine, serine, alanine, and cysteine. Moreover, Eagle had established the most effective concentrations of all of these constituents. The medium was not completely defined because 5%– 10% horse or human serum was required. (The serum was vigorously dialyzed, a treatment that removed all small molecules.) Thus, overnight, the world had a simple medium—Eagle's medium—that allowed the study of cells and virus multiplication under well-defined conditions, and thousands of gallons are still used every day around the world. Although not completely comparable to bacterial cultures, it was at least true that continuously cultured cells provided a vastly better defined experimental mammalian "organism" than was available before. In the last several decades, molecular biologists have also used continuously cultured insect cells, and, of course, much more molecular genetic work using yeast has become very important.

Ted Puck and Simple Cloning of Mammalian Cells

Eagle's early success with HeLa and L cells required the transferring of about one-tenth to one-twentieth of a crowded culture to a new flask roughly once a week as the best way to "maintain" the culture. If the experimenter diluted the culture too far, it often did not grow in the defined medium.

Ted Puck, a physical chemist who had earlier worked in bacterial and bacteriophage genetics, wanted a simple procedure to start colonies with single cells in order to use cloned cells and viruses, as was standard practice in bacterial genetics. He reasoned that such a feat would be necessary to study the genetics of the cells. Puck knew of the success that Earle and coworkers had in growing single L cells in the greatly reduced volume of a capillary tube and subscribed to the idea that cells "cross-feed" each other in dense cultures. He and his graduate student Philip Marcus took the following tack for growing single cells (Puck and Marcus 1955). They first plated embryonic cells that would not, on their own, grow continuously, and X-irradiated a layer of these cells so that no growth at all was possible. On top of this "feeder" layer, they then plated single dispersed HeLa cells. Virtually 100% of the individual HeLa cells gave rise to growing colonies. Individual HeLa cell "clones" derived from single cells could then be selected.

Among the very important early experiments that Puck and his colleagues performed with his colony formation technique was to test the sensitivity to X rays for mammalian cells (Puck and Marcus 1956). The lethal dose for the formation of colonies from individual HeLa cells was 97 rads. It was commonplace before these experiments to believe that killing of human cells by X rays required ∼10 times that much radiation. Puck's finding was very influential in making scientists and the public more aware of the potential dangers to humans of radiation that, for example, might result from fallout of nuclear tests or from overuse of medical X rays.

Renato Dulbecco and Quantitative Animal Virology

Renato Dulbecco, after surviving World War II, emigrated to the United States from Italy in 1947 to join the laboratory of Salvador Luria at Indiana University, where he shared bench space with Luria's first graduate student, Jim Watson. Dulbecco studied bacteriophage multiplication and genetics, discovering among other things that lethally ultraviolet (UV)-irradiated bacteriophage could be reactivated by exposure to natural light during a subsequent infection. This work was among the earliest hints that cells possess the capacity to "cure" damaged DNA (Dulbecco 1950). After moving to The California Institute of Technology (Caltech) in 1949, Max Delbrück offered Dulbecco the proceeds of a private grant given to Caltech to study animal viruses (Dulbecco 1966). Dulbecco and his longtime collaborator Marguerite Vogt set out to make animal virology quantitative, i.e., analytically rigorous, like bacteriophage work. The first challenge was to develop an accurate technique to count infectious particles (Dulbecco 1952; Dulbecco and Vogt 1954). At this point, it had not yet been established whether a virus infection in an animal cell required one or more than one particle. This was an issue mainly because by electron microscopic count, there were always many more particles than apparent infectious doses in all animal virus preparations.

Dulbecco and Vogt developed the first plaque assay for an animal virus using primary cultures of chicken embryo fibroblasts infected with a diluted suspension of viruses (Western equine encephalitis). After allowing time for the virus to attach to the cells, they removed the inoculum and covered the infected cells with a nutrient agar medium. After ~2–3 d, the underlying cell layer could be stained with a dye that differentiated live cells (red) from dead cells (white). Individual loci of infection—plaques—could be counted. This technique, of course, was modeled after the plaque technique that Luria and Delbrück had used to study bacteriophage.

The fact that plaque number was linear with dilution proved that a single virus starts an infection. This is of great practical importance, for if an experimenter wished to ensure that every cell in a culture was simultaneously infected, the multiplicity of infection should be at least 5–10 pfu (plaque-forming units) per cell. These simple precepts allowed quantitative assessment of a virus replication cycle and brought animal virology into the quantitative age. This transformation was necessary to reliably study the details of virus-specific nucleic acid and protein synthesis in infected animal cells.

The work of Eagle, Puck, and Dulbecco (Fig. 4-6) laid the foundation for successful experiments on the biochemistry of nucleic acid and protein synthesis in animal cells that began in the 1960s and continue to this day.

SEARCHING FOR mRNA IN CULTURED ANIMAL CELLS

When I left the Eagle laboratory in late summer of 1960, it was already clear that human or mouse cell cultures were the "organisms" best suited for quantitative molecular studies of animal cells and cells infected by viruses. When I returned to the Massachusetts Institute of Technology (MIT) in June 1961, our laboratory launched a direct attack on defining types of animal cell

RNA. Initially, we wished to simply extract *all* of the cell RNA undamaged and then examine the properties of the potentially different kinds of RNA from a growing mammalian cell. HeLa cells offered homogeneous cell populations in which radioactive nucleosides and $^{32}PO_4$ could be incorporated into RNA. If we grew HeLa cells in the presence of a labeled RNA precursor for one or two generations and extracted the RNA, we expected to label all of the major RNA species. Influenced by phage and bacterial findings that mRNA rapidly turned over, we expected that by restricting exposure of the cells to a labeled precursor for a very short time, we might be able to specifically label mRNA— at the very least, we could compare the rapidly labeled RNA to the majority of cellular RNA.

At this time ($\sim$1961), the only truly informative techniques to examine RNA molecules were (1) sedimentation through gradients stabilized by sucrose, so-called rate zonal analysis, and (2) base distribution, most effectively studied using $^{32}PO_4$-labeled RNA. Both size (length) and shape (folding) affect sedimentation rate, but it was possible to compare sedimentation rates with ribosomal and virus molecules whose mass had been determined by equilibrium analytical sedimentation analysis. Throughout the years, these comparisons have given quite accurate size (length) estimates of long single-stranded RNAs. The use of $^{32}PO_4$ for base distribution analysis was copied from the success achieved by Elliot Volkin and Lazarus Astrachan, who in 1956–1957 detected the rapidly labeled, unstable bacteriophage mRNAs by base distribution analysis (Chapter 2).

Others were also beginning to study animal cell RNA metabolism in cultured cells. Robert Perry studied mouse L cells at the Institute for Cancer Research (now the Fox Chase Cancer Center) in Philadelphia, and Georgii Georgiev in Moscow used a transplantable mouse tumor cell line.

Klaus Scherrer, who was trained in Switzerland as a chemist, joined me as a postdoctoral fellow in the late summer of 1961. He and I began labeling and extracting HeLa cell RNA. Following the lead of Alfred Gierer, Gerhard Schramm, and Heinz Fraenkel-Conrat with plant viruses (see Chapter 2), I had earlier learned the technique of phenol extraction of HeLa cells to liberate infectious poliovirus RNA (Darnell et al. 1960). This earlier success suggested that we could get long, unbroken RNA out of HeLa cells. The single RNA molecule in a poliovirus particle had been estimated to be $\sim$7000 nucleotides long (Schaffer and Schwerdt 1959). (When polio RNA was completely sequenced in 1981, it was 7433 nucleotides long [Kitamura et al. 1981].) By this time, it had been shown that one break would render tobacco mosaic virus (TMV) noninfectious (Mundry 1959), and the same appeared to be true for polio RNA, indicating that the phenol-extracted poliovirus RNA molecules that I obtained from infected cells were full length, because they were infectious (Boeye 1959).

HeLa Cells Contain Very Long RNA Molecules

By the fall of 1961, Scherrer and I had established a workable procedure using the detergent sodium dodecyl sulfate (SDS) along with phenol at 60°C ("hot" phenol), as Eberhard Wecker had done in extracting infectious virus RNA from chicken fibroblasts (Wecker 1959). With this procedure, we recovered (free of protein and DNA) the great majority of either the total HeLa cell RNA (long labeled) or the most briefly labeled, most recently synthesized RNA.

From the very beginning, we felt confident that we knew the *approximate* size of the extracted RNA molecules by comparing them with the ∼7000-nucleotide poliovirus RNA, which always sedimented at 35S, whether isolated from purified virus or from infected cells (Fig. 4-7) (Darnell 1963).

When the HeLa cells were labeled for 24–48 h (HeLa cells double every 24 h), as expected, the total *labeled* RNA in sedimentation analysis matched the size and proportional amounts of the *majority* of the cell RNA detected by UV absorption (Fig. 4-8A). Ben Hall, while a graduate student with Paul Doty, found that rat liver rRNA sedimented at 28S and 18S. Analytical ultracentrifuge analysis suggested that the 28S rRNA was ∼4500 nucleotides long (Hall and Doty 1959).

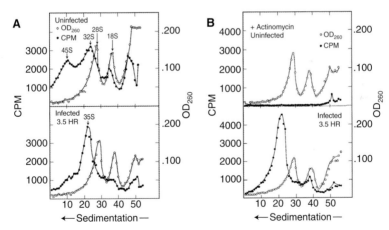

Figure 4-7. (A) HeLa cell RNA from growing and poliovirus-infected cells. (B) Cells, infected or uninfected, were treated or not with 5 μg/mL actinomycin. For all of the samples, RNA extraction and sedimentation analysis were similar to what is illustrated in Figure 4-8, but label was for 80 min with $^{32}PO_4$. Relative accumulation of 45S and 32S pre-rRNA (ribosomal RNA) is different from Figure 4-8 because the $^{32}PO_4$-labeled triphosphate pool requires 3–4 h to reach maximal, whereas uridine labeling, which labels UTP and CTP (by conversion of UTP to CTP) (in Fig. 4-8), is complete in 30–45 min. Actinomycin stops cellular RNA synthesis but does not affect polio RNA synthesis. The 35S peak is single-stranded polio RNA. (Modified from Darnell 1963. Presented June 1962 at the 27th Cold Spring Harbor Laboratory Symposia on Quantitive Biology.)

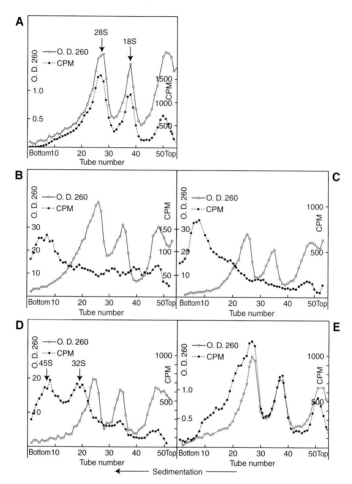

Figure 4-8. Comparison of "long-labeled" and "short-labeled" RNA from growing HeLa cells. Total cell RNA extracted from HeLa cells labeled with ^{14}C uridine for different times ([A] 24 h; [B] 5 min; [C] 30 min; [D] 60 min; [E] 4 h) was examined by zonal sedimentation analysis. To keep the relative incorporation into radioactive RNA comparable, smaller amounts of labeled uridine were used for longer label times. Used were (B) 10 μCi; (C) 5 μCi; (D) 2.5 μCi; and (E) 1 μCi. Unlabeled uridine of (A) 0.5 μCi plus 2 μM was used to ensure continued labeling. (Redrawn, with permission, from Scherrer and Darnell 1962 [©Elsevier].)

We also found, in agreement with Hall and Doty, that the majority of the HeLa cell RNA sedimented faster than bacterial rRNA (23S and 16S) and ultimately adopted 28S and 18S as the nominal size designations. These 28S and 18S HeLa cell RNAs, presumably rRNAs, appeared to comprise 80%–85% of the total, with the remainder, presumably, mostly tRNA in the 4S–5S range (Scherrer and Darnell 1962).

In contrast to the long-labeled RNA, the sedimentation profile of the very briefly (5 min) labeled HeLa RNA gave a surprising and striking result (Fig. 4-8B). The briefly labeled RNA appeared to have molecules of all sizes, ranging from a few hundred nucleotides to sizes much larger than the 28S RNA (or even the poliovirus RNA, as we soon found out). These rapidly labeled molecules did not correspond in size to the rRNAs, tRNAs, or any known cellular RNA.

When ^{3}H uridine was incorporated for 30 or 60 min (Fig. 4-8C,D), the sedimentation profile of the labeled RNA changed. First, a more discrete 45S species (estimated to contain >10,000 nucleotides) became the dominant labeled RNA, followed by a labeled 32S molecule just larger than 28S rRNA, and RNA that sedimented identically to 18S rRNA. By 4 h, the labeled 28S rRNA clearly exceeded the 32S RNA (Fig. 4-8E). Finally, by 24 h, when the cells had doubled (Fig. 4-8A), the labeled RNA sedimented identically to the rRNA and tRNA.

The most rapidly labeled HeLa cell RNA, some of which was larger than the 45S RNA, was all in the nucleus and clearly did not correspond to any of the major RNA molecules. Was any of the rapidly labeled RNA related to the major RNAs? As noted earlier (pp. 140–141), Henry Harris and others using autoradiography and cell fractionation had concluded that there was rapidly labeled, unstable nuclear RNA in cultured cells that was, they believed, completely unrelated to the stable cytoplasmic RNA of the cell. Perhaps the rapidly labeled heterodisperse RNA was an unstable type of RNA that might be related to mRNA. At this time, the only well-characterized unstable cellular RNA was bacterial mRNA. However, the heterodisperse sedimentation pattern of the most rapidly labeled HeLa cell RNA was fairly quickly overtaken by the appearance of the discrete 45S and 32S peaks. We decided to try to characterize these discrete-sized molecules first.

Ribosomal Precursor RNA: The First Recognized Case of RNA Processing

By the spring of 1962, we had determined the nucleotide distribution in the 45S, 32S, 28S, and 18S RNAs (Table 4-1). All were high in G + C content (60%–68%), unlike the total HeLa cell DNA, which was 42%–44% G + C (Darnell 1963). This analysis strongly suggested that the 45S and 32S RNAs were specific precursors to rRNA.

When actinomycin was added to a labeled culture after 30 min, quickly stopping all further RNA synthesis, two occurrences were noted: The 45S peak disappeared, and the 32S–28S and 18S peaks appeared (Fig. 4-9) (Scherrer et al. 1963; Darnell et al. 1964). Robert Perry had similar findings with

Table 4-1. Base composition (^{32}P distribution) of HeLa cell RNA molecules

Source	% of total counts as			
	A	U (T)	G	C
16S RNA	23.2	23.2	25.2	28.0
28S RNA	20.5	19.3	28.3	31.9
32S Rapidly labeled RNA	19.9	20.7	28.8	30.4
45S Rapidly labeled RNA	19.8	20.1	29.3	29.8
DNA, total	61		39	

The $(G + C)/(A + U)$ content 39%/61% of DNA was determined by Ed Simon using CsCl density gradients (Simon 1961). Later chemical analyses were usually closer to 42% G + C and 57%–58% A + U. (Modified from Darnell 1963.)

mouse L cells using pulse labeling followed by actinomycin treatment (Perry 1962). His original sucrose gradient analysis of briefly labeled RNA, however, did not show the discrete 45S or 32S peaks. He *did* find labeled molecules sedimenting faster than 28S, and after an actinomycin "chase," the large molecules decreased and labeled cytoplasmic rRNAs became the dominant species.

These actinomycin chase experiments, together with our base composition analysis, provided convincing evidence that 45S and 32S nuclear RNA molecules were related to rRNA (Scherrer et al. 1963). Animal cells apparently possessed a mechanism to convert a *primary nuclear RNA transcript* into a functioning cytoplasmic product. This "molecular carpentry" was a completely new wrinkle in RNA biochemistry. We used the term *RNA processing* to describe the apparent precursor-to-product conversion (Scherrer et al. 1963).

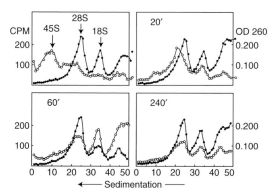

Figure 4-9. Fate of 45S RNA after synthesis. HeLa cell RNA was labeled in growing cells by incorporation of ^{14}C uridine for 30 min (*upper left*). Actinomycin, which halts RNA synthesis immediately, was then added for an additional 20, 60, or 240 min. Extracted RNA was analyzed by zonal sedimentation. (Open circles) counts per minute, cpm; (filled circles) OD$_{260}$. (Modified from Scherrer et al. 1963.)

Determination by sedimentation equilibrium of precise molecular weights (therefore lengths) of HeLa cell rRNA precursors and rRNAs was performed in 1967 and published in 1969 by Edwin H. McConkey and Johns W. Hopkins (1969). The 45S was ∼13,000 nucleotides, the 32S ∼7000, the 28S ∼5000, and the 18S ∼2000. Thus, it appeared likely that one 45S molecule gave rise to one ribosome, and the extra RNA did not accumulate, i.e., it was degraded.

Nature of the Most Briefly Labeled Heterodisperse RNA Remains Obscure

In addition to the apparent precursor role of 45S, when actinomycin was added after a brief label time, ∼30%–40% of the total briefly labeled RNA disappeared within 15 min. Was this material—the heterodisperse, most briefly labeled RNA that predominated the sedimentation profile before the accumulation of label in the 45S pre-rRNA (Darnell 1963; Scherrer et al. 1963)—conceivably related to mRNA?

During this time, Georgiev and colleagues and the Allfrey-Mirsky laboratories at The Rockefeller University had reported a very different approach to studying cellular RNA. They used cell fractionation procedures that aimed at a stepwise release of RNA from subcellular fractions of the isolated nucleus (Georgiev and Mantieva 1962; Sibatani et al. 1962) without regard to preserving the native size of extracted RNA molecules. Both of these groups reported labeled RNA fractions with a low G + C (40%–45%) content (i.e., DNA-like). However, neither the percentage of recovery of labeled RNA nor the size of the nuclear RNA in these experiments was clear. Certainly, no identification of an mRNA could be claimed by any of us at this point.

DISCOVERY OF POLYRIBOSOMES AND ANIMAL CELL mRNA

A technique for identifying mRNA was clearly needed to make sense of any of the labeled non-rRNA fractions that had been initially identified. This impasse was shortly broken in 1962–1963. Although early bacterial work had clearly shown association of mRNA with ribosomes, the numerology of this association was unsettled. Was there one ribosome per mRNA or more than one? From the size of a ribosome (measured in the EM [electron microscope]) and the estimated length of an mRNA encoding an ordinary protein, there was clearly room for at least several ribosomes on each mRNA (1 base ∼ 3 Å; 500 amino acids at 3 bases per amino acid = 1500 Å; ribosome diameter = 100 Å). The answer to this question of ribosomal density on mRNA came not from studies on bacteria but from studies with reticulocytes, which are devoted to making mainly hemoglobin, as well as from separate experiments with rat liver cells. (Fig. 3-7 in Chapter 3 shows multiple *E. coli* ribosomes translating mRNA as the RNA is being transcribed from DNA; this observation only came in 1970 [Miller et al. 1970].)

Reticulocyte Extracts Reveal Polyribosomes

In January and February of 1962, Jonathan Warner, an MIT graduate student at the time, and I added poliovirus RNA to *E. coli* ribosomes to determine whether poliovirus proteins would be made in the cell-free *E. coli* protein synthesis system of Marshall Nirenberg and Heinrich Matthaei. We *did* obtain some increase in amino acid incorporation dependent on polio RNA (Darnell 1963; Warner et al. 1963a). We then attempted, by sedimentation analysis after brief, labeled amino acid incorporation, to demonstrate functional ribosome association with the polio RNA. The poliovirus-RNA-dependent amino acid incorporation as well as labeled poliovirus RNA did, in fact, occur in structures, most of which sedimented faster than a single ribosome.

Warner then decided that it would be best to see whether an animal cell ribosome would do a better job of translating polio RNA. He therefore consulted Paul Knopf, who was working with reticulocytes one floor below us in Alex Rich's laboratory (recounted in Warner and Knopf 2002). This was on the same floor on which Howard Dintzis conducted his groundbreaking experiments showing the NH_2-to-COOH direction of synthesis of globin chains by reticulocytes (Chapter 3, pp. 120–121).

The reticulocyte cytoplasmic membrane can be easily dissolved, releasing the active protein-synthesizing complexes. Before Jon and Paul got to the stage of adding poliovirus RNA, they added labeled amino acids to the cells for 1 min, lysed the cells, and subjected the extract to sedimentation analysis. To their surprise, all of the nascent labeled protein (unfinished chains) was associated not with one ribosome but was present in what appeared to be a series of larger structures that they guessed were groups of ribosomes. Remarkably, in the sedimentation profile, they could count individual labeled nascent protein-containing units, the sizes of which suggested two to five associated ribosomes. This multiple ribosomal structure was confirmed by electron micrographs (Fig. 4-10). These groups of ribosomes were presumably associated with a single globin mRNA. A very brief ribonuclease treatment released the ribosomes that contained the labeled unfinished protein chains into particles equal in sedimentation rate to a single ribosome (Warner et al. 1963b).

The active unit of protein synthesis was a *polyribosome* (polysome). Obviously, several ribosomes could join sequentially and independently

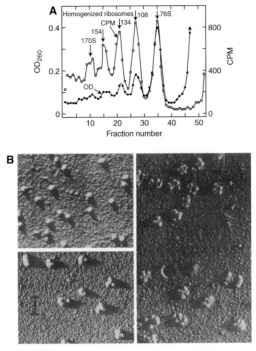

Figure 4-10. Discovery of polyribosomes in reticulocytes. (*A*) Jonathan Warner and Paul Knopf labeled reticulocytes with ³H leucine for 45 sec to label nascent protein, lysed the cells, and collected the ribosomes in a pellet and then resuspended them and subjected the resuspended material to sucrose gradient sedimentation. The radioactivity (open circles) was clearly associated with structures of different sizes (S values above peaks) that were suspected to be ribosome clusters (detected by UV absorbance, filled circles). (*B*) This was confirmed by electron microscopy (*upper left*, 76S peak; *lower left*, 134S peak; *right*, 170S peak). Vertical bar (*B*, *lower left*), 0.1 μm. (Reprinted, with permission, from Warner et al. 1963b.)

translate a single mRNA. Ribosome loading was faster than completion of a globin chain.

Alfred Gierer did very similar experiments with reticulocyte extracts, and Hans Noll and colleagues did similar experiments with rat liver extracts. All three laboratories discovered polyribosomes in cells within 1 or 2 months of one another (Gierer 1963; Warner et al. 1963b; Wettstein et al. 1963). Walter Gilbert (1963) also showed that multiple bacterial ribosomes could attach to individual long synthetic polynucleotides while making long peptides (Gilbert 1963). This was a case of a structure whose time for discovery had obviously arrived. Polyribosomes were the key to locating and examining HeLa cell mRNA.

Poliovirus Polyribosomes Help to Identify HeLa Cell mRNA in Polyribosomes

Sheldon Penman, a physicist who was instrumental in the characterization of the muon (an evanescent heavier relative of the electron), had decided to leave physics in the summer of 1962. Physics experimentation was becoming "impersonal" (i.e., was being performed by very large groups of people), and this did not suit Sheldon. Thus, fortunately for us, he decided to join our group in late summer of 1962. Penman, with Klaus Scherrer and me, made cytoplasmic cell extracts from HeLa cells that had incorporated amino acids for 1 min. When the extracts were examined by sedimentation analysis, polyribosomes of a wide variety of sizes marked by labeled amino acid incorporation were immediately identified (Fig. 4-11A) (Penman et al. 1963). When actinomycin was added for 3 or 7 h to stop all RNA synthesis, the polyribosomes declined 50% and 80%, respectively; we assumed that this was likely because of mRNA decay (Fig. 4-11B,C).

Although actinomycin stops transcription from DNA, it has no effect on the replication of RNA viruses such as poliovirus. When actinomycin-treated cells were infected with poliovirus for 2 h, all protein synthesis had ceased because of the coupled effect of virus infection (Darnell and Levintow 1960) plus actinomycin, and the polyribosomes had disappeared (Fig. 4-11D). After poliovirus RNA synthesis was well under way (3.75 h after infection) (Fig. 4-11E), polyribosomes larger than the average cellular polyribosomes appeared. These large structures contained the 35S poliovirus RNA. EM photos of the poliovirus polyribosomes showed up to 60 ribosomes, whereas the polyribosomes from uninfected HeLa cells contained ~5–20 ribosomes. Therefore, the size of the polyribosomes was apparently dictated by the size of the mRNA being translated (Fig. 4-11F–H).

HeLa Cell mRNA: Size and Base Composition

These experiments obviously indicated that the polyribosomes from uninfected cells were a semipurified source of cellular mRNA. Cellular polyribosomes were prepared from cells labeled for 40 min with $^{32}PO_4$, and labeled RNA was then extracted and examined for size and base composition (Fig. 4-12, top). The principle size range was broad, from 10S to 20S with a few larger or smaller molecules, all of which were distinctly not the size of rRNA. Most importantly, the G + C base composition of this briefly labeled polysomal RNA was not high like rRNA but was closer to the average in DNA (Fig. 4-12, bottom).

We concluded that we had identified mRNA in the cytoplasmic extracts of HeLa cells. Although the technique involved making polyribosomes and

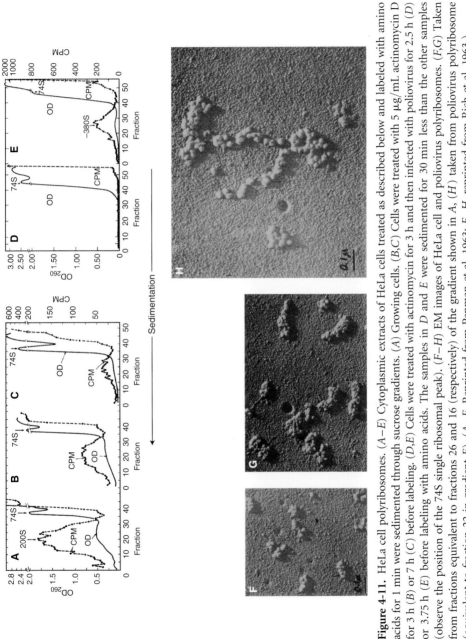

Figure 4-11. HeLa cell polyribosomes. (*A–E*) Cytoplasmic extracts of HeLa cells treated as described below and labeled with amino acids for 1 min were sedimented through sucrose gradients. (*A*) Growing cells. (*B,C*) Cells were treated with 5 μg/mL actinomycin D for 3 h (*B*) or 7 h (*C*) before labeling. (*D,E*) Cells were treated with actinomycin for 3 h and then infected with poliovirus for 2.5 h (*D*) or 3.75 h (*E*) before labeling with amino acids. The samples in *D* and *E* were sedimented for 30 min less than the other samples (observe the position of the 74S single ribosomal peak). (*F–H*) EM images of HeLa cell and poliovirus polyribosomes. (*F,G*) Taken from fractions equivalent to fractions 26 and 16 (respectively) of the gradient shown in *A*, (*H*) taken from poliovirus polyribosome (equivalent to fraction 22 in gradient *E*). (*A–E*, Reprinted from Penman et al. 1963; *F–H*, reprinted from Rich et al. 1963.)

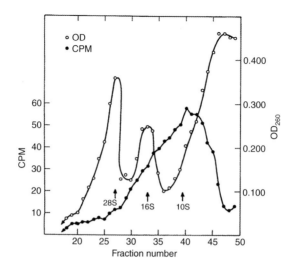

Material analyzed	——— Percent of nucleic acid as ———			
	A	U(T)	G	C
Polysomal RNA, pulse labeled, 6–12S	25.4	27.3	21.0	26.9
Polysomal RNA, pulse labeled, 12–22S	26.6	27.7	21.5	24.2
Ribosomal RNA, 28S	20.5	19.3	28.3	31.9
16S	23.2	23.2	25.2	28.0
DNA	29.1	27.5	22.0	21.4

Figure 4-12. Briefly labeled polysomal RNA has "DNA-like" base composition. (*Top*) Polysomal RNA from HeLa cells labeled for 40 min with $^{32}PO_4$ was collected, phenol-extracted, and analyzed for size by sucrose gradient sedimentation. (*Bottom*) The labeled RNA of two sizes ($\sim$6S–12S, 12S–22S) was analyzed for nucleotide distribution to compare with rRNAs and total HeLa cell DNA. (Reprinted from Penman et al. 1963.)

using short label times, it was finally possible to say that at last we had a way to study radiochemically pure mRNA. From this point forward, the definition of *cell mRNA* was the non-rRNA isolated from polyribosomes. This criterion became very useful in later discoveries of chemical additions to mRNA after its initial synthesis.

All of these results had sharpened the question of the meaning of large nuclear RNA, but certainly had not answered it: Was the putative mRNA of polyribosomes, with a size range of $\sim$500–3000 bases long, related to the rapidly labeled polydisperse nuclear RNA that was 2,000–30,000 bases long? For lack of a better name, we called the large, nonribosomal nuclear RNA "heterogeneous nuclear RNA" (hnRNA). Because we had already concluded that rRNA was most likely derived by RNA processing from a larger precursor molecule in the nucleus, the conjecture that processing of at least some hnRNA into mRNA occurred was at least not an outrageous idea. It would be 14 years later before this puzzle was completely answered.

THE PENMAN-HOLTZMAN NUCLEAR FRACTIONATION
TECHNIQUE: SEPARATION OF PRE-rRNA FROM hnRNA

In the spring of 1964, we moved our laboratory to the Biochemistry Department of the Albert Einstein College of Medicine in New York, where my former mentor Harry Eagle had established a strong Department of Cell Biology that featured cell biologists and virologists, all using cultured animal cells. Sheldon Penman was still getting accustomed to a molecular biology career; thus, he, Jon Warner, and several others came along also. Sheldon was funded by a special forward-looking NIH fellowship that encouraged physicists to come into biology.

We needed a fractionation technique that would allow a clear separation of the two major classes of nuclear RNA, the pre-rRNA and the heterodisperse hnRNA, so that each could be studied separately. Penman, with an assist from the cell biologist Eric Holtzman, came to the rescue by 1966, developing the nuclear fractionation procedure that became the standard for RNA studies in cultured cells (Holtzman et al. 1966; Penman et al. 1966).

Penman and Holtzman knew from earlier work that the nucleolus was the likely site of synthesis of rRNA and now presumably the 45S pre-rRNA primary transcript. The nucleolus in electron micrographs is a granular intranuclear structure whose role in rRNA synthesis had been suggested years before because of its dense basophilic staining properties. Furthermore, Robert Perry (Gaulden and Perry 1958; Perry 1960) had shown that microbeam UV irradiation of the nucleolus largely prevented the labeling of cytoplasmic RNA (presumably, rRNA). In addition, using growing cultured cells, Perry had collected autoradiographic data compatible with the nucleolus as the origin of rRNA (Perry 1961, 1962).

Also by this time, Don Brown and John Gurdon had shown that the anucleolate mutant of *Xenopus laevis* (African clawed toad) goes through early embryonic stages using maternal RNA. Fertilized *Xenopus* eggs can be incubated and go through early development. Early in embryogenesis, anucleolate mutants can be recognized and separated from heterozygotes (genetically lacking one-half of rRNA genes) and normals. Brown and Gurdon (1964) labeled developing embryos (see legend to Fig. 4-13) and examined RNA of anucleolate mutants and normal plus heterozygote embryos. In contrast to the normal and heterozygous embryos, the anucleolate mutants failed to label rRNA, clearly indicating the nucleolus as the site of ribosome synthesis.

In addition, Ferruccio M. Ritossa, Kim Atwood, and Sol Spiegelman showed that a *Drosophila* mutant called *bobbed* lacked the normal amount of DNA that would hybridize to rRNA, i.e., it lacked some rRNA genes (Ritossa et al. 1966). Moreover, the *Drosophila* DNA that encoded rRNA

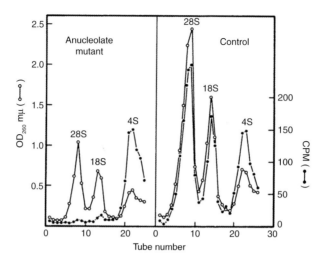

Figure 4-13. Failure of rRNA synthesis by the anucleolate mutant of *Xenopus laevis*. Early *Xenopus* embryos (a mixture of homozygous mutants, heterozygotes, and normals) were incubated in a sealed vessel with $^{14}CO_2$. After development, when the mutants could be physically recognized, RNA was prepared from each (mutant and the heterozygotes plus normals) and examined by sedimentation. No labeled rRNA was made by the mutants. (Reprinted, with permission, from Brown and Gurdon 1964.)

was in what morphologists called the "nucleolar organizer." All of these experiments clearly demonstrated that rRNA was copied from ribosomal DNA genes in the nucleolus.

By 1966, Penman and Holtzman succeeded in fractionating isolated HeLa cell nuclei into a nucleolar fraction that retained its characteristic appearance in the EM and an extranucleolar fraction (Penman et al. 1966). The procedure used DNase to destroy chromatin but left the nucleolus intact (Fig. 4-14A–F). When Penman examined the RNA of the nucleolar fraction by sedimentation analysis (Penman et al. 1966), it contained the previously identified 45S and 32S preribosomal components, now pure enough to observe as dominant UV-absorbing peaks (Fig. 4-14G). The extranucleolar fraction contained the heterodisperse RNA that had the same size distribution as the very briefly labeled nuclear RNA that Scherrer and I had observed in 1962. By the time he finished these experiments, Penman had returned to MIT, now as a Professor of Biology.

When Ruy Soeiro in my laboratory (Soeiro et al. 1966) examined the base distribution on these cleanly separated fractions, the preribosomal nucleolar 45S and 32S fractions, uncontaminated with hnRNA, were 65%–68% G + C (a composition similar to that of 28S and 18S rRNAs), as we had established earlier on molecules well separated by sucrose gradient sedimentation. Eight different samples (from ~2000 to perhaps 50,000 nucleotides long) of the extranucleolar

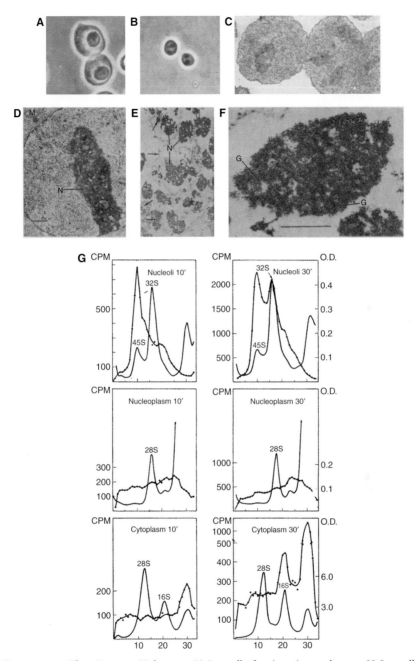

Figure 4-14. The Penman-Holtzman HeLa cell fractionation scheme. HeLa cells swollen in hypotonic buffer (*A*) were subjected to mild shear and then treated with detergent to prepare nuclei (*B*) where there was little or no cytoplasmic contamination as evidenced by electron microscopy (*C*). Nucleoli are visible in *C*, *D*, and *E*. (The nucleoli are labeled "N" in *D* and *E*.) The nuclei can be lysed in a high NaCl buffer plus RNase-free DNase; centrifugation leaves a particulate fraction that contains relatively intact nucleoli (*F*) and an extranucleolar fraction in the supernatant. (In *F*, the nucleolus is labeled "G," which refers to nucleolar granularity.) (*Legend continued on next page.*)

nucleoplasmic hnRNA fraction, now uncontaminated with pre-rRNA, were DNA-like (~42%–44% G + C), the same base distribution as in the previously analyzed polysomal mRNA fractions (500–3000 nucleotides) (Soeiro et al. 1968).

The base distribution of what we characterized as hnRNA agreed with the earlier reported nuclear subfractions of Georgiev. He suggested that his DNA-like nuclear fraction came from a subnuclear fraction that contained chromosomes (Georgiev and Mantieva 1962). After returning to Europe, Klaus Scherrer had also continued to work on large nuclear RNA molecules using a variety of eukaryotic cells in addition to HeLa cells. He and his colleagues established extraction procedures for duck erythroblasts, nongrowing cells that develop into the nucleated red blood cells characteristic of birds. Because these cells make relatively little pre-rRNA but do continue to make nuclear RNA, Scherrer and Marcaud (1965) obtained high-molecular-weight RNA, the base composition of which was very low in G + C content (nonribosomal in character). In addition, Scherrer with Nicole Granboulan (Granboulan and Scherrer 1969) made very important electron microscopic observations. The duck erythroblast and HeLa-cell nuclear RNA contained molecules several times longer than the 28S RNA. These molecules were observed in denaturing conditions, i.e., they were stretched linearly and ranged in size (compared to the ~5-kb 28S rRNA) of up to 50,000 bases. hnRNA did, indeed, include RNAs from a few thousand to many thousands of bases (Fig. 4-15).

All of these results concentrated interest in the possibility that RNA processing of hnRNA occurred during mRNA biosynthesis. Because of the size of the hnRNA, the idea of losing some RNA—either in whole or in part—during processing of large to smaller molecules might explain the long-discussed "unstable" nuclear RNA.

PROCESSING OF tRNA PRECURSORS TO tRNA

A quick mention of an illustration of RNA processing discovered subsequent to rRNA processing is in order. tRNAs, the ~70-nucleotide RNAs that play

---

Figure 4-14. (*Continued*) (*G*) RNA from cells labeled with ³H uridine for various times and examined by sedimentation analysis showed that the nucleoli contain the 45S and 32S pre-rRNA, observable for the first time by UV absorbance (smooth lines, OD_{260}) and uridine labeling (lines with dots, cpm). The synthesis of the pre-rRNA in the nucleolus was proven in this experiment. The extranucleolar fraction does not contain pre-rRNA but does contain the high-molecular-weight non-rRNA. (*A* and *B*, Reprinted, with permission, from Penman et al. 1966; *C–F*, reprinted, with permission, from Holtzman et al. 1966 [©Elsevier]; *G*, reprinted, with permission, from Penman et al. 1966 [©AAAS].)

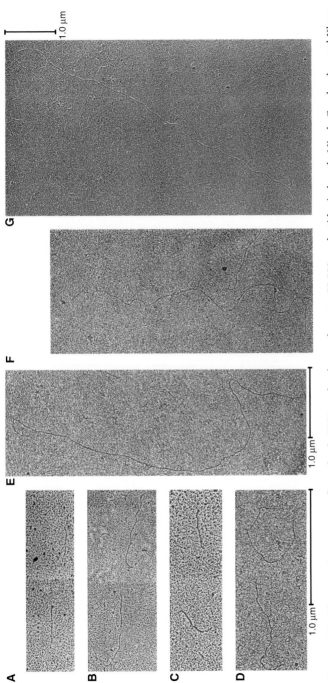

Figure 4-15. Electron microscopy confirms that hnRNA molecules can be up to 50,000 nucleotides in length. Nicole Granboulan and Klaus Scherrer prepared nuclear RNA molecules from a number of sources (including duck erythroblasts, HeLa cells, and silk worm embryos) and separated different size classes by sucrose gradient sedimentation. rRNA from *E. coli* (16S and 23S) and HeLa cells (18S and 28S) furnished size markers (length in micrometers, μm). The number of nucleotides in each unknown could be computed from their estimated equilibrium sedimentation mass (e.g., 45S between 13 and 14 kb). (*Left, A*) 16S (0.37 and 0.39 μm) and (*B*) 23S (0.82 and 0.83 μm) *E. coli* rRNAs. (*C*) 18S (0.58 μm) and (*D*) 28S (1.17 and 1.19 μm) HeLa cell rRNAs. (*E,F*) Two HeLa cell molecules >45S (6.93 and 5.57 μm in *E* and *F*, respectively). (*G*) Duck erythroblast molecule from an ∼50S region (7.15 μm). (Reprinted, with permission, from Granboulan and Scherrer 1969 [©Blackwell Publishing Ltd.].)

such a crucial part in translation, were known by the late 1960s to have many chemical additions. Methyl (—CH₃) groups are added to both the nucleic acid bases and the 2′-OH groups of riboses in tRNAs and to a greater extent in eukaryotes than in bacteria. The tRNAs are ∼70–80 nucleotides long, but radioactive label from methyl-labeled methionine (the universal methyl donor) appeared first in molecules ∼150 nucleotides long. When actinomycin was added to stop RNA synthesis, the ∼150-nucleotide methylated molecules disappeared, to be replaced by smaller tRNA-sized molecules. Such experiments were first performed in Roy Burdon's laboratory (Burdon et al. 1967) and at about the same time in my laboratory by Deborah Bernhardt (Darnell 1968; Bernhardt and Darnell 1969). Bernhardt (now Mowshowitz) also showed that starvation for methionine retarded the size conversion and that the longer molecules could be converted to the smaller size by incubation with cell extracts (Mowshowitz 1970). Clearly, the synthesis of tRNA was a second case of RNA processing. Later work indeed showed that the purified tRNA genes encoded the longer ∼150-nucleotide primary transcripts that are processed to tRNA. By the late 1960s, the idea of RNA processing had quite a head of steam.

SEARCHING FOR EVIDENCE OF hnRNA-TO-mRNA PROCESSING

If hnRNA was processed to mRNA, it was necessary to show a flow of some specific sequence(s) from longer to shorter molecules. At the time (mid 1960s), following individual cellular gene sequences was not possible. However, pursuing the sequence relatedness between hnRNA and mRNA offered the only route to understanding molecular details of the mammalian genome. The long RNA, hnRNA, was, after all, copied from long stretches of mammalian DNA, and if mRNAs were buried in these copies, the long molecules afforded the chance to learn something regarding the coding potential in mammalian DNA.

Sequence Complexity in Vertebrate DNA: Reassociation Rate of Denatured DNA

The only specific sequence comparison available was RNA:DNA hybridization. The general nature of sequence classes in vertebrate DNAs was uncovered during this time in the laboratory of Roy J. Britten. In 1968, Britten and David Kohne (Britten and Kohne 1968) described different classes of repetitive sequences in humans by observing rates of DNA:DNA hybridization. Samples of vertebrate (including human) DNA, broken into fragments (∼200 nucleotides long) and denatured, were allowed to reanneal. Hybridized

DNA was then selected at intervals. The most rapidly reassociating DNA and successive samples of intermediate-reassociating DNA were collected. Repeating the dissociation–reassociation with separated classes showed clearly that the most highly repeated sequences reassociated hundreds to thousands of times faster than the most slowly associating DNA. More than 50% of vertebrate DNA was highly repetitious. The presence of highly repetitious sequences that proved to be present in hnRNA and mRNA (to a much lesser extent) made attempts to determine whether sequences present in coding regions of mRNA were also present in hnRNA impossible. Attempted competition hybridization experiments done in several laboratories proved nothing other than that there were more repetitious sequences in total hnRNA than in mRNA. Essentially, the only mRNA sequences that hybridized to DNA in such experiments were copied from the repetitious DNA. Attempts to show convincing specific sequence relatedness between hnRNA and mRNA were unavailing. Certainly, it was not possible to prove that the coding regions of mRNA lay within hnRNA by these techniques (Soeiro et al. 1968; Greenberg and Perry 1971; Holland et al. 1980).

SPECIFIC SEQUENCES IN mRNA AND hnRNA: DNA VIRUSES IN THE NUCLEUS AND POLY(A)

There were, however, viruses whose DNA entered the nucleus to presumably produce mRNA during replication. Moreover, virus DNA could be purified in large amounts and used to measure viral mRNA. By the late 1960s and early 1970s, several laboratories had begun to investigate viral-specific RNA production from such viruses. At just about the same time (1971), another specific cellular sequence, a 3′-terminal poly(A) shared by most mRNAs, was discovered. These two quite different paths of research produced the first sequence-specific links between mRNA and large nuclear RNA.

SV40 and Herpes Large Nuclear RNA

Among the first experiments relating specific nuclear RNA to polyribosomal mRNA came from studying two DNA viruses, herpes simplex virus in cultured human cells and the mouse virus SV40, whose DNA genome becomes integrated in cell chromosomes. Work with human adenovirus followed soon after and proved to be the most informative.

Ed Wagner, a graduate student with Bernard Roizman at the University of Chicago, studied herpes simplex virus. RNA:DNA hybridization showed that some of the virus-specific RNA in the cell nucleus was much larger (20S–80S) than the virus-specific mRNA from polyribosomes (10S–20S) (Wagner and Roizman 1969). By competition hybridization, the unlabeled shorter poly-somal virus mRNA sequences were capable of blocking 50%–75% of the hybridization of the very large (>50S) herpes-specific RNA. Thus, the large RNA did contain sequences that were the same as those found in polyribosomes, but the experiments hardly proved that the polysomal molecules were derived from the larger nuclear molecule.

SV40 virus is one of a group of small DNA viruses that can enter the cell and have its DNA inserted into the chromosome of the host cell. The protein products of the virus genes then can *transform* the cell, i.e., make it competent to grow continuously in culture and form tumors when injected into animals. That this group of small DNA viruses were tumor viruses had been discovered by Sarah Stewart and Bernice Eddy at the NIH in 1958 (Stewart et al. 1958). The insertion of the virus DNA from one of these viruses, polyoma virus, into the chromosomes of transformed cells was proven in Renato Dulbecco's laboratory (Sambrook et al. 1968). Tom Benjamin, a graduate student with Dulbecco, showed by DNA:RNA hybridization that transformed cells with the inserted viral DNA made viral-specific mRNAs (Benjamin 1966).

Objections had been raised in some quarters that maybe all large, DNA-like nuclear RNA was somehow aggregated, e.g., by accidental base-pairing, and that was why it was "larger." Uno Lindberg in my laboratory used centrifugal separation of RNA from SV40-transformed cells in dimethyl sulfoxide (DMSO), which prevents RNA aggregation (Strauss et al. 1968), to counter this argument. He extracted RNA from the nucleus and polyribosomes of SV40-transformed cells and sedimented it in sucrose gradients using 98% DMSO. The SV40-specific RNA in the nucleus was larger than the virus-specific polysomal RNA (Lindberg and Darnell 1970). Susumu Tonegawa, as a postdoctoral fellow in Renato Dulbecco's laboratory, did an even more extensive experiment than Lindberg's, also showing that the nuclear SV40-specific RNA was larger than the cytoplasmic SV40-specific RNA (Fig. 4-16). Tonegawa's experiments also demonstrated that the larger nuclear RNA was labeled before most of the smaller cytoplasmic RNA (Tonegawa et al. 1971).

All of these early experiments with DNA virus-specific RNA were compatible with the idea that large nuclear viral-specific RNA originating from either a viral genome or a viral genome in a chromosomal site could be "processed" into polysomal mRNA. But at this point, no experiments had directly proved the elusive point of obligatory processing of a specific large RNA that was processed into a specific smaller molecule for any viral mRNA, let alone for any cellular mRNA.

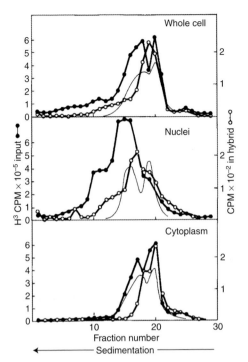

Figure 4-16. SV40-specific RNA in transformed cells: Some nuclear molecules are longer than cytoplasmic molecules. SV40-transformed cells were labeled with ^{3}H uridine for 90 min. RNA was then prepared from whole cells, nuclei, and cytoplasm and separated on sucrose-DMSO gradients, and fractions were hybridized to excess SV40 DNA. (The thin continuous lines are UV absorption.) (Reprinted from Tonegawa et al. 1971.)

POLYSOMAL mRNA IN MAMMALIAN CELLS HAS A POLYADENYLIC ACID ("POLY[A]") TAIL

The availability of briefly labeled polysomal mRNA allowed the discovery of the first distinctive cellular biochemical evidence that linked cellular hnRNA to mRNA. This was not a sequence copied from a gene. It was polyadenylic acid (poly[A]), a string of As added to hnRNA by a nuclear enzyme after it was copied from DNA. The poly(A) that originated on hnRNA then remained with the processed mRNA following arrival in the cytoplasm. Poly(A) proved eventually to be found on virtually all eukaryotic mRNAs.

Discovery of Poly(A)

This story begins with the discovery of poly(A) itself. In 1960 (before the discovery of a true RNA polymerase), Mary Edmonds and Richard Abrams reported an activity in lymphocyte extracts that would synthesize stretches of poly(A) from ATP without the presence of DNA (Edmonds and Abrams 1960). This activity had no known role at this time. In 1966, George Brawerman found stretches of poly(A) in the (unfractionated) cytoplasmic RNA of rat liver (Hadjivassiliou and Brawerman 1966).

In the late 1960s, Brian McAuslan, an Australian virologist, and his student Joe Kates made the very surprising discovery that vaccinia, a large DNA-containing virus, actually carries within the virus particle various virus-encoded enzymes (Kates and McAuslan 1967). When the vaccinia particle is exposed to a mild detergent, the outer lipid-containing coat of the virus is dissolved, as happens when the virus enters a cell. This uncoating releases a virus RNA polymerase that then copies the vaccinia DNA to make vaccinia mRNA. Kates then subsequently found that enzymes in the vaccinia particles also produced poly(A) that was associated with the vaccinia mRNA. But the poly(A) was not made if vaccinia mRNA synthesis was blocked with actinomycin (Kates and Beeson 1970). Kates reached the tentative conclusion that the poly(A) was copied from the vaccinia DNA.

Poly(A): Part of Polysomal mRNA

At this point, the knowledge that polyribosomes from briefly labeled cultured cells were a source of radioactively labeled mRNA (Penman et al. 1963) became extremely useful. In 1971, both the Edmonds and Brawerman laboratories, as well as our group, reported in the same issue of the *Proceedings of the National Academy of Sciences, USA* that adenosine-labeled mRNA from isolated polyribosomes of cultured cells (HeLa and mouse sarcoma 180 cells) contained a segment of labeled adenylic acid-rich RNA (Darnell et al. 1971a; Edmonds et al. 1971; Lee et al. 1971).

Both Edmonds and Brawerman had been following poly(A) for some years, and their laboratories both found poly(A) in mRNA by first preparing briefly adenosine-labeled polyribosomes, extracting the RNA, digesting it with pancreatic RNase (which does not digest poly[A]), and examining the digest by techniques that retained poly(A). I discovered poly(A) in polysomal mRNA because I was digesting ^{3}H-uridine- or ^{3}H-adenosine-labeled polysomal RNA with pancreatic RNase during the course of RNA:DNA hybridization experiments. The ^{3}H-uridine-labeled mRNA, before any interaction with DNA, was digested $> 99\%$, but there was an indigestible core ($\sim 4\%-5\%$ of the total) in the ^{3}H-adenosine-labeled mRNA. Furthermore, there was more of the adenylic acid-rich fraction in smaller mRNA than in larger mRNA, consistent with a constant-sized adenylic acid-rich sequence in each mRNA.

Actually, unbeknownst to Mary Edmonds, George Brawerman, or me, by this time Kates had also found poly(A) in cellular mRNA. I had reviewed his papers on the synthesis of vaccinia mRNA and poly(A), both of which were blocked by actinomycin. At that time, he did not suggest that cell mRNA contained poly(A). But after submitting the vaccinia virus papers, Kates located the RNase-resistant poly(A) on cellular polysomal mRNA, as each of our laboratories had done independently. (This latter finding is included in his Cold Spring Harbor paper given in June 1970 but published in the spring of 1971 [Kates 1971].) Thus, after more than a decade, this distinctive and previously mysterious stretch of eukaryotic RNA had a home—it was part of polysomal mRNA.

Poly(A) Is Added to hnRNA in the Cell Nucleus

The newly labeled poly(A) segment, when released from the rest of the RNA by digestion with pancreatic RNase, was ~200 nucleotides in length. That made it easy to assay quite specifically.

Of great importance, a poly(A) segment of the same size was also present in hnRNA (Fig. 4-17) (Darnell et al. 1971a; Edmonds et al. 1971). When HeLa cells were labeled for very brief times (45 sec), >95% of the labeled poly(A) was in the nucleus as part of hnRNA (Jelinek et al. 1973). After a few minutes, newly labeled poly(A) appeared in polyribosomal mRNA. But only after ~40 min was there more labeled poly(A) in polysomal RNA

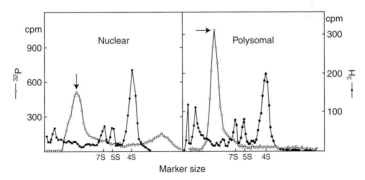

Figure 4-17. Poly(A) from newly synthesized nuclear and polysomal mRNA (arrows) is the same size. Poly(A) (^{32}P label) was prepared by nuclease digestion of phenol-extracted nuclear RNA or polysomal mRNA and selection by specific binding to and eluting from oligothymidylate-cellulose columns. The poly(A) (open circles) and size markers (^{3}H uridine [closed circles]) from low-molecular-weight cytoplasmic RNA were mixed and subjected to electrophoretic separation on polyacrylamide gels. (Modified from Edmonds et al. 1971.)

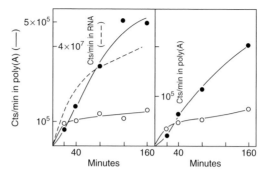

Figure 4-18. Time course of labeled poly(A) appearance in hnRNA and polysomal RNA. Two experiments (*left, right*) in which HeLa cells were labeled with ^{3}H adenosine and samples taken at indicated times for poly(A) assay in hnRNA (open circles) and polysomal RNA (filled circles). In the experiment at *left*, total labeled RNA is plotted (dashed line). (Redrawn, with permission, from Jelinek et al. 1973 [©Elsevier].)

than in the nuclear RNA (Fig. 4-18). One of the first biochemical tasks was to determine whether the poly(A) was copied from DNA or was perhaps added after transcription. In fact, very short actinomycin treatment (2 min) reduced total adenosine incorporation in the ensuing 2 min by ∼85%, but poly(A) formation on nuclear RNA was reduced by only ∼25%, consistent with the possibility of poly(A) addition by an enzyme different from RNA polymerase immediately after transcription was complete (Darnell et al. 1971b).

Several biochemical reports soon proved that the poly(A) was on the 3′-OH end of both hnRNA and rRNA. For example, pancreatic RNase released a fragment that contained only one adenosine residue (the 3′ residue) per 200 AMP (adenosine 5′ phosphate) residues (Mendecki et al. 1972; Molloy et al. 1972a,b; Nakazato et al. 1973). The conclusion was hence quite firm that poly(A) was added as a posttranscriptional event at the 3′ end of some hnRNA molecules and was present in the great majority of mRNA. Whether actual chain termination and poly(A) addition occurred or whether RNA chain cleavage and poly(A) addition occurred was not known. (As discussed in Chapter 5, the latter is the case.)

Based on all of these experiments, a putative model of at least one method of hnRNA processing to yield mRNA seemed very logical, at least for the formation of sequences from the 3′ end of hnRNA (Fig. 4-19).

Soon after the discovery of poly(A) in cultured human and mouse cells, poly(A) was found in many other vertebrate and invertebrate mRNAs. Although the unit of poly(A) was shorter in yeast cells (∼50 nucleotides), it was definitely present in polyribosomal mRNA (McLaughlin et al. 1973).

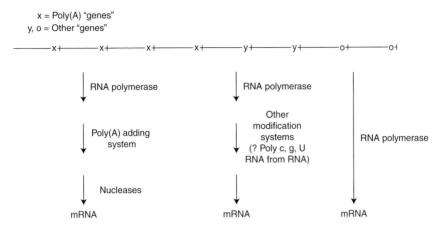

Figure 4-19. Diagram from 1972 summarizing ideas on processing of hnRNA to mRNA as poly(A) studies progressed. (*Top left*) At sites marked x, cleavage of hnRNA could free a 3' end to allow formation of poly(A)-terminated mRNA. (*Top center*) Another possible recognition system (y) for making an mRNA is envisioned. (*Top right*, o) As in bacteria, RNA polymerase would simply copy an mRNA molecule that would find its way into the cytoplasm. The temper of the times is captured in the discussion, "The excitement at the present time is generated by the apparent certainty that at least one mechanism of mRNA biogenesis in eukaryotes has at least been described—and this mechanism is very different from mRNA biogenesis in bacteria." (Redrawn from Darnell et al. 1972.)

USE OF POLY(A) IN GENE TECHNOLOGY: EASY MRNA PURIFICATION

The discovery of poly(A) in polysomal mRNA conveyed an instant technological benefit. Because of the presence of the poly(A), mRNA molecules could be separated from rRNA and tRNA by binding to immobilized polyribouridylic acid (polyU) or polydeoxyribothymidylic acid (poly[dT]). Both poly(U) and poly(dT), when attached to appropriate solid materials (e.g., treated paper or the commercial gel Sepharose), allowed the collection of mRNAs free of rRNA and tRNA.

A very important immediate use of such physical mRNA selection came through the use of an enzyme also discovered in 1971 by Howard Temin and Satoshi Mizutani (1971) and separately by David Baltimore (1971). By this time, several RNA-containing animal viruses (e.g., vesicular stomatitis virus [VSV] and reoviruses) in addition to the DNA virus vaccinia had been discovered to carry their own mRNA-producing enzymes (Kates and McAuslan 1967). All of these viral enzymes made RNA.

Temin and Baltimore took the adventuresome leap to determine whether the RNA tumor viruses that become established in tumor cells could copy their RNA into DNA. Tumor viruses somehow exert continuing genetic control in transformed cells, and the synthesis of DNA seemed possible, especially after polyoma and SV40 DNA were shown to become part of the cell DNA (Sambrook et al. 1968). The lipid envelope of the virus particles was removed with a mild detergent (as McAuslan and Kates had done with vaccinia virus), added labeled deoxynucleoside triphosphates, and the virus cores made DNA. Moreover, the initial single-stranded DNA copy was eventually found to be converted inside cells into a double-stranded molecule that becomes integrated into the cell chromosomes, causing cell tumorigenic transformation.

The enzyme in the RNA tumor virus particle was named *reverse transcriptase*. A whole field of cancer biology was opened by these remarkable results. The integration of viral DNA into the DNA of transformed cells convinced scientists that changes in cell DNA were very likely an underlying cause of cancer. Baltimore and Temin shared the 1975 Nobel Prize in Medicine or Physiology with Dulbecco.

The practical utility of copying mRNA into DNA—so-called *cDNA* or *complementary DNA*—made reverse transcriptase one of the most widely used and valuable reagents in the recombinant DNA revolution that began by 1972. Preparation of cDNA with reverse transcriptase is now a part of standard molecular biology repertoire (Cohen et al. 1973).

As discussed shortly, the formation of cDNA and subsequent molecular cloning of specific cDNA molecules also provided the first supply of specific DNA with which individual cell mRNAs could be measured by DNA:RNA hybridization.

METHYLATED STRUCTURES ON mRNA 5′ ENDS

Although the poly(A) structures of some hnRNAs and of virtually all mRNAs were a common characteristic of the 3′ ends of these molecules, nothing had been discovered in the early 1970s regarding the 5′ end of either hnRNA or mRNA. In bacteria, the 5′ end of an mRNA is usually pppAp or pppGp, whichever starts the RNA chain. But no one had found these nucleoside polyphosphates on eukaryotic nuclear or cytoplasmic RNA. The nature of the 5′ ends of eukaryotic mRNA was ready for attack.

By 1973–1974, chemical "decorations," especially methyl groups, of both the bases and sugars of tRNA and rRNA were well recognized. When it became possible because of the poly(A) tail to purify mRNA away from rRNA and tRNA, methylation of mRNA was uncovered. This led to the important discovery of a universal methylated structure on the 5′ end of all eukaryotic mRNA.

Methylation in mRNA

The solution to this important question of mRNA methylation began when Bob Perry and Dawn Kelley, his assistant, labeled growing mouse L cells with ^{3}H-methyl-labeled methionine (the methyl donor through S-adenosylmethionine) (Perry and Kelley 1974). They selected the poly(A)-containing mRNA, and it was methyl labeled.

When the methyl-labeled poly(A)-selected RNA was subjected to alkaline degradation to determine which nucleotides were labeled, the majority of the labeling was in mononucleotides (Fig. 4-20). Later that year, these mononucleotides were identified mainly as N-6-methyl adenylic acid (Desrosiers et al. 1974). But perhaps the most novel finding in Perry and Kelley's first experiment was an alkali-resistant methyl-labeled structure that was more highly charged than individual nucleotides. Perry and Kelley (1974) suggested that

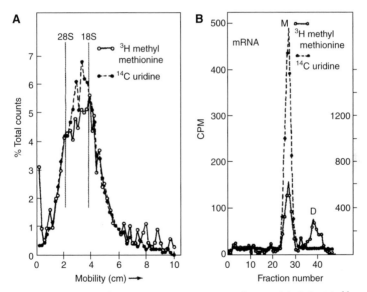

Figure 4-20. Methylation of mRNA. Growing mouse cells were labeled with ^{14}C uridine and ^{3}H methyl methionine (methyl donor). Poly(A)-containing RNA (mRNA) was prepared from the cytoplasm. (*A*) Electrophoresis of the RNA through polyacrylamide gels showed that the mRNA molecules of all sizes were labeled and therefore methylated. (*B*) The labeled mRNA was hydrolyzed by alkali treatment and analyzed by column chromatography. Methylation occurred to nucleotides within the mRNA chain that eluted in the M peak. The D peak was labeled by —CH$_3$-labeled methionine, but not by uridine. (*B*) (*Left* y axis) ^{3}H methyl methionine, cpm (open circles); (*right* y axis) ^{14}C uridine, cpm (closed circles). The nature of the D peak was not proven. (Modified, with permission, from Perry and Kelley 1974 [©Elsevier].)

this structure consisted of dinucleotides because of its greater charge but did not identify its chemical nature (D peak in Fig. 4-20B).

At about the same time, several groups of virologists studying animal viruses also found evidence of methyl-labeled viral mRNAs. Bernard Moss and Cha-Mer Wei were studying mRNAs transcribed from vaccinia cores (Wei and Moss 1974). Aaron Shatkin and colleagues were studying reovirus mRNAs also produced by viral enzymes that were part of the reovirus particle (Shatkin 1974). In Japan, Kin-ichiro Miura and colleagues had found methylation in the mRNA of the polyhedrosis virus of insects (Miura et al. 1974).

Within the virion, all of these viruses contain not only polymerases that make mRNA by copying their genomes but also methylases that use S-adenosylmethionine to methylate virus mRNA. In examining the nature of the methyl labeling of the viral mRNAs, alkaline hydrolysis showed that some bases within the RNA chains were labeled, but more importantly, the virus mRNA also contained a more highly charged, alkali-resistant, methyl-labeled structure.

The m^7GpppN "Cap" on Viral and Cell mRNAs

Both the Moss and Shatkin groups went to work on the chemistry of this structure, and when the dust settled, they had discovered something very important (Furuichi et al. 1975a; Wei and Moss 1975). The 5' ends (the beginning end) of the mRNAs had the structure m^7GpppNmpNp; the formation of this structure by a series of enzymes is diagrammed in Figure 4-21.

Shatkin's and Moss's discoveries had suddenly provided a new, well-characterized chemical structure that marked the 5' end of animal virus mRNAs. Although vaccinia and reovirus RNAs had been used to characterize the "cap" structures, the same methylated 5'-end marker was soon found in polysomal poly(A)-containing mRNAs of uninfected human cells by several laboratories (Furuichi et al. 1975b; Wei et al. 1975). Thus, mammalian cell and virus mRNAs are chemically marked at both ends. In fact, all animal and plant mRNAs and mRNAs from single-cell eukaryotes have a methylated cap structure at their 5' ends and a poly(A) structure at their 3' ends (Shatkin 1976).

This cap structure proved to be very important in helping to locate start sites for eukaryotic RNA polymerase in later studies. Although the specific biochemical steps outlined in Figure 4-21 were not completed for several years, it is implied in the diagram that the cap is formed from a 5'pppN nucleotide, the first nucleotide in a primary transcript.

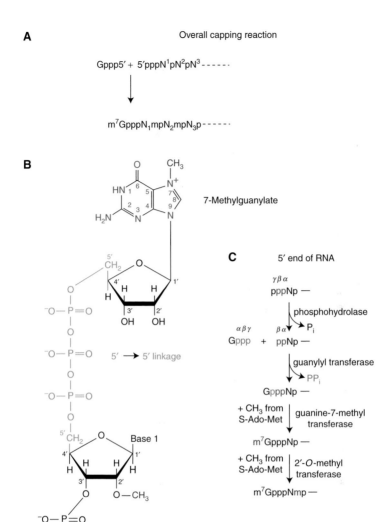

A

Overall capping reaction

$$Gppp5' + 5'pppN^1pN^2pN^3 - - - - - -$$

$$m^7GpppN_1mpN_2mpN_3p - - - - - -$$

B

7-Methylguanylate

$5' \rightarrow 5'$ linkage

Base 1

Base 2

C

5' end of RNA

$\gamma \beta \alpha$
pppNp —

phosphohydrolase

$\alpha \beta \gamma$ $\beta \alpha$ P_i
Gppp   +   ppNp —

guanylyl transferase

PP$_i$

GpppNp —

+ CH$_3$ from | guanine-7-methyl
S-Ado-Met | transferase

$m^7GpppNp$ —

+ CH$_3$ from | 2'-O-methyl
S-Ado-Met | transferase

$m^7GpppNmp$ —

Figure 4-21. Capping reaction of eukaryotic mRNAs. (*A*) In the overall reaction, the 5'pppN of the first nucleotide laid down by RNA polymerase II is combined with an incoming 5'pppG (GTP) to form a linked structure with three phosphates. During this process, methyl groups are added to the guanine of the incoming GTP and to the ribose of the first and second nucleotides of the mRNA (the second ribose is not always methyl-ated in mammalian cells and never, e.g., in yeast). (*B*) Structure of the completed cap. (*C*) Shown is the stepwise formation of the cap and the enzyme. The phosphates that are retained in the final linkage are indicated. (See Shatkin 1976; Venkatesan and Moss 1982.)

SOME LONG HNRNAS HAVE BOTH AN M⁷G CAP
AND A POLY(A) TAIL

During the summer of 1974, Aaron Shatkin, whom I had first met at the NIH in 1960 and frequently saw at virology meetings, told me of the important results concerning the cap structure. We discussed using this new marker for the 5' ends of RNA to perhaps connect HeLa cell mRNA and hnRNA. My laboratory assistant, Marianne Salditt-Georgieff, provided Shatkin and coworkers with methyl-labeled, polysomal, poly(A)-containing mRNA and also poly(A)-containing hnRNA from the extranucleolar portion of HeLa cell nuclei. In short order, we had established that mRNA and poly(A)-containing hnRNA both had the, by now, well-characterized 5' caps as well as, of course, the 3' poly(A) (Furuichi et al. 1975b). The great majority (if not all) of hnRNA molecules had a cap, but a substantial fraction (about two-thirds) lacked the ~200-nucleotide poly(A) tail. Because we could label the RNA molecules with $^{32}PO_4$, we got a very accurate count of the nucleotides in different sizes of mRNA and the *poly(A)-containing hnRNA*: caps have five Ps per molecule, poly(A) has ~200 Ps per molecule, and the rest of the Ps are in the internal bases in the RNA chains (Fig. 4-22).

These results settled any question on the size range of poly(A)-containing hnRNA: It varied from several thousand to >15,000 nucleotides (average of 8,000 to 10,000 nucleotides), whereas the mRNA was 500–3000 nucleotides with an average of ~1200 nucleotides (Salditt-Georgieff et al. 1976).

It seemed to us that if mRNA resulted from processing the hnRNA, we must postulate some event(s) that used both ends of the hnRNA and possibly internal sections as well. Recall that colinearity was a prized principle of bacterial genetics and biochemistry (Sarabhai et al. 1964; Yanofsky et al. 1964), and we were greatly influenced by that dictum in thinking about the possible generation of mRNA from hnRNA. Because both ends on every mRNA are modified, this might indicate both capping at a 5'-cut site or polyadenylation at a 3'-cut site within the hnRNA. We already knew that poly(A) was added on the 3' end of hnRNA; thus, why not also on an internal cut site where a cap might also be added (see Fig. 4-19)? In addition, Shatkin had told me of experiments by a colleague of his, Amiya Banerjee, on mRNA generation by VSV (vesicular stomatitus virus) that were interpreted to allow an internal cleavage of virus RNA followed by addition of a cap (first published in Banerjee and Rhodes 1976; for review, see Banerjee et al. 1977). Thus, it seemed possible to us that cleavage of hnRNA might yield either a 5' end to be capped or a 3' end to be polyadenylated.[2]

---

[2] Finally, three decades later, Banerjee and colleagues (Koonin and Moss 2010; Ogino et al. 2010) have defined the chemical mechanism of VSV cap formation that appears to require a 5'pppNp to attach a cap (m⁷G). Thus, the early results with VSV caps proved to be a red herring.

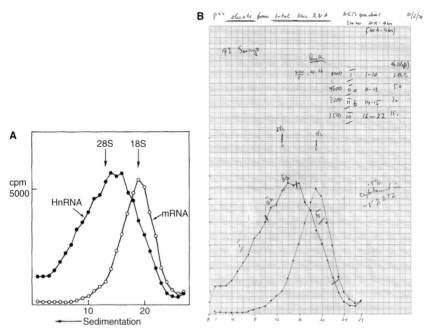

Figure 4-22. Sizes of capped-polyadenylated HeLa cell hnRNA and mRNA. (*A*) ^{32}PO$_4$-labeled HeLa cell hnRNA and polysomal mRNA were separated on a sucrose gradient. Samples were analyzed for total radioactivity, radioactivity in poly(A) (~200 nucleotides long), and radioactivity in caps (5 nucleotides). Assuming one cap and one poly(A) per molecule, these analyses allowed an accurate size of the hnRNA and mRNA molecules: hnRNA >28S = ~10,000, 18S–28S = 4000, and mRNA = 1000 nucleotides. (*B*) Actual page from laboratory notebook of Marianne Salditt-Georgieff, who performed the analyses. (*A*, Reprinted from Salditt-Georgieff et al. 1976 [©Elsevier]; *B*, reprinted from Darnell 2002, graph from Marianne Salditt-Georgieff's notebook.)

A postdoctoral fellow, Uri Lavi, came into my office one morning in 1975 to offer an alternative that did not use the majority of an hnRNA in making an mRNA. Uri suggested, "Jim, maybe you cut out the middle and join both ends." I laughed him off. As we will see below, I was not wise enough to take the suggestion seriously.

A Potential Globin mRNA Precursor

In 1975–1976, there was additional strong evidence that at least one cellular mRNA was derived from a larger nuclear RNA. Globin mRNA from reticulocytes had been copied into cDNA using reverse transcriptase and cloned to make large amounts of double-stranded copies of globin cDNA. Using this synthetic globin DNA, it was possible to hybridize to the unlabeled cDNA copies, the labeled newly formed RNA either from erythroleukemia cells that had been induced to make globin or from embryonic liver that contains red blood cell precursors that normally make globin. Using this strategy, a nuclear globin-containing molecule that was larger than the cytoplasmic mRNA was found in several laboratories (Fig. 4-23) (Curtis and Weissmann 1976; Ross 1976). These investigators cautiously concluded that this ~14S molecule was most likely a precursor to globin mRNA.

Negative but Faulty Experiment Argues against Processing to Make Ovalbumin mRNA

The resistance to accepting hnRNA → mRNA processing was still strong in some quarters. For example, a negative result using a different DNA:RNA hybridization protocol to search for a precursor to ovalbumin mRNA was reported, and the experiment gained some currency. Bob Schimke and Stan

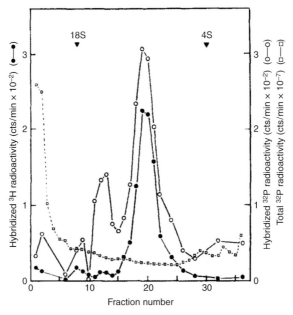

Figure 4-23. Possible β-globin mRNA precursor. Cultured erythroleukemia cells can be induced (by treatment with DMSO) to make globin mRNA. Induced cells were labeled for 24 h with ^{3}H uridine and for 20 min with ^{32}PO$_4$. β-globin cDNA was used to purify labeled globin molecules, and the sizes of the newly synthesized (^{32}P labeled) and the 24-h-labeled (H^3) molecules were determined by sucrose gradient sedimentation. The 24-h-labeled globin-specific RNA (closed circles) sedimented at 9S, the size of the globin mRNA. In the RNA labeled for 20 min with ^{32}PO$_4$ (open circles), there was both a 9S and a larger ~14S–15S globin-specific RNA. The total ^{32}P-labeled RNA (open squares) was also determined. (Modified, with permission, from Curtis and Weissmann 1976 [©Elsevier].)

McKnight used a radioactively *labeled* cDNA copy of ovalbumin to hunt for any large, *unlabeled* molecules in the RNA from ovalbumin-producing oviduct cells (McKnight and Schimke 1974). Total RNA from oviduct tissue was used in the experiment to test for any large or small unlabeled RNA that could drive the labeled cDNA into a hybrid. The ability to form hybrids with the labeled ovalbumin cDNA by total RNA larger than the ovalbumin mRNA was compared with the hybridization of unlabeled cloned ovalbumin cDNA in an attempt to quantitate the level of any large ovalbumin RNA precursor. This experiment was *not* done by actually labeling the RNA produced in the oviduct cells. With this approach, the only unlabeled RNA found in the oviduct cells to which the labeled cDNA would hybridize was the already finished product, the size of ovalbumin mRNA. The problem with the experiment was its lack of sensitivity, because the shorter cytoplasmic ovalbumin mRNA had been accumulating for days and vastly outnumbered any potential nuclear RNA precursor molecules in the oviduct tissue. If any nuclear precursors were promptly processed, one would not have expected to detect many molecules of it. As we will see later (p. 197), this experiment led to a wrong conclusion. Nevertheless, some investigators at the time took the experiment to mean that there was in general no connection between the larger hnRNA and mRNA despite extensive evidence by this time against aggregation of large nuclear RNA.

SPECIFIC PROCESSING OF mRNA FROM A LARGE NUCLEAR PRECURSOR: "LATE" ADENOVIRUS mRNAs ARE SPLICED

The processing of large nuclear RNA into mRNA was finally solved by studies on human adenovirus, a DNA virus that uses cellular enzymes and processing machinery to generate its mRNAs. Work leading to this conclusion began in 1970.

Properties of Adenovirus-Specific Nuclear RNA and mRNA

Our laboratory began using adenovirus as the DNA virus of choice to study nuclear RNA processing because ~50% of pulse-labeled nuclear RNA late in adenovirus infection was adenovirus specific, even though the total rate of incorporation of labeled uridine was still equal to that in growing cells. This meant that late in adenovirus infection, virtually the entire RNA synthetic machinery produced adenovirus mRNA.

In early 1971, Lennart Philipson, a prominent Swedish virologist, student of adenovirus, and an old friend, joined our laboratory at Columbia University for a sabbatical year. It is conventional for sabbatical visitors to take it easy. However, Lennart did not come to New York to sit and think; he came to do experiments. He already knew of our recently completed demonstration that nuclear SV40-specific RNA was longer than polysomal SV40-specific mRNA.

When he arrived, we immediately told him about finding poly(A) in HeLa cell mRNA and explained that the poly(A) was most likely added in the nucleus during mRNA processing. Len and I had two immediate questions to answer: Did the late adenovirus mRNA have poly(A) and what was the size of adenovirus-specific nuclear RNA relative to the cytoplasmic mRNA?

Both of these questions were quite promptly answered. First, adenovirus-specific polysomal RNA and nuclear RNA (both early and late after infection) were selected by hybridization to adenovirus DNA. Both RNAs had poly(A) that was the same size as that in cellular mRNA (Philipson et al. 1971). Second, the nuclear adenovirus-specific RNA late in infection was as large or larger than the 45S pre-rRNA (13,000 nucleotides), but the cytoplasmic polysomal adenovirus-specific mRNA was in the 2000–4000-nucleotide-size range (Wall et al. 1972).

By competition hybridization, all of the adenovirus sequences in the polysomal mRNA were also shown to be present in the large nuclear RNA, i.e., the nuclear RNA could compete (block) hybridization of polysomal adenovirus sequences (Fig. 4-24). However, adenovirus polysomal mRNA could not

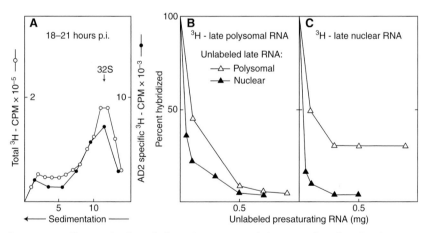

Figure 4-24. Characterization of adenovirus-2 RNAs. (*A*) HeLa cells infected with Type 2 adenovirus for 18 h ("late" infection) were labeled with ^{3}H uridine for 2 h and separated into cytoplasmic (polyribosomal) and nuclear fractions. RNA was extracted and the nuclear RNA was subjected to sedimentation (*A*). (The 32S marker is ∼6700 nucleotides.) A portion of every fraction was sampled for total labeled RNA (open circles) and a portion was hybridized to adenovirus-2 DNA on filters (filled circles). (*B,C*) Competition hybridization between RNA fractions was performed by exposing filters to unlabeled RNAs as indicated, washing away unbound RNA and then hybridizing labeled RNA as indicated. (Redrawn, with permission, from Wall et al. 1972 [©Elsevier].)

block a substantial fraction (~30%) of the nuclear adenovirus sequences. If processing of large adenovirus-specific molecules was occurring, most of the sequences that were transcribed exited to the cytoplasm, but at least 30% did not.

Thus, we had an easy-to-study virus model for hnRNA-to-mRNA conversion that possibly mimicked what appeared to be going on with cellular RNA molecules. We concluded in 1971, "Detailed biochemical knowledge concerning the nuclear origin and processing of mRNA may therefore first become available from studies on [adeno] virus infected cells" (Philipson et al. 1971).

Mapping Sites of Origin of Adenovirus mRNAs

In the years between 1971 and 1975, there was great progress in learning the locations on the adenovirus DNA genome from which virus mRNA originated both early and late in infection. Several scientists at Cold Spring Harbor Laboratory—particularly Phil Sharp, Sarah Jane Flint, and Phil Gallimore—as well as several students in Lennart Philipson's laboratory in Uppsala, Sweden, were especially prominent in this effort (Lindberg et al. 1972; Mulder et al. 1975; Philipson et al. 1975; Sharp et al. 1975). DNA restriction enzymes had been discovered by this point (for review, see Arber 1974). Using different restriction enzymes to separate specific virus segments, the sites on the 36,000-nucleotide adenovirus genome that were complementary to mRNAs, both early and late in adenovirus infection, were determined (Fig. 4-25).

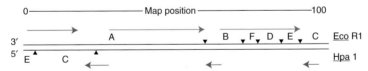

Figure 4-25. Early mapping of mRNA sites on the adenovirus genome. This figure shows the approximate sites and the direction of transcription of adenovirus-specific cytoplasmic RNA early and late in infection, as determined by sites to which virus mRNA bound. The data required determination of the polarity of the two separated strands of the whole adenovirus molecule (Tibbets and Pettersson 1974), followed by separation of and determination of the polarity of each restriction fragment. Hybridization of cytoplasmic RNA (early and late in infection) to each strand of each fragment then allowed positioning of rightward-reading (r) and leftward-reading (l) mRNAs on the map. Early (green) mRNAs were found to have complementarity to r or l strands, but most late adenovirus RNA (red) came from a long stretch of the rightward-reading strand. Furthermore, the amount of late adenovirus RNA greatly exceeded the amount of adenovirus early RNA. (See Philipson et al. 1975; Sharp et al. 1975.)

Richard Roberts, a biochemist at Cold Spring Harbor, and colleagues had a big hand in preparing and defining the specificity of all of the necessary restriction enzymes to produce this detailed map (Mulder et al. 1975). The convention was to describe map positions in percentages, with 0 at the left end and 100 at the right end. One percent was equal to ~360 nucleotides.

The sophistication of the map reached the level of separating the strands of various restriction fragments of the adenovirus DNA so that the strand that bound the mRNA could be detected (Tibbets and Pettersson 1974; Pettersson et al. 1976). In this way, "rightward"- and "leftward"-reading mRNAs could be distinguished. The direction of transcription was determined for mRNAs both early and late in virus infection. A critical conclusion concerning the mRNA late in infection was obvious. A great deal of late mRNA was transcribed from long regions of the rightward-reading strand of DNA. The region covered by these rightward-reading mRNAs was at least 25–30 kb long (with some gaps). Perhaps this entire stretch corresponded to the large adenovirus-specific nuclear RNA that we had found in 1971–1972 (Philipson et al. 1971; Wall et al. 1972).

Mapping the Primary Transcript Precisely

For the first time, it seemed possible to study a well-mapped, single, large nuclear molecule. In 1974–1975, we decided to map the late adenovirus *nuclear* RNA more precisely to determine whether the 20–25-kb stretch encompassing the sites from which the several late mRNAs arose was transcribed as *one primary transcript*. If the entire region did produce one long transcript, we expected to learn much regarding processing. A second major impetus for this effort was to locate a "start site" for a specific long RNA. At this point, no start site for any viral or cell mRNA had been determined. Furthermore, there were no prospects in 1974–1975 of locating such a start site except through molecular techniques. Of course, knowledge of a start site would allow biochemists to begin to study sequences required for an RNA polymerase to start transcription.

Nascent Chain Transcript Mapping

The logic behind our first experiments was to label *nascent* RNAs—those molecules in the process of being synthesized but not yet finished (Fig. 4-26A). If there was a single precursor RNA, then all of the nascent adenovirus-specific nuclear RNA would have a common 5′ end, beginning at the transcription start site. A growing set of nascent RNAs that all started at the same place would, if labeled on their growing 3′ ends, have the

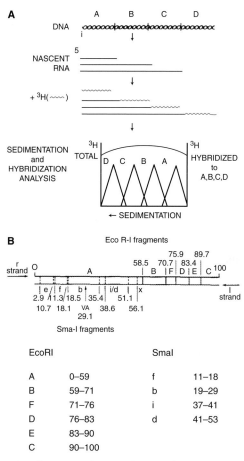

Figure 4-26. (*A*) Logic of nascent chain analysis to find a transcription start site. (*B*) Restriction maps of adenovirus DNA with enzymes EcoRI and SmaI. Various fragments were used to hybridize nascent late-adenovirus-labeled RNA (see Fig. 4-27). (*A*, Reprinted from Bachenheimer and Darnell 1975; *B*, reprinted from Weber et al. 1977.)

following property: The shortest labeled molecules in this set would hybridize closest to the start site, and progressively longer labeled molecules in the set would all project back to a common start site. Thus, the question became, how could we achieve labeling of nucleotides near the growing ends of nascent RNA? We knew from a number of estimates that RNA polymerase rates in making hnRNA or pre-rRNA were in the range of 25–50 nucleotides/sec (Greenberg and Penman 1966 [for Pol I]; Sehgal et al. 1976 [for Pol II]). In addition, it took at least 5 min (5–10 min of ^{3}H-uridine labeling) for the 45S ribosomal precursor (13,000 nucleotides long) to become a dominant nuclear RNA peak of completed molecules (Scherrer and Darnell 1962). Thus, label

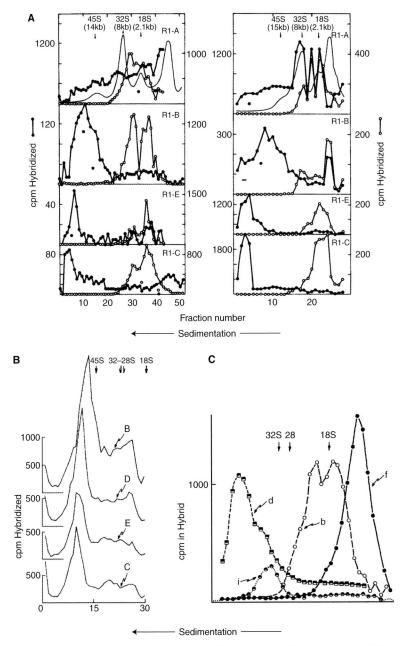

Figure 4-27. Nascent chain analysis detects >25-kb primary late adenovirus RNA tran-
script; the transcription start site is in the SmaF fragment. (*A*) In vivo–labeled (^{3}H uridine)
RNA from nuclei of cells infected for 18 h (*left*, 1-min label; *right*, 2-min label) separated
on a sucrose gradient and hybridized to EcoRI restriction fragments (filled circles).
^{32}P-labeled (2 h) cytoplasmic RNA from late-infected cells (open circles) was used as a marker.
Smooth line in both "R1-A" boxes is OD$_{260}$ showing rRNA. (*Legend continued on facing page.*)

times shorter than several minutes should result in labeling mainly nascent RNA of the putative adenovirus late primary transcript of ~25,000 bases. (Fig. 4-26B shows the adenovirus genome divided by two restriction enzymes, EcoRI and SmaI. The rightward-reading strand r is marked, and if transcription started near the left end and proceeded to the right, the longer and larger ³H molecules would hybridize to EcoRI fragments in the order ABFDEC, so named by size, with A being the largest.)

In the first experiments, Steve Bachenheimer and I exposed cells late in adenovirus infection to very brief label times (³H uridine for 1 or 2 min) (Fig. 4-27A) (Bachenheimer and Darnell 1975). In addition, Jeffrey Weber, Warren Jelinek, and I took isolated nuclei from adenovirus-infected cells late in infection and allowed the nuclei to incorporate labeled nucleotides (Weber et al. 1977). Several others (Ron Evans, Nigel Fraser, Ed Ziff, and Michael Wilson) joined later in using isolated nuclei (Evans et al. 1977). In isolated nuclei, any Pol II that has already started copying DNA elongates nascent RNA chains only by ~500 nucleotides, thus labeling only the ends of long nascent and still-growing RNA chains (Fig. 4-27B,C) (Cox 1976; Weber et al. 1977). With these two approaches, we were reasonably certain to be studying labeled nascent RNA chains at their growing ends.

After extraction and centrifugation, samples of the short and progressively longer separated molecules, labeled presumably only on their 3′ ends, were hybridized to an ordered set of adenovirus DNA restriction fragments. Regardless of which procedure we used to label the RNA, the longest labeled adenovirus-specific molecules hybridized to a region from 90 to 100 on the DNA map (e.g., the EcoRI C fragment), i.e., near the very end of the adenovirus genome (Fig. 4-27). Increasingly shorter nascent, labeled molecules hybridized to DNA fragments at map points 80, then 70, 60, etc., all the way back to the map position between 10 and 20. The shortest abundant hybridized RNA bound to SmaI F (Fig. 4-27C), the boundaries of which are 11.3−18.1 on the restriction map (Fig. 4-26B), a total of ~2500 nucleotides. Because the SmaF-specific RNA sedimented at ~10S, or ~1000 nucleotides, the transcription start site appeared to lie within the SmaF fragment. We concluded that the great majority of rightward-reading adenovirus late

Figure 4-27. (*Continued*) (*B,C*) Isolated nuclei from cells infected for 15 h with adenovirus were exposed to ³H-UTP for 2 min to label nascent chains; RNA was then extracted and sedimented (in two different experiments, *B* and *C*) to size-fractionate RNA (each of ~30 fractions). Size-separated RNA was hybridized to the indicated DNA fragments on the restriction map. (*B*) EcoRI fragments are shown in uppercase letters, (*C*) SmaI fragments are shown in lowercase letters. Chain growth begins in SmaF fragments (shortest labeled chains) and proceeds across the entire adenovirus genome. (Modified from Bachenheimer and Darnell 1975; Evans et al. 1977; Weber et al. 1977.)

transcription started at about position 15–16 and continued almost to position 100 on the map (Figs. 4-26B, 4-27C).

UV Transcription Mapping

We also used a second mapping technique called *UV transcription mapping*. Walter Sauerbier and colleagues developed this technique while studying bacteriophage RNA synthesis (Sauerbier and Broutigan 1970). Exposure to UV light cross-links DNA, prematurely terminating RNA synthesis. Thus, the farther from a transcription start site an RNA sequence lies in a growing RNA chain, the more frequently its synthesis is interrupted by UV irradiation. The interruption of synthesis is measured by incorporation of radioactivity into finished molecules in untreated (and UV-treated) cells. Progressively, sites closer to the start site are more and more resistant to interruption by UV light. Sauerbier and his student Perry Hackett proved the validity of this technique in animal cells using the effects of UV on the synthesis of 45S pre-rRNA, which contains sequences for both 28S and 18S rRNA. Formation of the complete 45S pre-rRNA is the most sensitive to UV interruption because a UV hit anywhere in an rRNA gene stops 45S pre-rRNA synthesis. The 28S–32S is the next most sensitive, and the 18S is the least sensitive (Fig. 4-28A). Therefore, the 28S–32S sequences are farther from the start site of the 45S pre-rRNA, and the 18S sequence is closer to the start site (Hackett and Sauerbier 1974, 1975). When the sequencing of ribosomal genes was eventually performed, the UV transcription mapping was proven to be correct.

Seth Goldberg, Joe Nevins, and I (Darnell et al. 1978; Goldberg et al. 1978) used this technique to study the products of the large RNA transcript late in adenovirus infection: Adenovirus nuclear RNA sequences that hybridized near the right-hand end of the genome (90–100) were the most sensitive to inhibition by UV light (Fig. 4-28B). As regions to the left on the genomic map were tested, they were found to be progressively less sensitive. Very importantly, the cytoplasmic adenovirus mRNA, which, of course, is much shorter than the adenovirus nuclear RNA, was affected similarly: The mRNAs known from hybridization experiments to contain sequences from the right-hand end of the genome were the most sensitive, the mRNAs that hybridized to the more leftward fragments were progressively less sensitive, and the UV inhibitions of complete nuclear and cytoplasmic sequences were indistinguishable (Fig. 4-28B). This last point strongly supported the obligatory derivation of mRNA from the long nuclear RNA transcript that began at approximately position 20 on the adenovirus map. These experiments were finished in the fall of 1976 and early spring of 1977 and sent off for publication. We were confident that we had proven (or, at the very

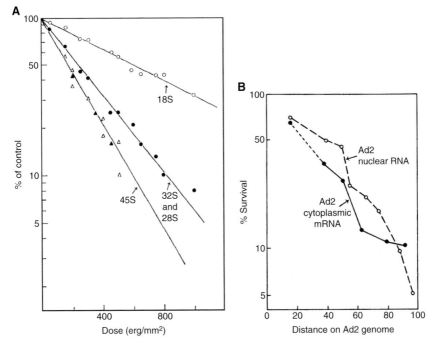

Figure 4-28. UV transcription mapping of late adenovirus transcripts indicates processing of mRNA from nuclear RNA. (*A*) UV transcription method of Sauerbier and Broutigan (1970) was used to show the organization of a pre-rRNA gene that encodes 45S rRNA and is processed to 32S, 28S, and 18S rRNAs. Each labeled RNA species was measured as a function of UV dose. (*B*) UV targeting of late adenovirus nuclear transcription and mRNA formation as a function of location on the adenovirus (Ad2) EcoRI restriction map. ^{3}H-uridine-labeled adenovirus-specific RNA on different restriction fragments across the genome was assayed in the nucleus or cytoplasm after a 30-min label before or after UV irradiation. Survival of adenovirus RNA is plotted. (See text for description.) (*A*, Modified, with permission, from Hackett and Sauerbier 1975 [©Elsevier]; *B*, modified from Goldberg et al. 1977, 1978).

least, strongly suggested) that the late adenovirus mRNA molecules were derived from the processing of a *single* large nuclear RNA precursor.

   Finally, in the spring of 1977, we did similar UV-irradiation experiments on the reduction of labeling of total uninfected HeLa cell hnRNA and mRNA (Goldberg et al. 1977; Darnell et al. 1978). With increasing UV doses, first the labeling of the longest and then the labeling of the shorter and shorter hnRNAs was eliminated by UV, but the labeling of the *total* mRNA declined even beginning with the *lowest* UV doses. This meant that the size of an mRNA did not necessarily correspond to the relative size of its hnRNA precursor. It appeared that long precursors might give rise to mRNAs of various sizes.

The UV mapping technique not only strongly indicated a single large late adenovirus transcript to be the precursor to late adenovirus mRNAs, but it also provided very strong evidence that the long hnRNA transcripts of HeLa cell DNA were the likely precursors of cellular mRNA.

By this time (spring of 1977), both we (Sommer et al. 1976) and the Cold Spring Harbor group (Gelinas and Roberts 1977) had established that the late adenovirus mRNA had a cap on the 5′ end and poly(A) on the 3′ end, as did the hnRNA and mRNA from uninfected cells. Moreover, we had found caps on the adenovirus-specific nuclear RNA that we believed had to start on a site between 15 and 20 on the genome. Rich Gelinas from Rich Roberts's laboratory at Cold Spring Harbor had joined us for our laboratory meetings twice during 1976–1977. He had told us that all of the caps of late mRNAs were attached to an identical 11-nucleotide sequence (Gelinas and Roberts 1977). (Dan Klessig at Cold Spring Harbor was in the midst of completing an analysis on two specific highly purified late adenovirus mRNAs that would prove the identity of the 11-nucleotide cap structure on different mRNAs [Klessig 1977].)

From our pulse-labeled (nascent chain) analysis and UV transcription mapping, it seemed to us that it was proven that the large nuclear RNA of both adenovirus and the cell had to be processed into mRNA, saving disparate parts of the primary transcript. And we had known since the early 1970s (Wall et al. 1972) that at least 30%–40% of the sequences in the late adenovirus primary transcript did not exit the cytoplasm. Something unusual was surely occurring in making both adenovirus and cellular mRNAs. Before we put two and two together as to how the ends of the transcript were saved with much RNA being discarded, we were let in on the secret in late April 1977, just before the Cold Spring Harbor Symposium in early June, by a visit from Richard Roberts of Cold Spring Harbor. But I'll maintain the suspense for another page or two.

Discovery of RNA–RNA Splicing in Adenovirus mRNA Processing

Two of the scientists in the adenovirus group at Cold Spring Harbor—Louise Chow and Tom Broker—along with Phil Sharp (by this time at MIT), Sharp's postdoctoral fellow Susan Berget, and a technical assistant, Claire Moore, performed the experiments that unlocked the mystery of how the large capped and polyadenylated adenovirus nuclear RNA gives rise to smaller capped and polyadenylated adenovirus mRNA. The technique that finally solved the mystery was electron microscopy of individual long, single-stranded adenovirus DNA molecules hybridized to adenovirus mRNA.

The techniques for spreading DNA molecules for electron microscopic examination and then observing sequence identity in hybridized DNA molecules originated in the laboratory of Norman Davidson at Caltech. DNA molecules with stretches of homology and nonhomology, when denatured and allowed to renature, could be viewed as double strands where there was complementarity and single strands where there was not complementarity (Davis et al. 1971).

R Loops

The so-called R-loop adaptation of the DNA homology method was discovered by Ray White in David Hogness's laboratory at Stanford (Fig. 4-29). To standardize the technique, White joined Ron Davis, also at Stanford, who earlier had worked with Davidson to establish EM observations on DNA hybrids (Thomas et al. 1976). *Drosophila* rRNA exposed to a cloned double-stranded

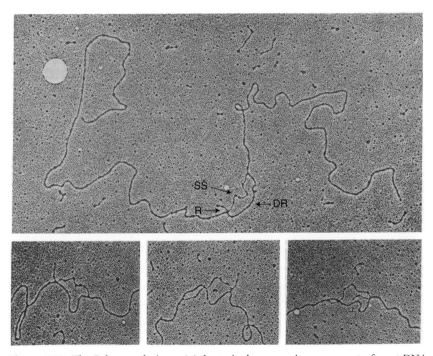

Figure 4-29. The R-loop technique. A λ bacteriophage carrying a segment of yeast DNA that encoded a fragment of 26S rRNA was exposed to conditions (47°C, 70% formamide) known to allow added full-length 26S rRNA to displace one strand of DNA and hybridize to the coding strand. The two ends of the λ DNA are obvious. The loop approximately in the middle is the result of the full-length rRNA inserted as a partial hybrid. The projection (R) at the end of the loop is the portion of the 26S rRNA not encoded in the recombinant λ phage. There is a characteristic difference in appearance of a displaced single strand (SS) compared to the thicker RNA:DNA duplex (DR). (*Bottom*) Three other R loops are shown that illustrate the reproducibility. (Reprinted, with permission, from Thomas et al. 1976.)

segment of *Drosophila* DNA that contained a piece of an rRNA gene displaced a loop—a so-called R loop—of DNA approximately the length of the encoded piece of rRNA (White and Hogness 1977). The conditions for RNA displacement of one strand of DNA (high formamide concentrations) were based on the known fact that RNA:DNA hybrids are more stable than DNA:DNA hybrids (Black and Knight 1953).

All three of the scientists who performed the adenovirus EM hybridization studies—Broker, Chow, and Sharp—had been in Norman Davidson's laboratory in the early 1970s. The R-loop method was perfect to show the exact position on a DNA molecule of an RNA:DNA hybrid. The accuracy was ±50 nucleotides.

The three scientists all came to Cold Spring Harbor Laboratory—Sharp in 1970, and Chow and Broker in 1975—and all worked on problems concerning adenovirus mRNA formation. Sharp had had a leading role in the hybridization experiments that located the sites and specific strand on the DNA to which different adenovirus mRNAs hybridized (Sharp et al. 1975). By 1975, Sharp had left for MIT, but a large group of other investigators including Chow and Broker continued to work on the details of how adenovirus mRNAs were produced.

Late Adenovirus mRNAs All Contain "Leader" Sequences from Upstream

Both the Cold Spring Harbor and MIT groups spread adenovirus DNA fragments together with late adenovirus mRNAs for EM analysis. In their first experiments, both groups used DNA from near the center of the adenovirus map (e.g., from map position ~50–73). As expected, in the EM they observed R loops (RNA:DNA hybrids that displaced single-strand DNA) on the fragments that had earlier been determined to hybridize labeled mRNA. They also saw short nonhybridized projections from what they knew to be the 3' end of the mRNAs from the 5'-to-3' direction of the mRNA. These projections were consistent with the ~200-nucleotide poly(A) that was known to be present in the viral mRNAs. *But* they also saw a short, single-strand projection at the 5' end of the mRNA (Fig. 4-30A). Of course, they knew of our evidence that the long nuclear adenovirus-specific RNA began at ~15–20 on the adenovirus map and was very likely the precursor to cytoplasmic adenovirus mRNA. The Cold Spring Harbor group also knew about the 5'-capped 11-nucleotide sequence present on all of the late mRNAs that possibly came from upstream. Both groups guessed that the 5' projection must be derived from some adenovirus site in an upstream region, perhaps at ~20 on the map.

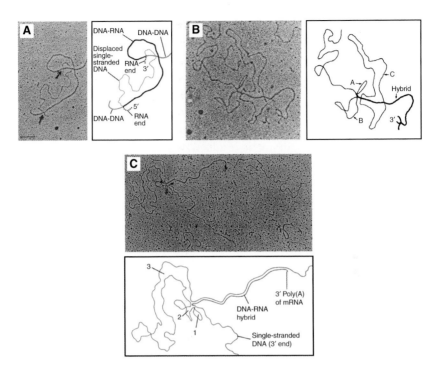

Figure 4-30. Adenovirus mRNA–DNA hybrids suggest composite nature of mRNA. (*A*) When a double-stranded DNA from 50–73 map units of adenovirus was exposed under R-loop conditions to a late adenovirus mRNA (the hexon mRNA), a long hybridized region (RNA:DNA [purple line]) formed a single-strand R loop (DNA [blue line]). There were also nonhybridized projections at both the 3′ and 5′ ends of the mRNA (red lines). (*B,C*) Two examples in which single strands of adenovirus DNA hybridized to two different late adenovirus mRNAs (hexon, *B*; 100K, *C*). These are long stretches representing the body—the coding region—of the mRNA (labeled DNA:RNA hybrid). A second single-stranded DNA fragment containing sequences at the left end of the genome produced short regions of hybridization that produce three loops (labeled A, B, C in *B* and 1, 2, and 3 in *C*). The short regions of hybridization are located at approximately 16, 20, and 27 on the adenovirus map. The short regions in the mRNA were termed leaders. (*A,B, left,* Reprinted, with permission, from Berget et al. 1977; *A,B, right,* modified, with permission, from Berget et al. 1977; *C*, reprinted from Darnell et al., 1990 [photo courtesy of L.T. Chow and T.R. Broker]; see Chow et al. 1977.)

When they added a single-stranded DNA that included this upstream region, to their great surprise, they got not one but three additional short regions of hybridization to the 5′ projection of not one but several different late adenovirus mRNAs (Fig. 4-30B,C). It appeared that, regularly, there was a union of RNA sequences copied from three upstream locations, called "leaders," with the "body" of the mRNAs that hybridized to downstream regions (Berget et al. 1977, 1978; Chow et al. 1977; Broker et al. 1978).

From the very convincing EM results, both groups suggested that some mechanism, that Sharp termed *splicing*, existed in cells to bring together disparate parts of long nuclear RNA and eliminate the intervening sequences. In the process, the capped 5' end and three "leader" sequences appeared to have been joined to a downstream poly(A)-terminated portion from the same long RNA molecule. Thus, the final step in RNA processing discarded a large portion of the nuclear RNA transcript in order to fashion an mRNA. This was an absolutely revolutionary idea. These dramatic results were presented by Tom Broker and Phil Sharp for the first time to a large audience one evening at the June 1977 Cold Spring Harbor meeting and published separately in journals several months later (Berget et al. 1977, 1978; Chow et al. 1977; Broker et al. 1978).

RNA SPLICING IS GENERALIZED: "GENES IN PIECES"

Once it was clear to look in the EM for hybrid molecules of cloned genomic DNA and mRNA transcribed from that genomic DNA, it was easy to confirm the likelihood of splicing based on the adenovirus mRNA model. Word had leaked out regarding the composite structure of the late adenovirus mRNAs. At the same June 1977 Cold Spring Harbor meeting, Heiner Westphal, using the EM technique, showed that early adenovirus mRNAs were also spliced but had only one leader sequence (Westphal and Lai 1978). George Khoury's group at the NIH (Aloni et al. 1978) and Ming-Ta Hsu, a young colleague of mine at The Rockefeller University, both described spliced mRNAs that were produced during SV40 infection (Hsu and Ford 1978).

By this time, laboratories all over the world not only had purified cDNA copies of particular mRNAs but also genomic DNA clones for specific mRNAs (see Maniatis et al. 1978). Richard Flavell, a British scientist working at this time in Amsterdam, presented originally puzzling results at the June 1977 meeting (Flavell et al. 1978). Some restriction enzymes cut globin genomic DNA but did not cut globin cDNA, which included the entire globin mRNA sequence. Recall that potential globin precursor RNA had already been identified (Curtiss and Weissmann 1976; Ross 1976). The conclusion reached at the meeting was that the puzzling restriction enzyme sites in the genomic DNA were in the globin-gene–encoding part of a primary transcript that was eliminated by splicing and therefore not present in the finished globin mRNA. Phil Leder actually presented electron micrographs at the meeting, showing loops created by the hybridization of globin mRNA and globin genomic DNA, but originally interpreted them to be adjacent β-globin genes (Leder et al. 1978). This was speedily corrected by Jake Maizel by the R-loop method, together with Leder's group, after the meeting (Tilghman et al. 1978). The formation of globin mRNA clearly also depended on splicing.

Pierre Chambon, who gave the traditional summary closing talk at the meeting (the nominal subject of which was "Chromatin"), phoned his laboratory in Strasbourg, France, with the news. Very soon after he returned home, he and his colleagues showed that chicken ovalbumin mRNA hybridized to ovalbumin genomic DNA at multiple separate sites in the genomic ovalbumin DNA (Breathnach et al. 1977; Garapin et al. 1978). This was not only a satisfying result that broadened the conclusion of splicing to another cellular gene, but was to our laboratory a fitting debunking of Schimke's earlier claim (McKnight and Schimke 1974) of "no processing" in ovalbumin mRNAs.

The astounding fact was that the sequences in the DNA of animals and animal viruses that end up in mRNA to encode protein were not contiguous in the genomic DNA. "Genes in pieces" became the operative phrase. All of a sudden, the world was ready to adopt the idea that the hnRNA → mRNA conversion was now expected in making most mRNAs, at least in animal cells and their viruses.

Our laboratory and Shatkin's laboratory had known since early 1975 (Salditt-Georgieff et al. 1976) that capped, polyadenylated HeLa-cell nuclear RNA averaged a length of 8,000 to 10,000 nucleotides with many molecules >15,000 nucleotides, compared to an ~1000-nucleotide average for capped, polyadenylated mRNA (Fig. 4-22). But no one, including Sharp and coworkers, Chow and Broker and coworkers, or certainly us, guessed the answer to the puzzle of how mRNA was finally produced. Enzymatic cutting and ligating DNA segments was known for some time by 1977, but no such RNA–RNA chemistry—cutting and rejoining selected pieces—had ever been suspected. Not until the EM pictures of the late adenovirus mRNAs were produced was the possibility of such a process revealed.

Because of their explanatory power and the total surprise of the results, the EM experiments with adenovirus rank among the most informative biological experiments ever performed. The 1993 Nobel Prize in Physiology or Medicine went to Phil Sharp and Rich Roberts, who did not perform the EM experiments at Cold Spring Harbor but was singled out from the large Cold Spring Harbor group to share the prize.

Differential Poly(A) and Splicing Choices Lead to Posttranscriptional Gene Regulation

A most important point, implicit but not yet mentioned, was inherent in the formation of the late adenovirus mRNA and was proven more thoroughly after the Cold Spring Harbor June 1977 meeting. The adenovirus major late transcript is cut, not terminated, at five 3' sites to add poly(A) (Fig. 4-31) (Nevins and Darnell 1978a). Transcription continues in an equimolar fashion

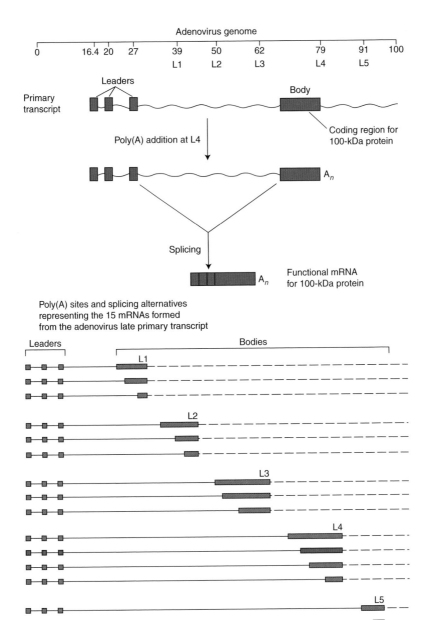

Figure 4-31. Differential poly(A) and splicing choices in formation of adenovirus late mRNAs. The combined efforts of many investigators produced detailed map coordinates of the five poly(A) sites (L1–L5) and of the splice junctions in the multiple spliced forms of late adenovirus mRNAs. All of the mRNAs have the same three leader sequences, but different splice site choices are made within each of the five poly(A)-terminated groups. The diagram illustrates the initial choice of the L4 poly(A) site followed by one of four possible splice sites. The poly(A) is added to a cut site because transcription goes equally all of the way to at least 98.2 on the map, regardless of which 3′ end is chosen (Nevins and Chen-Kiang 1981). (Reprinted from Darnell et al. 1986.)

all the way to at least ~98.2 on the physical map (Fraser et al. 1979). We later proved that globin gene transcription also proceeded past the poly(A) site (Falck-Pedersen et al. 1986). Nested within the "body" of each of the five poly(A)-terminated mRNA sets are several possible splice choices (Chow and Broker 1978). All of the late adenovirus mRNAs get the three tripartite leaders, but the site of splicing of these leaders to the body region on each poly(A)-terminated set varies, so that each mRNA has more or less of the downstream coding region. Thus, not only is there differential poly(A) choice, but differential splicing results in many different proteins being encoded by a single primary transcript.

Because long poly(A)-terminated adenovirus nuclear RNA not yet spliced was found, we concluded that the choice of one of the five poly(A) sites was made first and then one of several protein-coding regions was chosen (Nevins and Darnell 1978a,b). Each primary transcript can produce only one mRNA, but it can be any one of more than a dozen different mRNAs (Fig. 4-31). It is now proven that most cellular pre-mRNA transcripts in humans also can be processed to produce different final mRNA products (Wang et al. 2008). Both alternative poly(A) choices and splice choices exist in the great majority of human pre-mRNAs. This situation offers the opportunity of *posttranscriptional* gene regulation, which, in fact, occurs (discussed in Chapter 5).

There is no doubt that the development of methods for DNA sequencing, also in 1977, the subsequent era of individual gene sequencing, and finally high-throughput sequencing have dramatically changed the landscape of biological research. *But* if we had not learned about RNA processing of large nuclear RNA *before* the sequencing onslaught, there would have been mass confusion in trying to understand the initial results of genomic sequencing. The shibboleth "everything in its time and place" seems perfectly applicable here.

DNA Sequences Confirm a Different Gene Structure in Animals: Introns and Exons

Efficient DNA sequencing techniques were described in 1977 by Allan Maxam and Walter Gilbert at Harvard (Maxam and Gilbert 1977) and by Fred Sanger's laboratory in Cambridge, England (Sanger et al. 1977). In short order, sequences of portions of genomic DNA clones and cDNA clones from specific genes confirmed that not only did each mammalian mRNA sequence encode only one protein, but that the genomic DNA had coding sequences interspersed with noncoding regions. This added final confirmation to the EM pictures and the hnRNA → mRNA pathway defined by the earlier labeling and UV mapping experiments.

Walter Gilbert coined the term *intron* for the "intervening sequence" between coding regions that were referred to as *exons*, meaning that they were "exported" to the cytoplasm (Gilbert 1978). Thus, the storage of information is fundamentally different in eukaryotes compared to bacteria. Any hope of successful recovery and use of the information in our DNA *demands* special processing mechanisms for making mRNA. The most fundamental tenet of information storage in bacteria is colinearity; one can line up the triplets in the gene with the amino acids in the protein (Sarabhai et al. 1964; Yanofsky et al. 1964). There is still colinearity, of course, in eukaryotic *mRNA* once the introns have been removed. But, strictly speaking, colinearity between genes and proteins does not exist in most genes of animals and plants. All eukaryotes have at least some spliced mRNAs.

A final, perhaps obvious, point can stand emphasis. This process of discarding a large portion of many, perhaps most, hnRNA molecules explains the early evidence for "instability" in nuclear RNA, which was noticed in the late 1950s and proved for HeLa cell hnRNA as a separated fraction in the 1960s (Soeiro et al. 1968). Further to this point, less than one-half of all hnRNA molecules in cultured cells have the ~200-nucleotide poly(A) (Salditt-Georgieff and Darnell 1982). Thus, the nonpolyadenylated fraction (which is capped) presumably is also rapidly destroyed in the nucleus. Recent experiments on nuclear RNA, discussed in Chapter 5, have identified long, nuclear, noncoding RNA molecules that do not yield mRNA but most likely have other functions. There is a great deal of RNA synthesis of short and long RNA that appears to be simply turned over in the nucleus regardless of whether it has a function. Thus, the early choice of the term *hnRNA* for the totality of nonribosomal nuclear RNA seems appropriate; some long nuclear RNA is pre-mRNA, but a large fraction is not.

The Eukaryotic/Bacterial Divide

By 1982, it was strikingly clear that the storage and recovery of information from vertebrate DNA (and, as it turned out later, *all* eukaryotic DNA) was fundamentally different from that in bacteria. It had been recognized quite early from the average size of mammalian cell mRNAs (with a mean of ~1200 bases) (Penman et al. 1963) and then from the size identification of many specific mRNAs that, in general, each mammalian mRNA only encodes a single polypeptide. Of course, operons in bacteria are copied into strands of mRNAs, most of which encode multiple proteins. But quite early (1958–1962), a perspicacious geneticist, Guido Pontecorvo (1963), had spoken out against a universal adoption of the "operon" model. He pointed out that in eukaryotes such as fungi (e.g., *Aspergillus nidulans*), the several genes in pathways devoted to synthesis of amino acids and also those for the synthesis of nucleic acid bases most often were not clustered. Thus, some individuals, at least, were on guard from the early days of "information storage and retrieval" that bacterial and eukaryotic genomes had a different design.

IMPACT OF THE DISCOVERY OF SPLICING ON RNA CHEMISTRY

After the surprise and excitement of 1977 subsided a bit and splicing was recognized to be widespread throughout eukaryotes, two lines of research on RNA arose that are still actively ongoing: (1) the biochemistry/molecular biology of pre-mRNA processing, both differential splicing and differential poly(A) choice, and (2) the potential evolutionary implications of splicing. Both of these topics are revisited, respectively, in Chapters 5 and 6 to summarize recent results. In this section, a brief discussion is given of a few important developments that quickly followed the discovery of splicing.

The "Spliceosome": Nuclear Machinery That Performs pre-mRNA Splicing

A fortuitous discovery stemming from an unfortunate medical condition greatly contributed to solving how cells splice nuclear RNA. Patients with lupus erythematosus often have antibodies against nuclear components. Michael Lerner (whose father was a physician) knew of this, and in Joan Steitz's laboratory at Yale, where Lerner was a student, two particular antibodies were discovered that proved to be very useful in determining the components involved in splicing. Five small nuclear RNAs (snRNAS) were precipitated with these antibodies. One of these RNAs, termed U1 (because the molecule was rich in uridylate), had already been uncovered and sequenced by Harris Busch, Ramachandra Reddy, and colleagues (Reddy et al. 1974), although no function for it had yet been uncovered. An antibody termed Sm (an abbreviation of the patient's name) precipitated not only U1 but also four other small RNAs: U2, U4, U5, and U6. Each of these RNAs was assumed to associate with at least one common protein. Eventually, these snRNAs were isolated in association with proteins as snRNPs. Many of the proteins were shared among all of the snRNPs (Hinterberger et al. 1983; for review, see Tycowski et al. 2006). The first snRNP to be extensively studied was U1, the most abundant of the snRNPs ($\sim 10^6$ per human cell nucleus). By this time, enough sequencing around the spliced boundaries had been performed to suggest consensus sequences for marking exon–intron boundaries (Fig. 4-32) (for review, see Breathnach and Chambon 1981). The suggestion

Figure 4-32. Intron–exon junction sequences. Early sequencing of cDNAs and genomic sequences (the first 85 or 90 sequences) showed that strong conservation was likely at intron borders but there was essentially no nucleotide preference in the exon regions. (Redrawn, with permission, from Breathnach and Chambon 1981.)

was first made in 1980 that perhaps these abundant small nuclear particles might all be involved in splicing (Lerner et al. 1980). Support for this view came when the U1 snRNP was found to be able to bind to a 5′ exon–intron junction in the precursor to globin mRNA (Mount et al. 1983).

To establish a functional test for the snRNPs, an in vitro splicing system was needed. This was achieved in 1983 with a crude nuclear extract and snRNP antibodies (Fig. 4-33) (Padgett et al. 1983a). In the in vitro splicing reaction, the addition of Sm or the U1-specific antibody blocked the reaction (Padgett et al. 1983b). It was clear that it would be possible to learn details of the splicing reaction in a cell-free system. And the snRNP particles would be at the center of this research.

An in vitro splicing system was also established in John Abelson's laboratory. He and his colleagues showed that the components required and, indeed, the partially spliced primary transcripts were contained in, a 35S–40S complex (i.e., the size of a single ribosomal subunit) that was named a *spliceosome*, a name that stuck (Brody and Abelson 1985).

The stepwise involvement of the different spliceosomal particles is now understood in detail (see Chapter 5 and Tycowski et al. 2006). In addition, what became clear in the early 1980s was that RNA by itself had unsuspected chemical abilities. Studies in the intervening years showed that RNA sequences within the snRNPs quite likely play a direct part in RNA splicing to produce mRNAs. A summary of recent progress on splicing is given in Chapter 5.

RNA-Directed RNA Cleavage and Transesterification

Momentous discoveries by Tom Cech and Sid Altman in the early 1980s reshaped thinking regarding the chemical capacities of RNA. These two investigators found that RNA, *in the total absence of protein*, can have enzymatic activity.

RNase P: An RNA-Containing Enzyme

Altman and coworkers purified a bacterial ribonuclease (called RNase P) that cleaves several nucleotides from the 5′ end of newly made precursors of

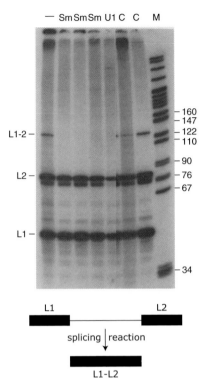

Figure 4-33. In vitro splicing is inhibited by antisera against U1 snRNP. A ^{32}P-labeled RNA containing the first two exons (leaders, L1, L2) of adenovirus was incubated (with or without antiserum, as indicated at the *top* of the figure) together with a nuclear extract from HeLa cells (uninfected) capable of splicing. The assay for splicing consisted of transcribing a ^{32}P-labeled RNA from a DNA clone that contained the L1 and L2 exons separated by an intron, as diagrammed at the *bottom* of the figure. The labeled RNA was hybridized with a complementary single-stranded DNA in which L1 is spliced to L2. After RNase treatment and gel electrophoresis, the spliced RNA gave the L1 · L2 band, and unspliced RNA gave the separate L1 and L2 bands. With no antiserum (column 1) or control antiserum (columns 6 and 7), some of the substrate L1 and L2 exons were linked to form L1 · L2 (column 1). Both the Sm antisera (which can precipitate all of the snRNPs: U1, U2, U4, U5, and U6) and the U1 antiserum (which precipitates only the U1 snRNP specifically) blocked the splicing reaction. The columns marked C are control sera with no affinity for snRNPs. (*Right* margin lists size markers for RNA.) (Reprinted, with permission, from Padgett et al. 1983b [©Elsevier].)

bacterial tRNA (and also residues from some other small RNAs). This pure enzyme preparation contained a small RNA molecule, i.e., the "enzyme" is a complex between a protein and a small RNA molecule (Stark et al. 1978). Proteases destroyed the enzymatic capacity of the enzyme, but, surprisingly, so did nuclease treatment (with either pancreatic ribonuclease or micrococcal nuclease) (Fig. 4-34A). The Altman group, in collaboration with Norman

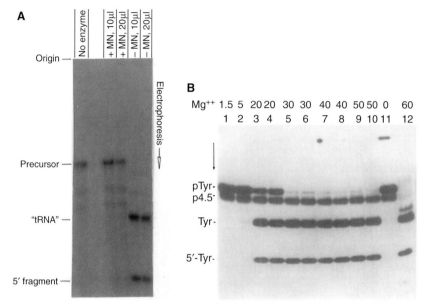

Figure 4-34. Enzymatic activity of the P1 RNase from *E. coli* resides in the RNA moiety. (*A*) Purified RNase P1 was exposed to micrococcal nuclease (+MN, columns 3 and 4). The enzyme not treated with micrococcal nuclease (columns 5 and 6) cleaved a radioactive tRNA precursor, releasing the tRNA and a 5′ fragment. (*B*) The RNA moiety of RNase P1, free of protein, was tested for its ability to digest two substrates, pre-tRNA-Tyr (pTyr) and pre-4.5S RNA (p4.5), as a function of increasing Mg^{++} ions (concentrations are millimolar, mM). When Mg^{++} concentration was 30 mM or above, cleavage of pTyr occurred. (*A*, Reprinted, with permission, from Stark et al. 1978; *B*, reprinted, with permission, from Guerrier-Takada et al. 1983 [©Elsevier].)

Pace's group, showed that the pure RNA component of P1 (obtained from either *E. coli* or *Bacillus subtilis*), in the presence of a high concentration of Mg^{++}, had precise ribonuclease activity (Guerrier-Takada et al. 1983). Normally in the enzyme, the protein simply keeps the RNA folded into its active form, and this RNA structure is most likely mimicked in the high Mg^{++} concentration (Fig. 4-34B). Altman's experiments were the first to show that RNAs are directly involved in enzyme biochemistry (for a thorough historical review of this important result by an early colleague of Altman's, see McClain et al. 2010).

Self-Splicing RNA

In 1979, Joe Gall and his colleague Martha Wild found that *Tetrahymena thermophila* encoded a pre-rRNA containing large and small rRNAs. But processing of the pre-rRNA was very uncharacteristic: There was an intron in the sequence for the large rRNA (Wild and Gall 1979). This protozoan has ribosomes with the conventional large and small RNAs; thus, this intron must be removed in the processing of *Tetrahymena* rRNA.

Studying the *Tetrahymena* reaction, Tom Cech and Arthur Zaug, Cech's technician and colleague, made an astounding discovery (Zaug and Cech 1980). Because the isolated *Tetrahymena* nuclei continued to synthesize RNA, they produced labeled RNA in nuclei and extracted the protein-free labeled RNA to use as a substrate to search for a specific processing enzyme. To their total amazement, Cech and coworkers found that when they incubated the labeled pre-rRNA by itself in the absence of protein, the intron was excised and the two flanking pieces of rRNA were joined (Cech et al. 1981). The reaction required Mg^{++} and a moderate salt concentration (75 mM NH_4SO_4). Significantly, ATP, UTP, or CTP had no effect on the reaction, but GTP strongly promoted the reaction. The GTP, however, was not a source of energy because guanosine served equally well (Fig. 4-35A).

The Cech group proved that the 2′ OH of the guanosine appeared to act as a nucleophile to attack the phosphodiester bond at the 5′ (UpA) intron–exon junction (Fig. 4-35B). The guanine-containing residue (GMP is shown in Fig. 4-35) becomes attached by its 3′ hydroxyl to the phosphate on the pA residue of the intron, preserving the energy of the nucleotide linkage (a phosphoanhydride energy-rich linkage). The now-freed 3′ OH on the uridylate residue (U_{OH}) of the 5′ exon attacks the nucleotide bond at the 3′ intron–exon boundary (GpU), with a resulting second transfer event that unites the two exons (UpU) and again preserves the energy originally present in the nucleotide linkage. Thus, there was a transfer of attachment of one portion of the original polynucleotide chain to another site along the original chain with no loss or gain of energy. Cech referred to this reaction as a *transesterification*, a term used earlier to describe the transfer of DNA nucleotide linkage during recombination. The Cech group went ahead to show that the now-freed intron, at first released as a linear molecule, underwent a second transesterification (Fig. 4-35B) directed by the 3′ hydroxyl of the G residue at the end of the intron (Fig. 4-35B). This G_{OH} attacked a UpA linkage in the intron, resulting in a transesterification that formed a circle and released the first 15 nucleotides of the intron. To emphasize again, all of this chemistry occurred with no protein present and with no outside source of energy (for review, see Cech 1983).

These original experiments of the Cech group and succeeding brilliant experiments proved that chemically active groups in RNA are just as available, although they are not as numerous as chemically active groups in protein. RNA is now known to be folded into three-dimensional shapes that promote not only intron excisions but RNA synthetic reactions as well (see Chapter 6).

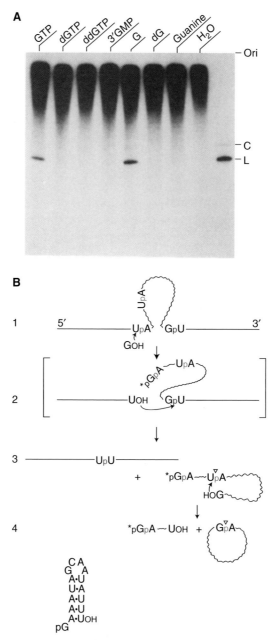

Figure 4-35. Transesterification reaction in removal of the *Tetrahymena* rRNA intervening sequence. (*A*) In vitro protein-free removal of the *Tetrahymena* intron is scored in this experiment by the release of the linear intron (L). Compounds added to pure RNA in appropriate buffer (Mg^{++} ions and [NH$_4$]$_2$SO$_4$) are given above each lane. The requirement for a guanine-containing 5′ ribonucleotide (GTP) or the nucleoside guanosine (G) (not a deoxyguanosine or deoxy-GTP) to effect the intron release is clear. (*Legend continued on facing page.*)

It is widely believed, but not proven, that in pre-mRNA splicing facilitated by spliceosomes, the essential chemistry of transesterification is performed by RNA present in the spliceosomes that align the correct intron–exon junctions and promote RNA-catalyzed splicing. Perhaps the only role of proteins in the spliceosome is to act as a scaffold (discussed in Chapter 5; for review, see Pyle 2010).

Splicing of pre-tRNA

First discovered in 1967–1968 (p. 169) and as noted above (p. 202), primary transcripts of tRNA are longer than mature tRNA, and methylation of precursors was probably required in pre-tRNA processing. Cell extracts could convert the pre-tRNA into tRNA. After pre-mRNA splicing was discovered, sequencing of several tRNAs in yeast (*Saccharomyces cerevisiae*) showed an intron (intervening sequence) between two "half" units that were joined to make mature tRNA. John Abelson and colleagues found that this processing used protein enzymes (i.e., they lack RNA) that removed the intervening sequence as a linear molecule. The "front half" (the 5′ end) of the pre-tRNA was left with a 3′ phosphate that was required to join in ligation to a 5′ OH that was left on the "back half." These reactions, now well understood, are completely different from the transesterification reactions in spliceosomes or in *Tetrahymena* rRNA intron removal (Knapp et al. 1979; Peebles et al. 1979).

RNA POLYMERASES AND GENERAL TRANSCRIPTION FACTORS OF EUKARYOTIC CELLS

In this chapter thus far, the major types of RNA in eukaryotic cells have been described and the idea developed of a "primary transcript"—an initial RNA

Figure 4-35. (*Continued*) (*B*) Diagrams summarize the events that occur during the two transesterifications, showing the origins of the phosphates (red, purple, and blue) involved in the interaction. (1) The substrate with the intron (A∼G) in place is bounded by a 5′ U and a 3′ U that are united after splicing. (2) The 2′ OH of the attacking guanosine (GOH) breaks the UpA bond and becomes attached to the 5′ phosphate (red) of the A residue at the intron boundary. (3) This allows the 3′ OH of the U residue of the upstream exon to attack the 5′ phosphate of the U residue that is the first nucleotide in the downstream exon, leading to a union by transesterification of the two uridine nucleotides in the two exons. (4) The released intron bearing the initiating guanosine residue undergoes a second transesterification that cuts off the first 15 nucleotides of the intron and cyclizes the remainder of the intron. The cyclization reaction is essentially a *ligase* reaction. (*A*, Reprinted, with permission, from Cech et al. 1981; *B*, redrawn, with permission, from Zaug et al. 1983 [©Macmillan].)

product that after processing becomes rRNA, tRNA, or mRNA. The discussion now turns to an obvious and most crucial topic to overall progress in the RNA field, the enzymes—the RNA polymerases—that make the primary transcripts. The initial discovery of the mammalian RNA polymerases occurred in 1969–1970, earlier than the discovery of splicing. Early progress on the nature of these enzymes and requirements for their activity occurred in parallel with studies that uncovered the details of pre-mRNA processing. But the events of 1977 and the recombinant DNA revolution that began in ∼1975 greatly aided in characterizing site-specific initiation by mammalian, and much later, yeast RNA polymerases.

As noted in Chapter 3 (p. 123), Sam Weiss, in 1960, found that the first discovered cell-free RNA polymerase activity in rat liver nuclei required nucleoside triphosphates (ATP, UTP, GTP, and CTP) for incorporation into RNA. However, until 1969, no purification of this enzymatic activity was successful from any eukaryotic source. An important reason for this difficulty was that eukaryotic DNA does not exist "naked" but is complexed with proteins (mostly histones, which are discussed in detail in Chapter 5). The polymerases are part of this complicated protein/DNA mixture and were initially difficult to liberate.

Three RNA Polymerases: A Graduate Student Strikes Gold

In the mid 1960s, Robert G. Roeder was an adventuresome graduate student in the laboratory of W.F. "Bill" Rutter, a biochemist (enzymologist) who had become interested in embryonic development. After the mRNA hypothesis was proven for bacteria in 1961, scientists had guessed/assumed that developmental events would depend on controlling specific mRNA production.

Roeder and Rutter started at the absolutely necessary starting point to study the molecular details of development: What enzyme(s) made RNA in developing systems? They succeeded in making protein extracts free of DNA from nuclei of sea urchin embryos and then also from rat liver nuclei. Following addition of ribonucleoside triphosphates and protein-free DNA to these extracts, they obtained incorporation of ribonucleotides into RNA (Roeder and Rutter 1969). The liberation in soluble form of RNA-polymerizing activity was the first necessary step for studying the RNA polymerases. Many other laboratories essentially followed the Roeder-Rutter procedures and also obtained soluble RNA polymerase activity.

When Roeder used column chromatography to begin enzyme purification from the original crude nuclear protein mixture, three different RNA-synthesizing activities were revealed. The enzymes were named polymerase I (Pol I), polymerase II (Pol II), and polymerase III (Pol III) for the order in which they came through the purification procedure (Fig. 4-36A). An

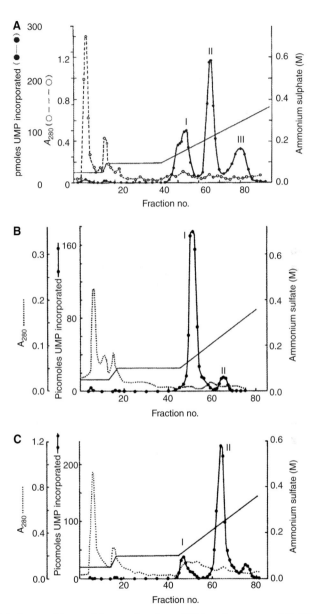

Figure 4-36. Sea urchin nuclei contain three RNA polymerases. (A) The nuclei of gastrula-stage sea urchins were broken by sonication, and the extracts were subjected to DEAE-Sephadex columns using an $(NH_4)_2SO_4$ gradient (solid line) to elute bound proteins. Fractions were tested for RNA polymerase activity by incorporation of labeled UTP into RNA after addition of unlabeled ATP, CTP, and GTP plus calf thymus DNA. Three activity peaks were found and named RNA polymerases I, II, and III (Roeder and Rutter 1969). Total protein was monitored by absorbance at 280 nm (open circles). (*Legend continued on next page.*)

important break occurred when the poison α-amanitin, from *Amanita phalloides*, a mushroom popularly known as the "death cap," was found by Roeder, Rutter, and colleagues to have selective action on the different RNA polymerases (Lindell et al. 1970). At low concentrations, α-amanitin completely blocked the partially purified Pol II but had little effect on Pol I or Pol III. The amanitin-resistant Pol I was found in association with nucleoli (Fig. 4-36B) and presumably was responsible for pre-rRNA synthesis (Roeder and Rutter 1970). In contrast, Pol II was outside of the nucleolus (in the extranucleolar fraction) (Fig. 4-36C) and was responsible for the synthesis of the DNA-like RNA corresponding to hnRNA in the extranucleolar fraction (Blatti et al. 1971). Pol III was later shown to make the smaller RNAs, tRNA and 5S RNA, the short rRNA that is not transcribed from rRNA genes (Weinmann and Roeder 1974). Thus, at this point (1970), three major classes of primary transcription units—pre-rRNA, hnRNA, and the small RNAs—and three polymerases, one for each class, had been identified in animal cells. (Chapter 5 presents a summary of the action of the three RNA polymerases in the synthesis of more recently discovered RNAs.)

Different animal cells and tissues other than those from sea urchin and rat liver were soon examined, and the existence of the three polymerases in all eukaryotic cells was quite well established by the early 1970s (for review, see Roeder 1976). In all of these cases, the activity of partially purified polymerases depended on a free end of DNA in the template, often a "nick" (a cut) in the template DNA. This was surely not physiologically meaningful initiation. However, each of the polymerases was capable of starting new chains at such nicks or at the free ends of linear DNA molecules. The requirements for correct initiation were still in the future.

Coaxing Pol III and Pol II to Get Started Correctly In Vitro

Although these results definitely marked an important step forward in learning how information was pried out of DNA, there was still a long way to go. Getting the eukaryotic RNA polymerases to make proper RNA initiations, however, awaited knowledge of where specific start sites in DNA existed. By the late 1970s, proper initiation in bacteria had been found to depend on

---

Figure 4-36. (*Continued*) (*B,C*) Preparations from rat liver nuclei were centrifuged to separate nucleolar (*B*) and extranucleolar ("chromosomal") (*C*) fractions. Proteins in each fraction were chromatographed on DEAE-Sephadex as in *A*. Pol I was nucleolar and resisted α-amanitin, and Pol II was "chromosomal" and very sensitive to α-amanitin. (*A*, Reprinted, with permission, from Roeder and Rutter 1969 [©Macmillan]; *B,C*, reprinted, with permission, from Roeder and Rutter 1970.)

positive-acting protein factors to guide a polymerase to a specific site; initiation was not simply the release of transcription from repression. Thus, it was anticipated that accessory factors would also be required to get eukaryotic polymerases successfully started at a correct site in DNA.

In animal cells, there was no hope of classical genetics giving a direct helping hand in finding specific factors active in transcription. Biochemistry was therefore going to have to carry the load. The goal for Roeder and colleagues became the production of authentic RNA products in purified systems and the identification of the proteins required for correct initiation. They began with Pol II and Pol III, leaving aside the Pol I and pre-rRNA synthesis in the nucleolus.

Pol II was presumably the enzyme ultimately responsible for mRNA, and its regulation seemed most important to development. Pol III, which made the shortest RNAs, was thought to be the easiest to coax into making an authentic RNA product. Roeder and colleagues planned to use naked DNA plus purified polymerases plus cell extracts in which any required accessory factors might exist. If they could get correct initiation with purified RNA polymerases, there was some hope of ultimately purifying the accessory, presumably regulatory, factors. More than 30 years later, Roeder and dozens of laboratories around the world are still describing the inventory of proteins required in regulated eukaryotic RNA synthesis (summarized in Chapter 5).

Pol III Template

The first problem in mimicking specific gene transcription in vitro was to choose a template DNA to copy. In the early 1970s, the start sites for short RNAs were the only sites that promised some success. Don Brown had purified from *X. laevis* oocytes a repeating string of 5S genes encoding only the ribosomal 5S RNA molecule (Fig. 4-37) (Brown et al. 1971; see also Brown

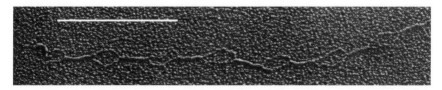

Figure 4-37. Tandem repeats of 5S DNA from *X. laevis*. In the course of purifying the ~20,000 tandemly arranged 5S genes in *X. laevis*, Donald D. Brown and colleagues took this electron micrograph. Under conditions of spreading on the EM grid, the spacer regions (rich in As and Ts) between the 5S genes (rich in Gs and Cs) denatured, and the length of the nondenatured gene corresponded to a length expected for the ~150-nucleotide 5S RNA, suggesting that the 5S RNA might be transcribed directly, not as a precursor molecule. (Reprinted, with permission, from Brown et al. 1971.)

and Sugimoto 1973). Pol III, however, could not copy this purified template correctly. Both strands were copied, not only the correct (template) strand that encoded 5S RNA but also its complementary strand. Even bacterial polymerase could do the same. Clearly, this was not specific copying (Parker and Roeder 1977; Parker et al. 1978). In contrast, Pol III added to isolated nuclei of frog cells or even of human cells, both of which were, of course, already making 5S RNA, greatly boosted the synthesis of 5S RNA by the isolated nuclei. It seemed likely that proteins already associated with the 5S DNA in the cell nucleus assisted Pol III to find a correct start site and make authentic 5S RNA. This proved to be the case.

Roeder and colleagues isolated DNA–protein complexes—"chromatin"—from nuclei and added Pol III (Parker and Roeder 1977). The synthesis of 5S RNA was greatly stimulated, generating the first eukaryotic cell-free RNA product of a cellular gene by an added polymerase. In this system, only purified Pol III worked (neither purified Pol I nor Pol II worked). Soon, crude nuclear protein extracts plus the purified 5S DNA template and purified Pol III were shown to be capable of making 5S RNA specifically (Ng et al. 1979). The 5S product in this reconstituted system was authentic. The 5S RNA in *Xenopus* (and in humans) is not modified at its 5′ end and retains its pppA residue that starts the RNA. The in vitro product had this marker of correct initiation. Thus, fractionation of crude nuclear protein extracts might be used to determine which protein(s) aided Pol III to find the correct start site and make 5S RNA.

Adenovirus as the First In Vitro Pol II Template

But the more absorbing challenge for Roeder, as well as the dozens of other laboratories now pursuing in vitro transcription in the late 1970s, was to determine how to get Pol II to correctly start a specific pre-mRNA, the precursor to a specific mRNA. Once again, it was adenovirus to the rescue. As adenovirus had done for the identification of the first high-molecular-weight nuclear precursor to mRNA, the adenovirus system provided the information that allowed the first Pol II start site to be located. By 1977, the mapping experiments of the late primary adenovirus RNA transcript, both biochemical and electron microscopic, had located the putative start site—the cap site—at ~16 map units from the left end of the adenovirus genome (Evans et al. 1977). Ed Ziff and Ron Evans proved, by DNA sequencing and direct RNA sequencing of the capped oligonucleotide from adenovirus mRNA, the precise position that is capped in the late adenovirus transcript (16.45 map units). This site was expected to be, but had not been proven to be, the presumed *transcription start site* (Ziff and Evans 1978).

In 1979, Tony Weil, Don Luse, and Jackie Segall in Roeder's laboratory took an adenovirus DNA fragment containing this region (Fig. 4-38A) and added ribonucleoside triphosphates and purified Pol II to a nuclear extract of the sort that had been successful in getting Pol III to make 5S DNA. The extract was from *uninfected* cells, thus containing only molecules that pre-existed before infection of the cell. This mixture produced a discrete product, starting at or very close to the expected position (the cap site at 16.45 map units) for the major late adenovirus transcript (Weil et al. 1979). Chain elongation extended rightward to the end of the chosen fragment, producing a "run-off" product (Fig. 4-38B,C). By choosing DNA fragments including different stretches beyond 16.45, different-sized run-off products were produced, thus mapping the start site quite exactly. Moreover, the run-off RNA products had a 5'-methylated G cap attached to the diagnostic undecanucleotide found in all late adenovirus mRNAs. Therefore, the crude cellular extract contained the necessary factors to allow purified Pol II to start transcription at the correct nucleotide and even had the enzymes that added the cap structure. Moreover, Pol II from rat liver and *X. laevis* nuclei produced the same in vitro products as the polymerase from HeLa cells. The reaction had re-created in the test tube a correct start for a precursor to mRNA.

The importance of this result cannot be overstated: The experiment supported the idea that Pol II does start at a particular site—the "cap" site—to generate long molecules that are to become mRNA precursors. The conclusion that the cap site on mRNA marks the start site for transcription of a gene had been widely guessed to be the case but had not been proven until this experiment. Many different eukaryotic genomic DNA samples that encoded different mRNA cap sites had been cloned by this time. Both the accessory proteins and the DNA sequences that guide the polymerase to correct start sites on many different genes were now open for examination. Extensive studies of sequences around the cap sites followed throughout the 1980s and continue to the present time, as we discuss in Chapter 5.

Recognition of the Complexity of General Transcription Factors

Roeder and colleagues began immediately using the tools of the biochemist to fractionate nuclear protein extracts and to test fractions for RNA synthesis beginning at the adenovirus major late promoter (the DNA around map position 16.45). The goal was to separate the specific protein factors required for RNA Pol II to correctly start RNA synthesis. At the same time, purified (cloned) 5S DNA, as well as DNA from many different tRNAs, was also available. All of these DNAs that encoded small RNAs allowed similar studies of proteins required for correct initiation by Pol III.

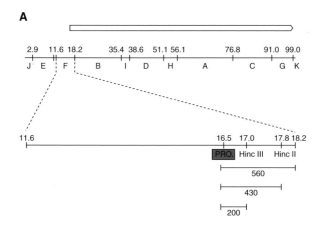

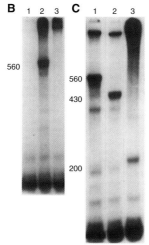

Figure 4-38. In vitro transcription by Pol II from the cap site of adenovirus. (*A*) Map of the adenovirus DNA in the region of the cap site thought to contain the major late transcription start site (labeled promoter, PRO). The SmaI F fragment was the first template used; in addition, this fragment was cleaved with restriction enzymes HindIII or HincII to yield additional shorter templates. (*B,C*) Electrophoretic migration of in vitro–labeled-RNA-transcribed products from the SmaI F fragment and from the shorter DNA templates identified a single starting region. (*B*) Transcription of the SmaI F fragment yielded a prominent product of ∼560 nucleotides (lane *2*) that was inhibited by α-amanitin (lane *3*). (Lane *1*) A control with no added DNA. (*C*) In vitro transcription of the SmaI F fragment (lane *1*) and the same fragment cleaved with HindIII (lane *2*) or HincII (lane *3*) resulted in shorter "run-off" products of ∼430 and ∼200 nucleotides, respectively. The in vitro product from SmaI F transcription labeled with ^{32}P-UTP was analyzed for oligonucleotide content and found to have the diagnostic capped 11-nucleotide fragment as well as the succeeding leader oligonucleotides that had been found (Ziff and Evans 1978) in all late adenovirus mRNAs made in cells. (Reprinted, with permission, from Weil et al. 1979 [©Elsevier].)

In less than 1 year, very significant results were obtained (Matsui et al. 1980; Segall et al. 1980). A detailed biochemical separation scheme was applied to the total nuclear protein extract that became widely used in the in vitro transcription field (Fig. 4-39). Reproducible separation occurred into at least four different fractions (still impure), each of which contained proteins required for correct Pol II RNA initiation and/or Pol III initiation. The separated fractions for Pol II and Pol III could not be substituted for each other, promising a different set of accessory transcription factors for each polymerase.

By this time, Pol I, II, and III had all been highly purified from cultured animal cells, and all were very different from the simpler bacterial enzymes. Instead of four subunits, there were at least 10 in mammalian polymerases, some similar in size in each of the three polymerases and some different (Roeder 1983).

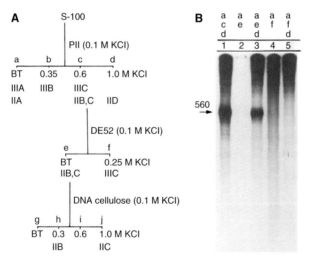

Figure 4-39. Flow diagram for purification of general transcription factors (GTFs). Robert Roeder and colleagues began the era of purification of GTFs (and later, many other factors) from eukaryotic cells using in vitro transcription of an adenovirus DNA template (SmaI F) (see Fig. 4-38), the transcription of which produced a characteristic 560-nucleotide "runoff" product. A 5S template from *X. laevis* was also used (results not shown; see Segall et al. 1980). (*A*) A nuclear extract (S-100; the supernatant after centrifugation at 100,000*g* for 1 h) was fractionated through several chromatographic steps of binding and elution (by designated KCl concentrations): P11 (phosphocellulose), DE52 (DEAE Sephadex), and DNA cellulose (salmon sperm DNA linked to cellulose). The fractions at each step are designated with letters (a–j); BT (breakthrough) indicates proteins that did not adsorb. (*B*) Adenovirus SmaF fragment transcribed by purified RNA Pol II and all four triphosphates (including [^{32}P]UTP) plus the designated fractions (shown above each lane) from the scheme in *A*. These initial results highlighted the importance of what were later purified as TFIIA, TFIIB, and TFIID. (Reprinted from Matsui et al. 1980.)

It had become increasingly clear from these early experiments that the bacterial and eukaryotic transcription initiation systems were quite different. Not only was the information in eukaryotic DNA stored differently (i.e., in pieces) than in bacteria, the proteins required to get the information out of DNA were also going to be different and would likely be much more complicated.

These first partially purified fractions were capable of allowing Pol II to start transcription on different mammalian promoters, e.g., on promoters for genes encoding β globin, histone, albumin, and other proteins produced in the liver (Luse et al. 1981). But such test tube reactions for correct initiation did not re-create regulated initiation. Globin mRNA initiation occurs only in red blood cell precursors, not in HeLa cells. Nevertheless, it was established that synthesis and elongation from a start site by Pol II required a complex mixture of proteins that became known as *general transcription factors* (GTFs). The GTFs interact with the polymerase and, as would be later discovered, have only one or two proteins that intimately contact DNA during initiation. It was another almost two decades before all of the more than 30 individual proteins in GTFs were defined and their individual roles in initiation of RNA synthesis more fully explained. Every initiation by Pol II to create a primary transcript, an mRNA precursor, requires GTFs. They are *necessary* but not *regulatory*.

IS TRANSCRIPTIONAL CONTROL THE MAIN CONTROL IN EUKARYOTES?

The Jacob-Monod hypothesis and later experiments in the 1960s proved that bacterial regulation occurs at the transcriptional level, i.e., when a bacterium makes a nutritional or other metabolic shift, it changes the rate of synthesis of particular mRNAs. A few cases of control of mRNA lifetime rather than control by changing the rate of synthesis are known in bacteria, but transcriptional control dominates.

But was the control of gene expression in animal cells vested mainly at the level of transcription? After all, the first transcription unit to be deciphered was the adenovirus late transcript, and this gave rise to more than a dozen different mRNAs (Fig. 4-31). Furthermore, most hnRNA molecules (capped and polyadenylated) were much longer than the mRNAs derived from them. Was posttranscriptional control—processing control—also important? Perhaps more important? Was it remotely possible that all pre-mRNAs were made at all times in mammalian cells—and could this explain the rapid synthesis and destruction of so much nuclear RNA? Although this seemed highly unlikely to most observers, it could not be peremptorily ruled out (for review, see Darnell 1982).

Importance of "Run-On" Transcription Assays

Before being able to measure individual mRNAs, the very important question of the dominant level of control in eukaryotes could not be answered. This situation changed with the arrival of recombinant DNA techniques. The cloning and amplification of cDNA complementary to virtually any desired mRNA became commonplace in the late 1970s and early 1980s. This amplified DNA furnished a reagent to perform the "run-on" transcription assay with isolated nuclei (Fig. 4-40). The use of isolated nuclei to study RNA synthesis began in the early 1960s. Even before the isolation of eukaryotic RNA

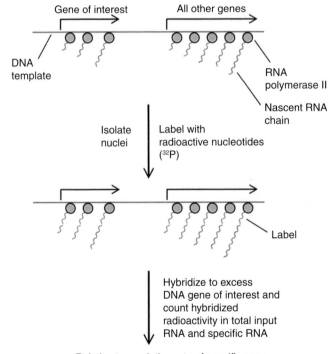

Relative transcription rate of specific gene

Figure 4-40. Run-on transcription assay. The RNA polymerases already engaged in RNA synthesis in isolated nuclei continue to elongate already-started, "nascent," chains by several hundred nucleotides (see text). This elongation can be as performed with a ^{32}P-labeled ribonucleoside triphosphate (often UTP) and the other three triphosphates. The resulting labeled RNA can be hybridized to a sample of purified DNA (most often cloned cDNA). The relative rate of transcription of a gene of interest (e.g., gene in the diagram) compared to total incorporation can be determined. The main enzyme responsible for total incorporation is Pol II because a low concentration of α-amanitin greatly decreases incorporation.

polymerases, C.C. Widnell and J.R. Tata showed that isolated nuclei could incorporate labeled nucleotides into RNAs (Widnell and Tata 1964).

Ronald Cox (1976) used the nuclei of chicken oviduct cells to investigate RNA synthesis by α-amanitin-sensitive polymerase (Pol II). More than 80% of the isolated nuclear RNA was inhibited by low-dose α-amanitin. Knowing that RNA grows from 5′ to 3′, the last nucleotide to be incorporated will not have a phosphate after alkali cleavage because the 5′ phosphate with which it entered the growing chain is left with the penultimate residue of the growing chain. Cox allowed isolated chick oviduct nuclei to incorporate ^{3}H-uridine-labeled UTP and then isolated the nuclear RNA. He then alkali-treated the labeled RNA and separated the UMP and uridine. The ratio was ~100 UMPs to 1 uridine. Because about one-quarter of the nucleotides in the RNA are UMPs, the average chain length synthesized by Pol II RNA in isolated nuclei was ~400 nucleotides. Thus, the already engaged Pol II in isolated nuclei could produce labeled nascent chains as a "run-on" product. The use of labeled triphosphates and cloned cDNAs afforded a direct measurement of the relative rate of synthesis of any specific gene.

There is still, more than three decades later, no substitute for run-on experiments to estimate the relative *rate* of transcription of a particular gene and its possible change under stimulated or inhibited conditions. The most common run-on assays in the 1970–1990s used full-length cDNAs; thus, these assays were not sensitive to premature termination or to events such as "paused" polymerases, which are discussed in Chapter 5. Of course, in cases of changing specific mRNA levels in cells, there may be both transcriptional and posttranscriptional controls, but the run-on assay is indispensable to measure relative transcription rates.

Differential Transcription: Hormone Stimulation and Tissue-Specific Control

One popular question under study in the late 1970s was the basis for the hormone-dependent increases of specific mRNAs. Two groups (Ringold et al. 1977; Young et al. 1977) showed that cells transformed by the estrogen-sensitive mouse mammary tumor virus (MMTV) incorporated ^{3}H uridine into MMTV-specific RNA at 10-fold-higher amounts when the cells were treated for 15 min with estrogen. But it is rare that cells produce enough specific RNA to allow a measurement in this fashion. The experiments nevertheless promised that the rate of transcription of individual genes could be measured and that estrogen stimulated specific transcription.

Furthermore, Stan McKnight and Richard Palmiter showed that hormone-dependent increases in transcription could be measured by a run-on assay.

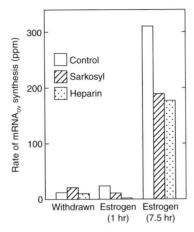

Figure 4-41. Estrogen stimulates transcription. Tissue fragments of chick oviducts were cultured for several hours without estrogen. The fragments were continued with no estrogen (withdrawn) or were given estrogen for 1 h or 7.5 h. Nuclei were prepared and incubated with four triphosphates including [^{32}P]UTP. Labeled RNA was isolated and hybridized to ovalbumin cDNA and hybridized RNA was measured. (The columns labeled "sarkosyl" and "heparin" were nuclei-treated with these surface active agents to ensure no new chain initiation.) (Redrawn from McKnight and Palmiter 1979.)

Ovalbumin mRNA transcription in oviduct tissue in which ovalbumin mRNA formation was already established was used. Oviduct tissue was removed and cultured briefly without estrogen. Then, tissue was stimulated with estrogen and McKnight and Palmiter (1979) used the nuclear run-on assay to show that the hormone stimulates oviduct cells to *transcribe* more ovalbumin and conalbumin RNA sequences (Fig. 4-41). Thus, a hormone that was known to enter the cell, together with something inside the cell, increased transcription.

Our laboratory used the run-on assay to study the relative transcription rates underlying apparent liver-specific gene expression. The mRNAs for a number of proteins were present in the liver (Derman et al. 1981; Powell et al. 1984) but not present or present in much lower levels in the kidney, brain, or spleen (Fig. 4-42). Using cDNAs for the liver-expressed mRNAs and RNA labeled by the run-on technique, we showed that the mRNAs expressed only (or mainly) in the liver were transcribed at a high rate in the liver and not in the other organs (Fig. 4-42). Other mRNAs encoding proteins present in all of the organs were transcribed at high levels in all of the organs and had high levels of mRNA in all of the organs (Derman et al. 1981; Powell et al. 1984).

Thus, by the early 1980s, it was a good bet that transcriptional regulation would be a *primary* mode of regulation in animal cells. Of course, these

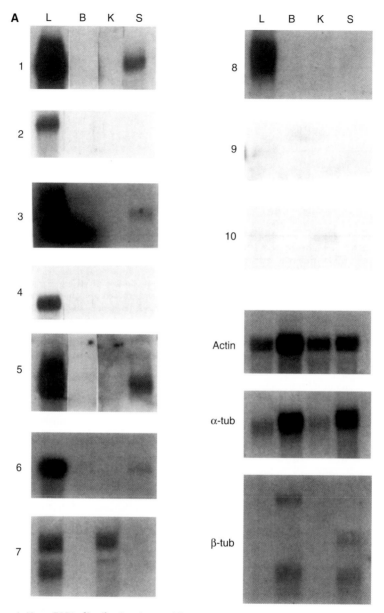

Figure 4-42. mRNA distribution in specific mouse tissues is controlled at the transcriptional level. (*A*) Cloned and ^{32}P-labeled cDNAs from mouse liver (1–10) were used to detect electrophoretically separated poly(A)-containing mRNA from liver, brain, kidney, and spleen using northern blots. The same RNA samples were used to detect actin (act), α-tub, and β-tub. (*B*) DNA from the same clones (unlabeled) as in *A* were used to hybridize ^{32}P-labeled RNA prepared from isolated nuclei from each of the four tissues. Autoradiograms were developed after 1 d (1×) or 10 d (10×). (Clone 11) A liver-expressed mRNA not shown in *A*, (pBR) the cloning vector; two tRNA clones were used as controls. (Reprinted, with permission, from Powell et al. 1984 [ⒸElsevier].) (*Fig. 4-42 continued on facing page.*)

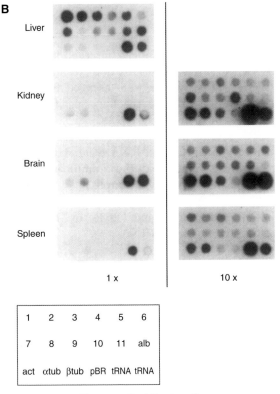

Figure 4-42. (*Continued*)

experiments did not mean that regulated poly(A) addition and regulated splicing do not also occur in deciding which mRNA might emerge from a particular transcript. But making the primary transcript is the first level of regulation. As we see in Chapter 5, posttranscriptional regulation of mRNA certainly does also occur for many mRNAs in eukaryotic cells.

CONCLUSION

At this point (early 1980s), we finally knew that the role in protein synthesis dictated by mRNA, and performed by tRNA and ribosomes, still held for all cells. But both the mechanical events of producing mRNA and the style in which information was stored in bacterial and eukaryotic DNA were very different. And although bacterial gene regulation seemed to have been largely solved by the early 1980s, we had scarcely scratched the surface in the details of how mammalian genes were regulated. The enzymes involved in RNA synthesis had been identified and purified. Correct initiation by mammalian polymerases plus a necessary group of accessory proteins—the GTFs—had been established, but how genes were chosen for transcription—how regulation occurred—was unsolved. However, more years of effort seemed clearly to promise success in learning the details of eukaryotic gene regulation. A summary of the detailed success plus additional startling discoveries of more classes of RNA involved in regulation is the aim of Chapter 5.

REFERENCES

Aloni Y, Bratosin S, Dhar R, Laub O, Horowitz M, Khoury G. 1978. Splicing of SV40 mRNAs: A novel mechanism for the regulation of gene expression in animal cells. *Cold Spring Harb Symp Quant Biol* **42:** 559–570.

Arber W. 1974. DNA modification and restriction. *Prog Nucleic Acid Res Mol Biol* **14:** 1–37.

Bachenheimer S, Darnell JE. 1975. Adenovirus 2 mRNA is transcribed as part of a high molecular weight precursor RNA. *Proc Natl Acad Sci* **72:** 4445–4449.

Baltimore D. 1971. RNA dependent DNA polymerase in virions of RNA tumor viruses. *Nature* **226:** 1209–1211.

Banerjee AK, Rhodes D. 1976. 5′ Terminal sequence of vesicular stomatitis virus mRNAs synthesized in vitro. *J Virol* **17:** 33–42.

Banerjee AK, Abraham G, Colonno RJ. 1977. Vesicular stomatitis virus: Mode of transcription. *J Gen Virol* **34:** 1–8.

Benjamin TL. 1966. Virus-specific RNA in cells productively infected or transformed by polyoma virus. *J Mol Biol* **16:** 359–373.

Berget SM, Moore C, Sharp PA. 1977. Spliced segments at the 5′ terminus of adenovirus 2 late mRNA. *Proc Natl Acad Sci* **74:** 3171–3175.

Berget SM, Berk AJ, Harrison T, Sharp PA. 1978. Spliced segments at the 5′ terminus of adenovirus-2 late mRNA: A role for heterogeneous nuclear RNA in mammalian cells. *Cold Spring Harb Symp Quant Biol* **42:** 523–530.

Bernhardt D, Darnell JE Jr. 1969. tRNA synthesis in HeLa cells: A precursor to tRNA and the effects of methionine starvation on tRNA synthesis. *J Mol Biol* **42:** 43–56.

Black FL, Knight CA. 1953. A comparison of some mutants of tobacco mosaic virus. *J Biol Chem* **202:** 51–57.

Blatti SP, Ingles CJ, Lindell TJ, Morris PW, Weaver RF, Weinberg F, Rutter WJ. 1971. Structure and regulatory properties of eukaryotic RNA polymerase. *Cold Spring Harb Symp Quant Biol* **35:** 649–657.

Boeye A. 1959. Induction of a mutation in poliovirus by nitrous acid. *Virology* **9:** 691–700.

Breathnach R, Chambon P. 1981. Organization and expression of eukaryotic split genes coding for proteins. *Annu Rev Biochem* **10:** 349–383.

Breathnach R, Mandel JL, Chambon P. 1977. Ovalbumin gene is split in chicken DNA. *Nature* **170:** 314–319.

Britten RJ, Davidson E. 1971. Repetitive and non-repetitive DNA sequences and a speculation on the origins of evolutionary novelty. *Q Rev Biol* **46:** 111–138.

Britten RJ, Kohne DE. 1968. Repeated sequences in DNA. Hundreds of thousands of copies of DNA sequences have been incorporated into the genomes of higher organisms. *Science* **161:** 529–540.

Brody E, Abelson J. 1985. The "spliceosome": Yeast pre-messenger RNA associated with a 40S complex in a splicing-dependent reaction. *Science* **228**: 963–967.

Broker TR, Chow LT, Down AR, Gelinas RE, Hassel JA, Klessig DF, Lewis JB, Roberts RJ, Zain BS. 1978. Displacement loops in adenovirus DNA-RNA hybrids. *Cold Spring Harb Symp Quant Biol* **42**: 531–554.

Brown DD, Gurdon JB. 1964. Absence of ribosomal RNA synthesis in the anucleolate mutant of *Xenopus laevis. Proc Natl Acad Sci* **51**: 139–146.

Brown DD, Sugimoto K. 1973. 5 S DNAs of *Xenopus laevis* and *Xenopus mulleri*: Evolution of a gene family. *J Mol Biol* **78**: 397–415.

Brown DD, Wensink PC, Jordan E. 1971. Purification and some characteristics of 5S DNA from *Xenopus laevis. Proc Natl Acad Sci* **68**: 3175–3179.

Burdon RH, Martin BT, Lal BM. 1967. Synthesis of low molecular weight ribonucleic acid in tumour cells. *J Mol Biol* **28**: 357–371.

Cech TR. 1983. RNA splicing: Three themes with variations. *Cell* **34**: 713–716.

Cech TR, Zaug AJ, Grabowski PJ. 1981. In vitro splicing of the ribosomal RNA precursor of *Tetrahymena*: Involvement of a guanosine nucleotide in the excision of the intervening sequence. *Cell* **27**: 487–496.

Chow LT, Broker TR. 1978. The spliced structures of adenovirus 2 fiber message and other late mRNAs. *Cell* **15**: 497–510.

Chow LT, Gelinas RE, Broker TR, Roberts RJ. 1977. An amazing sequence arrangement at the 5′ ends of adenovirus 2 messenger RNA. *Cell* **12**: 1–8.

Cohen SN, Chang AC, Boyer HW, Helling RB. 1973. Construction of biologically functional bacterial plasmids in vitro. *Proc Natl Acad Sci* **70**: 3240–3244.

Cox RF. 1976. Quantitation of elongating form A and B RNA polymerases in chick oviduct nuclei and effects of estradiol. *Cell* **7**: 455–465.

Curtis PJ, Weissmann C. 1976. Purification of globin mesenger RNA from dimethylsulfoxide-induced Friend cells and detection of a putative globin messenger RNA precursor. *J Mol Bol* **106**: 1061–1075.

Darnell JE Jr. 1963. Early events in poliovirus infection. *Cold Spring Harb Symp Quant Biol* **27**: 149–158.

Darnell JE Jr. 1968. Ribonucleic acids from animal cells. *Bacteriol Rev* **32**: 262–290.

Darnell JE Jr. 1982. Variety in the level of gene control in eukaryotic cells. *Nature* **297**: 365–371.

Darnell JE Jr. 2002. Special achievement in medical science award: the surprises of mammalian molecular cell biology. *Nat Med* **8**: 1068–1071.

Darnell JE Jr, Levintow L. 1960. Poliovirus protein: Source of amino acids and time course of synthesis. *J Biol Chem* **235**: 74–77.

Darnell JE Jr, Levintow L, Thoren M, Hooper L. 1960. The time course of synthesis of poliovirus RNA. *Virology* **13**: 271–279.

Darnell JE Jr, Penman S, Scherrer K, Becker Y. 1964. A description of various classes of RNA from HeLa cells. *Cold Spring Harb Symp Quant Biol* **28**: 211–214.

Darnell JE, Wall R, Tushinski RJ. 1971a. An adenylic acid-rich sequence in messenger RNA and its possible relationship to reiterated sites in DNA. *Proc Natl Acad Sci* **68**: 1321–1325.

Darnell JE, Philipson L, Wall R, Adesnik M. 1971b. Polyadenylic acid sequences: Role in conversion of nuclear RNA into messenger RNA. *Science* **174**: 507–510.

Darnell JE, Wall R, Adesnik M, Philipson L. 1972. The formation of mRNA in HeLa cells by post-transcriptional modification of nuclear RNA. In *Molecular genetics and developmental biology* (ed. M Sussman), pp. 201–224. Prentice-Hall, Englewood Cliffs, NJ.

Darnell JE, Evans R, Fraser N, Goldberg S, Nevins J. 1978. The definition of transcription units for mRNA. *Cold Spring Harb Symp Quant Biol* **42:** 515–522.

Darnell J, Lodish H, Baltimore D. 1986. *Molecular cell biology*. Scientific American/WH Freeman, New York.

Darnell J, Lodish H, Baltimore D. 1990. *Molecular cell biology*, 2nd ed. Scientific American/WH Freeman, New York.

Davis RW, Simon M, Davison N. 1971. Electron microscopic heteroduplex methods for mapping regions of base sequence homology in nucleic acids. In *Methods in enzymology* (ed. K Moldave, L Grossman), pp. 413–428. Academic Press, New York.

Derman E, Krauter K, Walling L, Weinberger C, Ray M, Darnell JE Jr. 1981. Transcriptional control in the production of liver-specific mRNAs. *Cell* **23:** 731–739.

Desrosiers R, Friderici K, Rottman F. 1974. Identification of methylated nucleosides in messenger RNA from Novikoff hepatoma cells. *Proc Natl Acad Sci* **71:** 3971–3975.

Dulbecco R. 1950. Experiments on the photoreactivation of bacteriophages inactivated by ultraviolet irradiation. *J Bacteriol* **59:** 329–347.

Dulbecco R. 1952. Production of plaques in monolayer tissue cultures by single particles of an animal virus. *Proc Natl Acad Sci* **38:** 747–752.

Dulbecco R. 1966. The plaque technique and the development of quantitative animal virology. In *Phage and the origins of molecular biology* (ed. J Carins et al.), pp. 287–291. Cold Spring Harbor Laboratory, Cold Spring Harbor, NY.

Dulbecco R, Vogt M. 1954. Plaque formation and isolation of pure lines with poliomyelitis viruses. *J Exp Med* **99:** 167–182.

Eagle H. 1954. Binding of penicillin in relation to its cytotaxic action. III. Binding of penicillin by mammalian cells in tissue culture (HeLa and L strains). *J Exp Med* **100:** 117–124.

Earle WR. 1943. Production of malignancy in vitro. IV. The mouse fibroblast cultures and changes seen in the living cells. *J Natl Cancer Inst* **4:** 165–212.

Edmonds M, Abrams R. 1960. Polynucleotide biosynthesis: Formation of a sequence of adenylate units from adenosine triphosphate by an enzyme from thymus nuclei. *J Biol Chem* **235:** 1142–1149.

Edmonds M, Vaughan MR, Nakazato H. 1971. Polyadenylic acid sequences in the heterogeneous nuclear RNA and rapidly-labeled polysomal RNA of HeLa cells: Possible evidence for a precursor relationship. *Proc Natl Acad Sci* **68:** 1336–1340.

Enders JF, Weller TH, Robbins FC. 1949. Cultivation of the Lansing strain of poliomyelities virus in cultures of various human embryonic tissues. *Science* **109:** 85–87.

Evans RM, Fraser N, Ziff E, Weber J, Darnell JE. 1977. The initiation sites for RNA transcription in Ad2 DNA. *Cell* **12:** 733–740.

Falck-Pedersen E, Logan J, Shenk T, Galli G, Darnell JE Jr. 1986. Adenovirus as a model for transcription-termination studies. In *DNA tumor viruses: Control of gene expression and replication* (ed. M Botchan et al.), pp. 267–273. Cold Spring Harbor Laboratory, Cold Spring Harbor, NY.

Flavell RA, Jeffreys AJ, Grosveld GC. 1978. Physical mapping of repetitive DNA sequences neighboring the rabbit β-globin gene. *Cold Spring Harb Symp Quant Biol* **42:** 1003–1010.

Fraser NW, Nevins JR, Ziff E, Darnell JE Jr. 1979. The major late adenovirus type 2 transcription unit: Termination is downstream from the last polyA site. *J Mol Biol* **129:** 643–656.

Furuichi Y, Morgan M, Muthukrishnan S, Shatkin AJ. 1975a. Reovirus messenger RNA contains a methylated, blocked 5'-terminal structure: m-7G(5')ppp(5')G-MpCp-. *Proc Natl Acad Sci* **72:** 362–366.

Furuichi Y, Morgan M, Shatkin AJ, Jelinek W, Georgieff MS, Darnell JE. 1975b. Methylated, blocked 5' termini in HeLa cell mRNA. *Proc Natl Acad Sci* **72:** 1904–1908.

Garapin AC, Cami B, Roskam W, Kourilsky P, Pennec JPL, Perrin F, Gerlinger P, Cochet M, Chambon P. 1978. Electron microscopy and restriction enzyme mapping reveal additional intervening sequences in the chicken ovalbumin split gene. *Cell* **14:** 629–639.

Gaulden ME, Perry RP. 1958. Influence of the nucleolus on mitosis as revealed by ultraviolet microbeam irradiation. *Proc Natl Acad Sci* **44:** 553–559.

Gelinas RE, Roberts RJ. 1977. One predominant 5'-undecanucleotide in adenovirus 2 late messenger RNAs. *Cell* **11:** 533–544.

Georgiev GP, Mantieva YL. 1962. The isolation of DNA-like RNA and ribosomal RNA from the nucleolo-chromosomal apparatus of mammalian cells. *Biochim Biophys Acta* **61:** 153–154.

Gey GO, Coffman WD, Kubicek MT. 1952. Tissue culture studies of the proliferative capacity of cervical carcinoma and normal epithelium. *Cancer Res* **12:** 264–265.

Gierer A. 1963. Function of aggregated reticulocyte ribosomes in protein synthesis. *J Mol Biol* **6:** 148–157.

Gilbert W. 1963. Polypeptide synthesis in *Escherichia coli*. I. Ribosomes and the active complex. *J Mol Biol* **6:** 374–388.

Gilbert W. 1978. Why genes in pieces? *Nature* **271:** 501. doi: 10.1038/271501a0.

Goldberg S, Schwartz H, Darnell JE Jr. 1977. Evidence from UV transcription mapping in HeLa cells that heterogenous nuclear RNA is the messenger RNA precursor. *Proc Natl Acad Sci* **74:** 4520–4523.

Goldberg S, Nevins J, Darnell JE. 1978. Evidence from UV transcription mapping that late adenovirus type 2 mRNA is derived from a large precursor molecule. *J Virol* **25:** 806–810.

Granboulan N, Scherrer K. 1969. Visualisation in the electron microsocpe and size of RNA from animal cells. *Eur J Biochem* **9:** 1–20.

Greenberg H, Penman S. 1966. Methylation and processing of ribosomal RNA in HeLa cells. *J Mol Biol* **21:** 527–535.

Greenberg JR, Perry RP. 1971. Hybridization properties of DNA sequences directing the synthesis of messenger RNA and heterogeneous nuclear RNA. *J Cell Biol* **50:** 774–786.

Guerrier-Takada C, Gardiner K, Marsh T, Pace N, Altman S. 1983. The RNA moiety of ribonuclease P is the catalytic subunit of the enzyme. *Cell* **35:** 849–857.

Hackett PB, Sauerbier W. 1974. Radiological mapping of the ribosomal RNA transcription unit in *E. coli*. *Nature* **251:** 639–641.

Hackett PB, Sauerbier W. 1975. The transcriptional organization of the ribosomal RNA genes in mouse L cells. *J Mol Biol* **91:** 235–256.

Hadjivassiliou A, Brawerman G. 1966. Polyadenylic acid in the cytoplasm of rat liver. *J Mol Biol* **20:** 1–7.

Hall BD, Doty P. 1959. The preparation and physical chemical properties of ribonucleic acid from microsomal particles. *J Mol Biol* **1:** 111–126.

Harris H. 1959. Turnover of nuclear and cytoplasmic ribonucleic acid in two types of animal cell, with some further observations on the nucleolus. *Biochem J* **73:** 362–369.

Harris H. 1964. Function of the short-lived ribonucleic acid in the cell nucleus. *Nature* **201:** 863–867.

Harris H, LaCour LF. 1963. Site of synthesis of cytoplasmic ribonucleic acid. *Nature* **200:** 227–229.

Harrison RN. 1909. The growth of nerve fibers. *Science* **30:** 158. doi: 10.1126/science. 30.761.158.

Harrison RN. 1910. The outgrowth of the nerve fiber as a model of protoplasmic movement. *J Exp Zool* **9:** 787–846.

Hinterberger M, Pettersson I, Steitz JA. 1983. Isolation of small nuclear ribonucleoproteins containing U1, U2, U4, U5, and U6 RNAs. *J Biol Chem* **258:** 2604–2613.

Holland CA, Mayrand S, Pederson T. 1980. Sequence complexity of nuclear and messenger RNA in HeLa cells. *J Mol Biol* **138:** 755–778.

Holtzman E, Smith I, Penman S. 1966. Electron microscopic studies of detergent-treated HeLa cell nuclei. *J Mol Biol* **17:** 131–135.

Hsu M-T, Ford J. 1978. A novel sequence arrangement of SV40 late RNA. *Cold Spring Harb Symp Quant Biol* **42:** 571–576.

Jelinek W, Adesnik M, Salditt M, Sheiness D, Wall R, Molloy G, Philipson L, Darnell JE. 1973. Further evidence of the nuclear origin and transfer to the cytoplasm of poly(A) sequences in mammalian cell RNA. *J Mol Biol* **75:** 515–532.

Kates JR. 1971. Transcription of the vaccinia virus genome and the occurrence of polyadenylic acid sequences in messenger RNA. *Cold Spring Harb Symp Quant Biol* **35:** 743–752.

Kates JR, Beeson J. 1970. Ribonucleic acid synthesis in vaccinia virus. II. Synthesis of polyriboadenylic acid. *J Mol Biol* **50:** 19–33.

Kates JR, McAuslan BR. 1967. Messenger RNA synthesis by a "coated" viral genome. *Proc Natl Acad Sci* **57:** 314–320.

Kitamura N, Semler BL, Rothberg PG, Larsen GR, Adler CJ, Dorner AJ, Emilio EA, Hanecak R, Lee JJ, van der Werf S, et al. 1981. Primary structure, gene organization and polypeptide expression of poliovirus RNA. *Nature* **291:** 547–553.

Klessig DR. 1977. Two adenovirus mRNAs have a common 5′ terminal leader sequence encoded at least 10 kb upstream from their main coding sequence. *Cell* **12:** 9–21.

Knapp G, Ogden RC, Peebles CL, Abelson J. 1979. Splicing of yeast tRNA precursors: Structure of the reaction intermediates. *Cell* **18:** 37–45.

Koonin EV, Moss B. 2010. Viruses know more than one way to don a cap. *Proc Natl Acad Sci* **107:** 3283–3284.

Landecker H. 2007. *Culturing life.* Harvard University Press, Cambridge, MA.

Leder P, Tilghman SM, Tiemeier DC, Polsky FI, Seidman JG, Edgell MH, Enquist LW, Leder A, Norman B. 1978. The cloning of mouse globin and surrounding gene sequences in bacteriophage λ. *Cold Spring Harb Symp Quant Biol* **42:** 915–920.

Lee SEY, Mondicki J, Brawerman G. 1971. A polynucleotide segment rich in adenylic acid in the rapidly-labeled polyribosomal component in mouse sarcoma 180 ascites cells. *Proc Natl Acad Sci* **68:** 1331–1335.

Lerner MR, Bouyle JA, Mount SM, Wolin SL, Steitz JA. 1980. Are snRNPs involved in splicing? *Nature* **283:** 220–224.

Lindberg U, Darnell JE. 1970. SV40-specific RNA in the nucleus and polyribosomes of transformed cells. *Proc Natl Acad Sci* **65:** 1089–1096.

Lindberg U, Persson T, Philipson L. 1972. Isolation and characterization of adenovirus messenger ribonucleic acid in productive infection. *J Virol* **10:** 909–919.

Lindell TJ, Weinberg F, Morris PW, Roeder RG, Rutter WJ. 1970. Specific inhibition of nuclear RNA polymerase II by α-amanitin. *Science* **170:** 447–449.

Lodish H, Baltimore D, Berk A, Zipursky SL, Matsudaira P, Darnell J. 1995. *Molecular cell biology*, 3rd ed. Scientific American/WH Freeman, New York.

Luse DS, Haynes JR, VanLeeuwen D, Schon EA, Cleary ML, Shapiro SG, Lingrel JB, Roeder RG. 1981. Transcription of the β-like globin genes and pseudogenes of the goat in a cell-free system. *Nucleic Acids Res* 9: 4339–4354.

Maniatis T, Hardison RC, Lacy E, Lauer J, O'Connell C, Quon D, Sim GK, Efstratiadis A. 1978. The isolation of structural genes from libraries of eucaryotic DNA. *Cell* 15: 687–701.

Matsui T, Segall J, Weil PA, Roeder RG. 1980. Multiple factors required for accurate initiation of transcription by purified RNA polymerase II. *J Biol Chem* 255: 11992–11996.

Maxam AM, Gilbert W. 1977. A new method for sequencing DNA. *Proc Natl Acad Sci* 74: 560–564.

McClain WH, Lai LB, Gopalan V. 2010. Trials, travails and triumphs: An account of RNA catalysis in RNAase P. *J Mol Biol* 397: 627–646.

McConkey EH, Hopkins JW. 1969. Molecular weights of some HeLa ribosomal RNAs. *J Mol Biol* 39: 545–550.

McKnight GS, Palmiter RD. 1979. Transcriptional regulation of ovalbumin and conalbumin genes by steroid hormones in duck oviduct. *J Biol Chem* 254: 9050–9058.

McKnight GS, Schimke RT. 1974. Ovalbumin messenger RNA evidence that the initial product of transcription is the same size as polysomal ovalbumin messenger. *Proc Natl Acad Sci* 71: 4327–4331.

McLaughlin CS, Warner JR, Edmonds M, Nakazato H, Vaughan MH. 1973. Polyadenylic acid sequences in yeast messenger ribonucleic acid. *J Biol Chem* 248: 1466–1471.

Mendecki J, Lee SY, Brawerman G. 1972. Characteristics of the polyadenylic acid segment associated with messenger ribonucleic acid in mouse sarcoma 180 ascites cells. *Biochemistry* 11: 792–798.

Miller OC, Hankalo BA, Thomas CA. 1970. Visualization of bacterial genes in action. *Science* 169: 392–395.

Miura K, Watanabe K, Sugiura M. 1974. 5′-Terminal nucleotide sequences of the double-stranded RNA of silkworm cytoplasmic polyhedrosis virus. *J Mol Biol* 86: 31–48.

Molley G, Thomas W, Darnell JE. 1972a. Occurrence of uridylate-rich oligonucleotide regions in heterogenous nuclear RNA of HeLa Cells. *Proc Natl Acad Sci* 69: 3684–3688.

Molloy GR, Sporn MB, Kelley DE, Perry RP. 1972b. Localization of polyadenylic acid sequences in messenger ribonucleic acid of mammalian cells. *Biochemistry* 11: 3256–3260.

Moore GP. 1984. The G-value paradox. *Bioscience* 34: 423–429.

Mount SM, Pettersson I, Hinterberger M, Karmas A, Steitz JA. 1983. The U1 small RNA-protein complex selectively binds a 5′ splice site. *Cell* 33: 509–518.

Mowshowitz DB. 1970. Transfer RNA synthesis in HeLa cells. II. Formation of tRNA from a precursor in vitro and formation of pseudouridine. *J Mol Biol* 50: 143–151.

Mulder C, Arrand JB, Delcino H, Keller W, Pettersson U, Roberts RJ, Sharp PA. 1975. Cleavage maps of DNA from adenovirus types 2 and 5 by restriction endonucleases EcoRI and HpaI. *Cold Spring Harb Symp Quant Biol* 39: 397–400.

Mundry K. 1959. The effect of nitrous acid on TMV: Mutation and selection. *Virology* 9: 722–725.

Nakazato H, Kopp DW, Edmonds M. 1973. Localization of the polyadenylate sequences in messenger ribonucleic acid and in the heterogeneous nuclear ribonucleic acid of HeLa cells. *J Biol Chem* 248: 1472–1476.

Nevins JR, Chen-Kiang S. 1981. Processing of adenovirus nuclear RNA to mRNA. *Adv Virus Res* 26: 1–35.

Nevins JR, Darnell JE. 1978a. Groups of adenovirus type 2 mRNAs derived from a large primary transcript. *J Virol* 25: 811–823.

Nevins JR, Darnell JE Jr. 1978b. Steps in the processing of Ad2 mRNA: Poly(A)$^+$ nuclear sequences are conserved and poly(A) addition precedes splicing. *Cell* **15:** 1477–1493.

Ng S-Y, Parker CS, Roeder RG. 1979. Transcription of cloned *Xenopus* 5S RNA genes by *X. laevis* RNA polymerase III in reconstituted systems. *Proc Natl Acad Sci* **76:** 136–140.

Nicholas JS. 1960. Ross Granville Harrison, 1870–1959. *Yale J Biol Med* **32:** 407–412.

Ogino T, Yadav SP, Banerjee AK. 2010. Histidine-mediated RNA transfer to GDP for unique mRNA capping by visticular stomatitis virus RNA polymerase. *Proc Natl Acad Sci* **107:** 3463–3468.

Padgett RA, Hardy SF, Sharp PA. 1983a. Splicing of adenovirus RNA in a cell-free transcription system. *Proc Natl Acad Sci* **80:** 5230–5234.

Padgett RA, Mount SM, Steitz JA, Sharp PA. 1983b. Splicing of messenger RNA precursors is inhibited by antisera to small nuclear ribonucleoprotein. *Cell* **35:** 101–107.

Parker CS, Roeder RG. 1977. Selective and accurate transcription of the *Xenopus laevis* 5S RNA genes in isolated chromatin by purified RNA polymerase III. *Proc Natl Acad Sci* **74:** 44–48.

Parker CS, Jaehning JA, Roeder RG. 1978. Faithful gene transcription by eukaryotic RNA polymerases in reconstructed systems. *Cold Spring Harb Symp Quant Biol* **42:** 577–587.

Peebles CL, Ogden RC, Knapp G, Abelson J. 1979. Splicing of yeast tRNA precursors: A two-stage reaction. *Cell* **18:** 27–35.

Penman S. 1966. RNA metabolism in the HeLa cell nucleus. *J Mol Biol* **17:** 117–130.

Penman S, Scherrer K, Becker Y, Darnell JE. 1963. Polyribosomes in normal and poliovirus infected HeLa cells and their relationship to messenger RNA. *Proc Natl Acad Sci* **49:** 654–662.

Penman S, Smith I, Holtzman E. 1966. Ribosomal RNA synthesis and processing in a particulate site in the HeLa cell nucleus. *Science* **154:** 786–789.

Perry RP. 1960. On the nucleolar and nuclear dependence of cytoplasmic RNA synthesis in HeLa cells. *Exp Cell Res* **20:** 216–220.

Perry RP. 1961. Kinetics of nucleoside incorporation into nuclear and cytoplasmic RNA. *J Biophys Biochem Cytol* **11:** 1–13.

Perry RP. 1962. The cellular sites of synthesis of ribosomal and 4S RNA. *Proc Natl Acad Sci* **48:** 2179–2186.

Perry RP, Kelley DE. 1974. Existence of methylated messenger RNA in mouse L cells. *Cell* **1:** 37–41.

Pettersson U, Tibbetts C, Philipson L. 1976. Hybridization maps of early and late messenger RNA sequences on the adenovirus type 2 genome. *J Mol Biol* **101:** 479–501.

Philipson L, Wall R, Glickman G, Darnell JE. 1971. Addition of polyadenylate sequences to virus-specific RNA during adenovirus replication. *Proc Natl Acad Sci* **68:** 2806–2809.

Philipson L, Pettersson U, Lindberg U, Tibbets C, Vennstrom B, Persson T. 1975. RNA synthesis and processing in adenovirus-infected cells. *Cold Spring Harb Symp Quant Biol* **39:** 447–456.

Pontecorvo G. 1963. The Leeuwenhoek lecture: Microbial genetics: Retrospect and prospect. *Proc Roy Soc B* **158:** 1–23.

Powell DJ, Friedman J, Oulette AJ, Krauter KS, Darnell JE Jr. 1984. Transcriptional and post-transcriptional control of specific messenger RNAs in adult and embryonic liver. *J Mol Biol* **179:** 21–35.

Puck TT, Marcus PI. 1955. A rapid method for viable cell titration and clone production with HeLa cells in tissue culture. *Proc Natl Acad Sci* **41:** 432–438.

Puck TT, Marcus PI. 1956. Action of X-rays on mammalian cells. *J Exp Med* **103**: 273–283.

Pyle AM. 2010. The tertiary structure of group II introns: Implications for biological function and evolution. *Crit Rev Biochem Mol Biol* **45**: 215–232.

Reddy R, Ro-Choi TS, Henning D, Busch H. 1974. Primary sequence of U-1 nuclear ribonucleic acid of Novikoff hepatoma ascites cells. *J Biol Chem* **249**: 6486–6494.

Rich A, Penman S, Becker Y, Darnell J, Hall C. 1963. Polyribosomes: Size in normal and polio-infected HeLa cells. *Science* **142**: 1658–1663.

Ringold GM, Yamamoto KR, Bishop JM, Varmus HE. 1977. Glucocorticoid-stimulated accumulation of mouse mammary tumor virus RNA: Increased rate of synthesis of viral RNA. *Proc Natl Acad Sci* **74**: 2879–2883.

Ritossa FM, Atwood KC, Spiegelman S. 1966. A molecular explanation of the bobbed mutants of *Drosophila* as partial deficiencies of "ribosomal" DNA. *Genetics* **54**: 819–834.

Roeder RG. 1976. Eukaryotic nuclear RNA polymerase. In *RNA polymerase* (ed. R Losick, M Chamberlin), pp. 285–329. Cold Spring Harbor Laboratory, Cold Spring Harbor, NY.

Roeder RG. 1983. Multiple forms of DNA-dependent RNA polymerases in *Xenopus laevis*: Properties, purification and subunit structure of class III RNA polymerases. *J Biol Chem* **258**: 1932–1941.

Roeder RG, Rutter WJ. 1969. Multiple forms of DNA-dependent RNA polymerase in eukaryotic organisms. *Nature* **224**: 234–237.

Roeder RG, Rutter WJ. 1970. Specific nucleolar and nucleoplasmic RNA polymerases. *Proc Natl Acad Sci* **65**: 675–682.

Rogers AW. 1979. *Techniques in autoradiography*. Elsevier/North Holland, The Netherlands.

Ross J. 1976. A precursor of globin messenger RNA. *J Mol Biol* **106**: 403–420.

Salditt-Georgieff M, Darnell JE Jr. 1982. Further evidence that the majority of primary nuclear RNA transcripts in mammalian cells do not contribute to mRNA. *Mol Cell Biol* **2**: 701–707.

Salditt-Georgieff M, Jelinek W, Darnell JE, Furuichi Y, Morgan M, Shatkin A. 1976. Methyl labeling of HeLa cell HnRNA: A comparison with mRNA. *Cell* **7**: 227–237.

Salk JE. 1953. Studies in human subjects on active immunization against poliomyelitis. *J Am Med Assoc* **151**: 1081–1098.

Sambrook J, Westphal H, Srinivasan PR, Dulbecco R. 1968. The integrated state of viral DNA in SV40-transformed cells. *Proc Natl Acad Sci* **60**: 1288–1295.

Sanford KK, Earle WR, Likely GD. 1948. The growth in vitro of single isolated tissue cells. *J Natl Cancer Inst* **9**: 229–246.

Sanger F, Nicklen S, Coulson AR. 1977. DNA sequencing with chain-terminating inhibitors. *Proc Natl Acad Sci* **74**: 5463–5467.

Sarabhai AS, Stretton AO, Brenner S, Bolle A. 1964. Co-linearity of the gene with the polypeptide chain. *Nature* **201**: 13–17.

Sauerbier W, Broutigan A. 1970. A simple method for isolating RNA from bacteria. *Biochem Biophys Acta* **199**: 36–40.

Schaffer FL, Schwerdt CE. 1959. The purification and properties of poliovirus. *Adv Virus Res* **6**: 159–204.

Scherer W, Syverton J, Gey G. 1953. Studies on the propagation in vitro of poliomyelitis viruses. *J Exp Med* **97**: 695–715.

Scherrer K, Darnell JE. 1962. Sedimentation characteristics of rapidly labeled RNA from HeLa cells. *Biochem Biophys Res Commun* **7**: 486–490.

Scherrer K, Marcaud L. 1965. Remarks on polycistron messenger RNA's in animal cells. *Bull Soc Chim Biol* **47**: 1697–1713.

Scherrer K, Latham H, Darnell JE. 1963. Demonstration of an unstable RNA and precursor to ribosomal RNA in HeLa cells. *Proc Natl Acad Sci* **49:** 240–248.

Segall J, Matsui T, Roeder RG. 1980. Multiple factors are required for the accurate transcription of purified genes by RNA polymerase III. *J Biol Chem* **255:** 11986–11991.

Sehgal PG, Derman E, Molloy GR, Tamm I, Darnell JE. 1976. 5,6-Dichloro-I-β-D-ribofurano-sylbenzimidazole inhibits initiation of nuclear heterogenous RNA chains in HeLa cells. *Science* **194:** 431–433.

Sharp PA, Gallimore P, Flint SJ. 1975. Mapping of adenovirus 2 RNA sequences in lytically infected cells and transformed cell lines. *Cold Spring Harb Symp Quant Biol* **39:** 457–474.

Shatkin AJ. 1974. Methylated messenger RNA synthesis in vitro by purified reovirus. *Proc Natl Acad Sci* **71:** 3204–3207.

Shatkin AJ. 1976. Capping of eucaryotic mRNAs. *Cell* **9:** 645–653.

Sibatani A, Kloet SRD, Allfrey VG, Mirsky AE. 1962. Isolation of a nuclear RNA fraction resembling DNA in its base composition. *Proc Natl Acad Sci* **48:** 471–477.

Simon EH. 1961. Transfer of DNA from parent to progeny in a tissue culture line of human carcinoma of the cervix (strain HeLa). *J Mol Biol* **3:** 101–109.

Skloot R. 2010. *The immortal life of Henrietta Lacks.* Crown Publishers, New York.

Smellie RMS. 1955. The metabolism of the nucleic acids. In *The nucleic acids* (ed. E Chargaff, JN Davidson), Vol. 2, pp. 393–434. Academic Press, New York.

Soeiro R, Birnboim HC, Darnell JE. 1966. Rapidly labeled HeLa cell nuclear RNA. II. Base composition and cellular localization of a heterogenous RNA fraction. *J Mol Biol* **19:** 362–372.

Soeiro R, Vaughan MH, Warner JR, Darnell JE Jr. 1968. The turnover of nuclear DNA-like RNA in HeLa cells. *J Cell Biol* **39:** 112–118.

Sommer S, Salditt-Georgieff M, Bachenheimer S, Furuichi Y, Morgan M, Shatkin A. 1976. The methylation of adenovirus-specific nuclear and cytoplasmic RNA. *Nucleic Acids Res* **3:** 749–766.

Stark BC, Kole R, Bowman EJ, Altman S. 1978. Ribonuclease P: An enzyme with an essential RNA component. *Proc Natl Acad Sci* **75:** 3717–3721.

Steward FC, Mapes MO, Kent AE, Holsten RD. 1964. Growth and development of cultured plant cells. *Science* **143:** 20–27.

Stewart SE, Eddy BE, Borgese N. 1958. Neoplasms in mice inoculated with a tumor agent carried in tissue culture. *J Natl Cancer Inst* **20:** 1223–1243.

Strauss JH Jr, Kelly RB, Sinsheimer RL. 1968. Denaturation of RNA with dimethyl sulfoxide. *Biopolymers* **6:** 793–807.

Temin H, Mizutani S. 1971. RNA dependent DNA polymerase in virions of Rous sarcoma virus. *Nature* **226:** 1211–1213.

Thomas M, White RL, Davis RW. 1976. Hybridization of RNA to double stranded DNA: R-loop formation. *Proc Natl Acad Sci* **73:** 2294–2298.

Tibbetts C, Pettersson U. 1974. Complementary strand-specific sequences from unique fragments of adenovirus type 2 DNA for hybridization-mapping experiments. *J Mol Biol* **88:** 767–784.

Tilghman SM, Tiemeier DC, Seidman JG, Peterlin BM, Sullivan M, Maizel JV, Leder P. 1978. Intervening sequence of DNA identified in the structural portion of a mouse β-globin gene. *Proc Natl Acad Sci* **75:** 725–729.

Tonegawa S, Walter G, Bernardini A, Dulbecco R. 1971. Transcription of the SV40 genome in transformed and lytic infection. *Cold Spring Harb Symp Quant Biol* **35:** 823–831.

Tycowski K, Kolev NG, Conrad NK, Fok V, Steitz JA. 2006. The ever-growing world of small nuclear ribonucleoproteins. In *The RNA world*, 3rd ed. (ed. RF Gesteland et al.), pp. 327–368. Cold Spring Harbor Laboratory Press, Cold Spring Harbor, NY.

Venkatesan S, Moss B. 1982. Eukaryotic mRNA capping enzyme-guanylate covalent intermediate. *Proc Natl Acad Sci* 79: 340–344.

Wagner EK, Roizman B. 1969. RNA synthesis in cells infected with herpes simplex virus. II. Evidence that a class of viral mRNA is derived from a high molecular weight precursor synthesized in the nucleus. *Proc Natl Acad Sci* 64: 626–633.

Wall R, Philipson L, Darnell JE. 1972. Processing of adenovirus specific nuclear RNA during virus replication. *Virology* 50: 27–34.

Wang J, Hevi S, Kurash JK, Lei H, Gay F, Bajko J, Su S, Sun W, Chang H, Xu G, et al. 2008. The lysine demethylase LSD1 (KDM1) is required for maintenance of global DNA methylation. *Nat Genet* 41: 125–129.

Warner JR, Knopf PM. 2002. The discovery of polyribosomes. *Trends Biochem Sci* 27: 376–380.

Warner J, Madden MJ, Darnell JE. 1963a. The interaction of poliovirus RNA with *E. coli* ribosomes. *Virology* 19: 393–399.

Warner JR, Knopf PM, Rich A. 1963b. A multiple ribosomal structure in protein synthesis. *Proc Natl Acad Sci* 49: 122–129.

Weber J, Jelinek W, Darnell JE Jr. 1977. The definition of a large viral transcription unit late in Ad2 infection of HeLa cells: Mapping of nascent RNA molecules labeled in isolated nuclei. *Cell* 10: 611–616.

Wecker E. 1959. The extraction of infectious virus nucleic acid with hot phenol. *Virology* 7: 241–243.

Wei C-M, Moss B. 1974. Methylation of newly synthesized viral messenger RNA by an enzyme in vaccinia virus. *Proc Natl Acad Sci* 71: 3014–3018.

Wei CM, Moss B. 1975. Methylated nucleotides block 5′-terminus of vaccinia virus messenger RNA. *Proc Natl Acad Sci* 72: 318–322.

Wei CM, Gershowitz A, Moss B. 1975. Methylated nucleotides block 5′ terminus of HeLa cell messenger RNA. *Cell* 4: 379–386.

Weil PW, Luse DS, Segall J, Roeder RG. 1979. Selective and accurate initiation of transcription at the Ad2 major late promoter in a soluble system dependent on purified RNA polymerase II and DNA. *Cell* 18: 469–484.

Weinmann R, Roeder RG. 1974. Role of DNA-dependent RNA polymerase III in the transcription of the tRNA and 5S RNA genes. *Proc Natl Acad Sci* 71: 1790–1794.

Westphal H, Lai S-P. 1978. Displacement loops in adenovirus DNA-RNA hybrids. *Cold Spring Harb Symp Quant Biol* 42: 555–558.

Wettstein FO, Staehelin T, Noll H. 1963. Ribosomal aggregate engaged in protein synthesis: Characterization of the ergosome. *Nature* 197: 430–435.

White RL, Hogness DS. 1977. R loop mapping of the 18S and 28S sequences in the long and short repeating units of *Drosophila melanogaster* rDNA. *Cell* 10: 177–192.

Widnell CC, Tata JR. 1964. A procedure for the isolation of enzymically active rat-liver nuclei. *Biochem J* 92: 313–317.

Wild MA, Gall JG. 1979. An intervening sequence in the gene coding for 25S ribosomal RNA of *Tetrahymena pigmentosa*. *Cell* 16: 565–573.

Witkowski JA. 1980. Dr. Carrel's immortal cells. *Med Hist* 24: 129–142.

Yanofsky C, Carlton BL, Guest JR, Helinski DR, Henning U. 1964. On the colinearity of the gene structure and protein structure. *Proc Natl Acad Sci* 51: 266–272.

Young HA, Shih TY, Scolnick EM, Parks WP. 1977. Steroid induction of mouse mammary tumor virus: Effect upon synthesis and degradation of viral RNA. *J Virol* **21:** 139–146.

Youngner JS, Ward EN, Salk JE. 1952. Studies on poliomyelitis viruses in cultures of monkey testicular tissue. II. Differences among strains in tissue culture infectivity with preliminary data and in the quantitative estimation of virus and antibody. *Am J Hyg* **55:** 301–322.

Zaug AJ, Cech TR. 1980. In vitro splicing of the ribosomal RNA precursor in nuclei of *Tetrahymena*. *Cell* **19:** 331–338.

Zaug AJ, Grabowski PJ, Cech TR. 1983. Autocatalytic cyclization of an excised intervening sequence RNA is a cleavage-ligation reaction. *Nature* **301:** 578–583.

Ziff EB, Evans RM. 1978. Coincidence of the promoter and capped 5′ terminus of RNA from the adenovirus 2 major late transcription unit. *Cell* **15:** 1463–1475.

5

Controlling mRNA

The Cell's Most Complicated Task

ALTHOUGH THE STUDY OF EUKARYOTIC RNAs from the early 1960s through the early 1980s produced many great advances and surprises—the recognition of ribosomal RNA (rRNA), transfer RNA (tRNA), and messenger RNA (mRNA) processing, including splicing; the description of three RNA polymerases together with proteins that allow correct RNA initiation; and the initial studies of RNA as an enzyme—things were just warming up. Since the early 1980s, an exquisitely detailed understanding, still not complete, of RNA biology and biochemistry has been achieved. This progress has been aided by spectacular advances in technology that have resulted in deep insight into how gene regulation in eukaryotes is managed, at both the transcriptional and posttranscriptional levels. During the thousands of experiments on these subjects in the 1990s and after 2000, many new species of RNA—both short and long noncoding RNAs—were discovered. Current research has provided general knowledge of the function of these new RNA species, and the specific importance of many individual representatives of these new molecules is accumulating rapidly. They operate at all levels of gene expression, affecting transcriptional initiation, elongation, and processing, as well as cytoplasmic translation and especially mRNA lifetime.

Granting the surprising importance of the newly discovered RNAs, there is still no doubt that the *primary* roles of RNA in cell function are those of the tRNA, rRNA, and mRNA. Furthermore, because all cells are equipped with the same tRNA and the same ribosomes (so far as we know), it is the control of the manufacture of mRNA—its synthesis and processing—that remains of primary importance in cell physiology.

To provide a useful overview of important recent discoveries regarding RNA, we must deal first with the elaborate events that determine the choice of and rate of transcription of specific mRNA precursors in eukaryotes, where so much progress has been made. After all, if a pre-mRNA is not transcribed, its differential processing in the nucleus or its translation rate and lifetime as a functional mRNA in the cytoplasm cannot be regulated (Table 5-1).

Table 5-1. The many ways that gene expression is regulated in eukaryotes

Nuclear events	Regulators and their actions

Pre-mRNA initiation	Positive acting activators are site-specific DNA-binding proteins that recruit
	  1. Coactivators (protein complexes)
	  2. Chromatin- (histone-) modifying complexes
	    a. Enzymes (e.g., acetyltransferases, methylases, demethylases, ubiquitin ligases)
	    b. ATP-dependent histone-remodeling and replacement complexes
	  3. General transcription factors (GTFs) and Pol II
	Negative acting repressors and corepressor complexes can
	  1. Block GTFs, Pol II, and positive-acting proteins
	  2. Be site-specific DNA-binding proteins or long noncoding RNAs that recruit corepressors (protein complexes) that contain enzymes (e.g., histone deacetylases, methylases, demethylases)

Pre-mRNA elongation	Elongation factors include
	  1. Promoter clearance proteins (e.g., phosphokinases)
	  2. Histone displacement/replacement

Pre-mRNA processing	Poly(A) choices 1. Protein complexes, including 3'-end cleavage proteins and poly(A) polymerase Splicing choices
Cytoplasmic events	

	1. Spliceosomes and general and cell-specific RNA-binding complexes
	mRNA translation frequency and mRNA lifetime
	  1. Quality-control complexes (nonsense-mediated decay)
	  2. RNA–protein complexes: 5'- and 3'-specific nucleases, small RNP complexes
	  3. mRNA–protein complexes and spatial control of expression in cytoplasm

Focal points at which regulation occurs, as opposed to biochemically necessary, but unregulated, steps are emphasized. For example, histone displacement–replacement factors are necessary during the elongation stage of pre-mRNA synthesis, but there is no evidence that these are specifically regulated.

THREE RECENT DECADES OF RNA
EXPLORATION: AN OVERVIEW

By the early 1980s, we knew that RNA polymerase II (Pol II) could, with the assistance of general transcription factors (GTFs), find a precise transcription start site (TSS) for mRNA precursors on *naked* DNA. Although this fundamental transcriptional machinery, i.e., the polymerase plus the GTFs, is necessary, it is *not* regulatory. The proteins that regulated transcriptional initiation, presumably by binding to specific DNA sites either to assist in in vitro transcription or, more importantly, to regulate transcription rate inside of the cell, were unknown. Furthermore, eukaryotic DNA in the cell nucleus was known since the 1950s to be associated with an equal weight of proteins, mainly histones. Such protein-associated DNA—*chromatin*—was widely regarded to be inaccessible to transcription.

Two main problems, therefore, existed for the transcription field circa 1980–1985. The first was how to define functional regulatory sites in DNA and identify the proteins, potentially both positive and negative acting, that bound the specific DNA sites in order to regulate transcriptional initiation. The second interrelated problem was to understand more exactly the roles of histones. Because specific transcription did, after all, occur, how was the presumed negative effect of histones circumvented?

Progress on these questions was spectacular (Table 5-1). The first step was the recognition of many hundreds of positive-acting transcription factors designated as *activators* that bind specific regulatory sites in DNA. Activators act alone or cooperate in groups. They recruit, by protein:protein association, assemblies of non-DNA-binding proteins called *coactivators*. These recruited protein complexes have a variety of functions including modifying histones.

Understanding details regarding activators and coactivators in *positively* promoting transcription initiation opened up the possibility of discovering negative-acting proteins widely described as *repressors*. The term *repressor*, originally coined for negative-acting transcriptional proteins in bacteria, is also in wide use in describing inhibitors of transcription in eukaryotes. However, in keeping with the much more complicated mechanisms of positive-acting transcriptional complexes in eukaryotes, the mechanism(s) of transcription prevention are also very different in the two cell types. In bacteria, repressors bind at (or near) the transcription start site and block the approach of the RNA polymerase. In eukaryotic cells, the negative-acting transcriptional proteins most often do not bind directly at TSSs. Some negative-acting eukaryotic proteins are site-specific DNA-binding proteins that are incompetent to drive transcription without a partner and act at least temporarily as a repressor. Most bind DNA at a distance from the TSS and recruit complexes of *corepressor* proteins. These complexes compete with, temper, or suppress positive-acting proteins at the level of transcriptional initiation.

The myriad interactions among these hundreds of accessory proteins are either positive acting—bringing the Pol II and the GTFs to a TSS to initiate transcription—or negative acting—preventing transcriptional initiation. Regulation of transcriptional initiation has by far the greatest impact on the number of any particular mRNA molecule that is present in a cell.

Even after initiation, however, Pol II and associated proteins require additional events (e.g., phosphorylation) to proceed to *elongation*. Furthermore, the growing group of proteins that accompanies the successful transcription machine includes many of the proteins required for pre-mRNA *processing* as well as complexes that remove histones and then restore them after passage.

Neither the initial elongation decision (overcoming polymerase pausing) nor the processing decisions (poly[A] site selection and splice site selection) are automatic. After elongation is sanctioned, the poly(A) polymerase complex and various components of the processing machine, often including the splicing machine (the *spliceosome*), are directed to their respective targets on a pre-mRNA in a regulated fashion because most pre-mRNAs, at least in mammals, are alternatively spliced and/or have more than one possible poly(A) site.

Once processing is complete, the mRNA is ushered into the cytoplasm, a necessary but perhaps unregulated event but one that still requires another coordinated set of protein complexes to ensure quality control, i.e., that nuclear processing was accurate. Some mRNAs are regulated by metabolic circumstances, e.g., iron concentration–specific nuclease complexes are involved in such mRNA turnover. Regulation resumes again in the cytoplasm. Complexes that include small RNAs and specific proteins control translation frequency and mRNA lifetime. The specificity of attack on some mRNAs is governed by base-pairing of an mRNA with a particular small RNA that is part of a cytoplasmic ribonucleoprotein complex. The base-pair complementarity is often in the 3'-untranslated sequences of the mRNA.

Much of this increased knowledge of mRNA production and function proceeded in the last three decades in a rather orderly fashion. Regulatory DNA-binding sites and cognate proteins were the main theme of the 1980s and early 1990s. Chromatin modification came center stage in the late 1990s and continues apace. Differential pre-mRNA processing was noted in the 1980s, but mechanisms of regulation at this important regulation point are much newer discoveries. Although the regulatory small RNAs were noticed early in the 1990s, learning of their impact on mRNA stability has been a feature mainly of the last decade. In addition to their cytoplasmic role, short RNAs can also participate in RNA processing (well documented in pre-rRNA processing) and in chromatin structure.

Research in the 1960s and early 1970s detected several classes of long, rapidly synthesized nuclear RNAs; some were precursors to rRNA and some to mRNA, but some were unclassified. New sequencing techniques have now identified specific long, nuclear, noncoding RNAs that are among the latest addition to the roster of defined RNA molecules that currently command attention.

Understanding the precise physical/biochemical role of some of the now identified proteins and RNAs engaged in regulation has also progressed remarkably.

The three-dimensional (3D) structures of dozens of regulatory proteins and many RNA molecules, including rRNAs and tRNAs (both on and off ribosomes), have been produced, and the understanding of their detailed roles in protein synthesis is quite advanced (Schmeing and Ramakrishnan 2009). Since the recognition in the early 1980s of the catalytic capacity of RNA, a detailed biochemical definition of the RNA/protein machines that add poly(A) and correctly steer splicing has been achieved, although not yet at the atomic level. In addition, some of the newest RNA categories and some of the proteins that mediate their effects on regulating already formed mRNAs have had their 3D structures solved. All of these areas of RNA biophysics/biochemistry are growing rapidly. They emphasize, perhaps as little else does, that after decades of darkness, RNA research is now in the spotlight.

This chapter ends by describing the explosive field of regulation of already formed mRNA by small RNAs and a brief discussion of the many (perhaps as many as 5000 in human cells) long RNAs that do not encode proteins but are synthesized by animal, plant, and yeast cells. At least in a limited number of well-studied instances, these different types of newly recognized RNA molecules are now strongly implicated in regulating the amount of functioning mRNA in any cell.

In earlier chapters, historical material was presented in some detail. The avalanche of new information since the early 1980s obviously precludes such attention to detail here. Perhaps all of this material is too new for "historical" treatment anyway. Specific descriptions of particular pioneering studies on RNA synthesis, processing, and regulatory functions are included, but emphasis is placed on providing references to reviews and to broad current conclusions.

CONSTANTLY IMPROVING TECHNOLOGY SINCE
1980 SPEEDS RESEARCH

It is a truism that the speed of advances in any field of science depends on relevant technological advances. The truism is especially clearly illustrated in the recent history of eukaryotic gene regulation. Ever more rapid DNA sequencing methods, both of whole genomes and of expressed RNA (copied into DNA), have produced for selected species (e.g., human, mouse, *Drosophila, Caenorhabditis,* and yeast) a total catalog of Pol II start sites, protein-coding sequences, intron–exon junctions, and poly(A) addition sites and have uncovered transcription units for an abundance of newly discovered noncoding RNAs.

Not only has nucleic acid technology galloped ahead, but protein identification has been enormously simplified by advances in mass spectroscopy (MS or "mass spec"). It is now trivial to identify single isolated proteins instantly by comparing their constituent peptides with whole-genome coding sequences—the *proteome*. In addition, sorting out mixtures of proteins in complexes has become possible. Analysis of protein mixtures, using secondary splitting of the initial peptide yield (so-called MS-MS, a double mass-spec run), often allows identification of all of the proteins in a complex.

This is only possible because of sophisticated computer matches of encoded peptides in sequence databases to peptides in the mixture of proteins in a complex. These 21st-century achievements of "global" analysis were not all available in the 1980s, but the techniques did become available piecemeal. Individual genes, their regulatory elements, and the proteins involved in regulation began to be described throughout the 1980s and 1990s.

As information regarding individual genes and proteins became increasingly available, it became imperative to test for their functionality. This was solved first and most completely in yeast, where transformation of selectable DNA constructs allowed the replacement of any gene by a mutant version. In mice, *Drosophila,* and plants, transgenic insertions (which, at first, were not necessarily at the correct chromosomal sites) allowed the "overexpression" of a gene or a mutant gene, thus providing clues to the organismal (or cellular) function of the gene product. In the late 1980s and early 1990s, techniques were developed for the precise replacement of individual genes in mice; these were first *knockouts* (removal of the gene and its product) and later *knockins* (replacement of a wild-type gene with a mutant or with a fused marker protein, detectable microscopically).

As described in the final section of this chapter, small RNAs (siRNAs [short-interfering RNAs, most often synthesized by machine]) can be introduced directly into cells, and genes encoding small RNAs (miRNAs [microRNAs]; shRNAs [short hairpin RNAs that are processed into miRNAs]) can be introduced into animals, to effect a decrease in a specific mRNA and its encoded protein. Thus, genetic and cell biological advances have kept pace with molecular advances and have afforded functional tests of newly recognized proteins and RNAs. Advancing knowledge and advancing technology surely go hand in hand. Which drives the other is a question for philosophers of science (Merton 1965, 1973).

REGULATING TRANSCRIPTIONAL INITIATION OF EUKARYOTIC PRE-mRNA

The ultimate details of regulation vary for each eukaryotic gene. Understanding these details obligatorily starts at transcriptional initiation. Locating the TSS and identifying the regulatory DNA sequences important for the transcriptional machinery to find the TSS is the first requisite for understanding the regulation of any gene.

Locating Promoters

In the absence of practical mutagenesis and selection—classical genetics approaches—to locate regulatory sites for transcription initiation, regulatory sites were found in eukaryotic DNA initially in animal cells by molecular genetics and biochemical experiments. Sequences near the TSSs, *promoter sequences*, were identified first in cell-free transcription systems on naked DNA and confirmed very soon after by introducing the same DNA (wild type or mutant) as part of a "reporter gene" into cultured animal cells. It was discovered in the 1970s that virus DNA (Graham and van der Eb 1973) and later recombinant viral or cell DNA (Wigler et al. 1977, 1979) could be taken up by mammalian cells and integrated into chromosomes. A bit later, this *transformation* (as it was first called) was accomplished in yeast (*Saccharomyces cerevisiae*) cells devoid of their cell walls (spheroplasts) (Hinnen et al. 1978).

Wild-type and mutant genomic sequences thought to govern transcription could be attached to coding sequences of a reporter protein to provide an in vivo test for promoter functionality. These introduced genes—now referred to as *transfected genes*—were integrated into the cell's chromosomes, although not at any particular site, and were associated with histones. Proteins encoded by the transfected DNA could be expressed, but often not in a regulated fashion. The transfected DNA was thought to be a reasonable initial guide to "in vivo" gene function.

The first promoter sequences identified in vertebrate cell DNAs by site-specific mutagenesis and in vitro transcription were located ~30 bp before (*upstream* of) the start sites of several different genes. These regions were rich in Ts and As and frequently had a TATA sequence located ~30 bp upstream of the TSS (Fig. 5-1) (for review, see Corden et al. 1980). Mutating the TATA sequences in these promoters blocked transcription in the cell-free initiation assay (originated by Weil et al. 1979; see Chapter 4, p. 213) and in transfected reporter genes. The first eukaryotic promoter sequences had therefore been identified. Transfecting yeast genes in the early 1980s also located promoters. *S. cerevisiae* DNA is AT rich (~62%), and tracts of AT-rich sequences were found before almost all promoter sites.

One of the protein complexes (discussed in detail later) that has been important in locating start sites contains the TATA-box-binding protein (TBP). All eukaryotes have this protein, and a TBP-like protein is found even in archaea (noneukaryotic single-celled organisms with transcriptional machinery similar to that of eukaryotes) (see Chapter 6). Furthermore, TBP (or a closely related protein) (see Reina and Hernandez 2007) is required for transcription by Pol III at its target promoters.

In vertebrate cells, first a few and then an increasing number of genes were found not to have a TATA box positioned upstream of the pre-mRNA start site. In the ensuing decades, as more sequences for the immediate upstream regions of transcribed loci ("genes") accumulated, the following

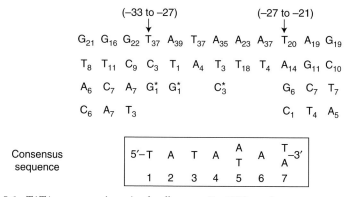

Figure 5-1. TATA sequence in animal cell genes. By 1980, perhaps as many as 50–60 sequences around the transcription start sites (TSSs; the "cap" sites) for vertebrate and invertebrate mRNAs were available. Comparison among 41 of these sequences revealed that many contained a conserved "TATA box" centered ~30 nucleotides upstream of the TSS. The subscript numbers indicate the frequency of a given nucleotide at the noted position upstream of the TSS. (Reprinted, with permission, from Corden et al. 1980 [©AAAS].)

picture of promoters emerged (Fig. 5-2). Jim Kadonaga, whose laboratory has been devoted to this problem, described several classes of promoters (Fig. 5-2A) (Juven-Gershon et al. 2008). Some (called *focused* promoters) cause transcription to begin at a precise site. In other regions, promoters do not direct Pol II to a specific single site but to a region that can be 50–100 nucleotides long; these are called *dispersed* promoters. One common sequence feature in vertebrate DNAs for dispersed promoters is CpG islands. These runs of repetitive CpG sequences lie upstream of many genes—particularly the "housekeeping" genes—that are transcribed in all cells.

In focused promoters, sequence motifs other than TATA are present in some but not all promoters. The focused promoters seem evolutionarily

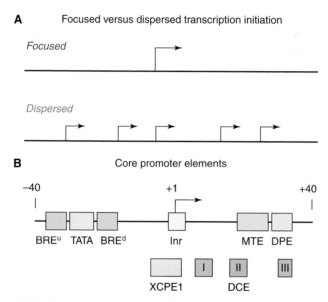

Figure 5-2. (*A*) In focused promoters, transcription initiates at a single site or in a cluster of sites in a narrow region of several nucleotides. Dispersed promoters are often found in CpG islands in vertebrates and show multiple weak start sites over a region of 50–100 nucleotides. (*B*) Conserved DNA recognition sequences of focused promoters in the vicinity (the "core") of various eukaryotic TSSs (+1) for Pol II. TATA-box genes often have poorly conserved upstream (BRE^u) and downstream (BRE^d) regions that bind the transcription factor for RNA Pol IIB (TFIIB), one of the general transcription factors that aligns Pol II on DNA. An initiator element (Inr) is found in some focused promoters with or without TATA element and contains an A residue, which is frequently the first transcribed nucleotide. For focused promoters lacking the TATA box, various sequences (remainder of boxed elements) have been proved through mutagenesis studies to bind proteins that assist transcriptional initiation of a limited number of genes. None of these elements appears to be conserved in dispersed promoters. (Redrawn, with permission, from Juven-Gershon et al. 2008 [©Elsevier].)

more ancient because single-celled organisms have mostly focused promoters. For example, yeast DNA, as noted, is very AT rich around TSSs, and most transcription begins at particular sites. Complete genome sequencing has now revealed that only about one-third of vertebrate pre-mRNA promoters have a focused TSS (Carninci et al. 2006).

Mutagenesis of many individual promoter elements has proved that promoters are required both in vitro and in vivo for getting transcription started—so-called *basal* transcription. However, the promoter alone is insufficient to direct regulated transcription.

Enhancers Allow Discovery of Transcriptional Activator Proteins

Regulatory sequences to which activating proteins bind were first uncovered by transfection experiments of reporter gene constructs into cultured mammalian cells. Early in the 1980s, it was discovered that high levels of transcription of a transfected DNA required sequences in addition to those of the promoter. Walter Schaffner and colleagues (Banerji et al. 1981) transplanted a segment of SV40 virus DNA that was known to be required for early steps in virus growth in front of a β-globin gene reporter. When transfected into HeLa cells, this construct expressed the β-globin sequences 200 times better than a construct without the SV40 sequences. In addition, the SV40 insert gave the same result regardless of the orientation ($5'-3'$ or $3'-5'$) and regardless of whether it was close to or as far as 1400 bp upstream of or 3000 bp downstream from the promoter region of the β-globin gene. Schaffner referred to these sequences as *enhancers*. Many laboratories quickly confirmed these results with other virus and cell DNA segments in many other cell types.

These results set off a frenzy of activity, especially in the hope of finding enhancers for genes in specialized cell types. However, constructs containing immediate upstream sequences from genes known to be expressed in a cell-specific manner (e.g., globin in reticulocytes and albumin in hepatocytes or heptoma cells) did not enhance expression in cells such as HeLa cells. The first specialized enhancer-driven transcription was accomplished with transfection of upstream sequences from chymotrypsin and insulin genes (which are expressed specifically in the pancreas); these sequences were introduced into DNA constructs and transfected into cancer cells derived from pancreas (Walker et al. 1983). Similarly, cells derived from liver cancer or lymphomas could transcribe DNA constructs containing enhancers from genes expressed only in liver (D'Onofrio et al. 1985) or in lymphocytes (Queen and Baltimore 1983). These results hinted that specialized cells would contain transcriptional activators not found in nonspecialized cells.

With recombinant DNA techniques, it was firmly established that enhancers can function close to or far from promoters and that the orientation of the promoter and enhancer are independent of each other. Enhancers were even found within introns and 3' to (after) the end of regions encoding mRNAs. The techniques also established the necessity of using specialized cells for genes that were naturally expressed in specific differentiated cell types (Gluzman and Shenk 1983). Among the important points established by these early transfection experiments was that the start sites in the experimentally introduced enhancer-reporter gene constructs were the same as those for transcription of the natural chromosomal genes.

Enhancers were obvious candidates as targets for site-specific DNA binding of regulatory proteins. Protein:DNA-binding assays were developed that, together with deletion and site-specific sequence mutagenesis, located the exact sequences (10–15 nucleotides long) responsible for activator binding and enhancer function. The function of enhancers, especially cell-specific enhancers, has remained an active research activity. The effects of enhancers in chromatic modifications is discussed later (see pp. 272–273) (Odom et al. 2007; Bulger and Groudine 2011).

TRANSCRIPTIONAL ACTIVATORS

In the mid 1980s, short, functional enhancer sequences were used to detect site-specific DNA-binding proteins; this led to the purification of these proteins and the subsequent cloning of their genes. These purified, presumably regulatory proteins somehow allowed RNA Pol II to locate and transcribe DNA in vitro and in vivo (e.g., during transfection of cells) and were designated *transcriptional activators*, or just *activators*.

There are more than 1000 transcriptional activator proteins encoded in the human genome. Even in the simplest well-studied eukaryotes (the yeasts *S. cerevisiae* and *Schizosaccharomyces pombe*) there are perhaps 200 activators. Only a fraction of the activators encoded in the human genome is present in each type of cell, and the restricted distribution of transcription factors is itself transcriptionally regulated (Xanthopoulos et al. 1991; St Johnston and Nüsslein-Volhard 1992).

Sequence comparisons and X-ray crystallography showing how activators contact DNA allow the divison of activating proteins into distinct families that show strong evolutionary conservation. It was originally thought that only a few structural classes of activators existed, and it remains true that some motifs have been repeatedly used. But by now, more than several dozen different DNA-binding motifs have been described. Likewise, no particular structural classes of proteins are devoted to limited biological purposes (e.g., cell growth or cell-specific gene expression). The enormous amount of information regarding transcriptional activators is collected at many different websites (e.g., TRANSFAC [see Wingender et al. 2000]; TrSDB [Hermoso et al. 2004]; DBD [Wilson et al. 2008]).

Activator Proteins Have Functional Domains

Activator proteins have several distinct functions, each of which is dependent on a different protein domain within the complete activator (Fig. 5-3). The different functional domains were identified by mutagenesis soon after the first activators were cloned. The order of domains differs among activators. All activators have a *DNA-binding domain* (DBD) (Giguere et al. 1986; Ma and Ptashne 1987). A separate section (sometimes several sections) of the activator protein acts as the *transcription activation domain* (TAD). Activators must interact with several different proteins/protein complexes in order to facilitate transcription. It is the TAD that allows these specific binding functions. Not infrequently, DNA-binding proteins lack a TAD but recruit a partner protein that has a TAD but lacks a DNA-binding domain (Fig. 5-3B). In such cases, the two necessary functions come in the form of two proteins. As an example, the Notch protein family (present in all animals) operates in this fashion (Weinmaster 2000). The Notch intracellular domain (NICD) is cleaved from a cell-surface receptor and enters the nucleus to bind DNA but requires a partner to stimulate transcription.

Some activators must receive signals from outside of the cell to function in transcriptional activation (Evans 1988; Darnell 1997). Such activators have *signal receptor domains*. When an activator receives an extracellular signal, the

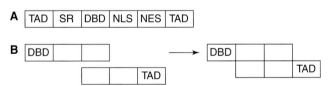

Figure 5-3. (*A*) Common domains in eukaryotic transcriptional activators include the transcription activation domain (TAD), signal receptor (SR), DNA-binding domain (DBD), nuclear localization signal (NLS), and nuclear export signal (NES). Among the hundreds of activators, there is no general rule for the order of domains. SR domains, which interpret information from outside of the cell, are most often phosphorylation sites that can either be activating or inhibitory. Some signaling molecules (e.g., steroids) can diffuse into cells and bind the SR domain. Methylation and acetylation can also influence a transcriptional activator. (*B*) Activators can come in two parts. Two different genes produce two proteins—one with a DBD and one with a TAD—that cooperate. The molecule with only a DBD can act as a repressor.

most common event, by far, is phosphorylation at a specific site in the SR domain. Phosphorylation activates some factors, whereas in other cases, it is inhibitory. Other enzymatic actions, such as regulated proteolysis, methylation, acetylation, or attachment of short proteins (e.g., ubiquitination, sumoylation), also occur to activators. These changes are not as frequent as phosphorylation but are now coming under intensive study. Finally, transcriptional activators must be imported into the nucleus (and in some cases, specifically exported from the nucleus). Relatively short amino acid stretches serve as nuclear localization signals (NLSs) or nuclear export signals (NESs) also exist in activators.

The great majority of eukaryotic site-specific DNA-binding proteins are activators, but some site-specific DNA-binding proteins act negatively on transcription—almost always by recruiting protein complexes that themselves are responsible for the negative action on transcription. Discussion of inhibitory or repressive proteins is presented below after positive regulation is described.

Examples of Transcription Activator Functions

Transcriptional activators have far too many molecular types and physiological roles in directing eukaryotic cell functions to describe them in detail. But a few examples are provided here to illustrate (1) how some important classes were first recognized in the 1980s and 1990s and (2) the central role that activators have in eukaryotic biology.

Transcriptional Activators and Response to Environmental Stress

All cells—bacteria and eukaryotes—possess *heat shock proteins*. New, as-yet-not-folded, proteins and partially unfolded proteins after a heat shock are protected from quick destruction by *unfolded protein complexes*, sentinels that help to destroy denatured proteins (Marcu et al. 2002; Schroder and Kauman 2005). These complexes contain heat shock proteins plus other components. Production of mRNAs for heat shock proteins in animals is increased by heat-activated DNA-binding transcriptional activators, *heat shock factors*, most completely studied in *Drosophila* cells, although all eukaryotes have such factors (Morimoto 1993). (The study of heat shock genes has also unearthed a phenomenon of "paused" RNA polymerase at start sites [Core and Lis 2008; Core et al. 2008].) This important phenomenon is discussed below in the section "Regulated Transcriptional Events after Initiation: Paused Polymerases and Completion of Pre-mRNA" on p. 274.

Animal cells are also sensitive to oxygen deprivation, *anoxia* or *hypoxia* (no or low oxygen tension), and survival depends on a transcriptional activator called hypoxia-induced factor (HIF). One subunit of this dimeric protein (HIF1α) is formed at all times but is normally rapidly destroyed. Hypoxia blocks the ongoing destruction and thus stabilizes HIFα. Accumulation of HIFα (with HIF1β) then occurs in the nucleus and takes part in activating genes that help cells to survive the anoxia (Wang and Semenza 1995).

DNA Damage and p53 Activation

Radiation damage and induced DNA repair mechanisms have been well studied for decades in all kinds of cells. In addition to DNA repair proteins, some transcriptional activators are also increased after DNA damage. p53, one of the most well studied of all mammalian transcriptional activators, has a critical role in mammalian DNA damage response. This protein was originally discovered because it is frequently inactivated in cancer (for review, see Vousden and Prives 2009). An important functional role of p53 is to assist in the decision of whether a damaged cell can be repaired or should enter the cell death pathway (see the section below, *C. elegans*: A Genetically Tractable Developmental Target on p. 311) and be removed from the body. p53 is normally synthesized and rapidly destroyed. Following DNA damage, however, p53 turnover ceases and a rise in total p53 protein occurs that then allows formation of transcriptionally active oligomers. Both the p53 and the HIF protein illustrate the important mechanism of activator control by protein stabilization.

Developmental and Cell-Specific Functions

Among the transcriptional activators most thoroughly studied for biological function are the products of genes required in early development. In the early 1980s, by producing and recognizing mutant *Drosophila* embryos unable to form the normal segments of the fly larva, Christiane (Jani) Nüsslein-Volhard and Eric Wieschaus discovered an array of genes required for normal segmental development (Nüsslein-Volhard and Wieschaus 1980). When these genes were cloned, nine of the 15 genes in the original set encoded DNA-binding transcription factors (Fig. 5-4). (Most were activators; some were repressors.) In addition to uncovering transcription factors, these early studies made historic inroads on the signaling pathways that cause production of or activation of the transcription factors themselves (St Johnston and Nüsslein-Volhard 1992). The expression of the developmental transcription factors and therefore the genes that they control is specifically distributed in different embryonic cells, in specific areas (stripes) of the developing embryo. These patterns of expression precisely divide groups of cells of the developing *Drosophila*

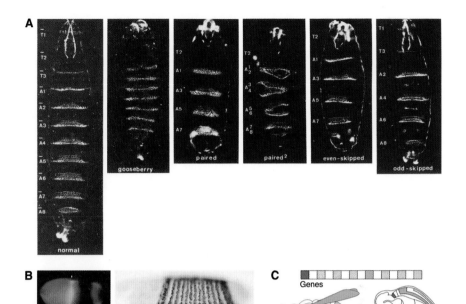

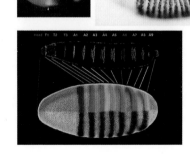

Figure 5-4. The Nüsslein-Volhard/Wieschaus screen illuminates *Drosophila* gene regulation. (*A*) In the first report in 1980 describing *Drosophila* mutants that could not form larval segments correctly, Nüsslein-Volhard and Wieschaus described 15 different mutant genes, nine of which turned out to encode site-specific DNA-binding transcription regulators. The other six were important in signal pathways that report to transcriptional regulators. The body defects of five mutants in activators are shown here. (The alternating denticle [transverse bands of light pinpoints] and naked bands [dark] in the cuticle of the embryo are clearly visible in light microscopy.) (*B*) After the genes for developmental mutants were cloned, antibodies against the various proteins were prepared. Developing larvae stained with these antibodies depict the time and location of the very precise expression in specific segments. The stripes show narrow bands (∼ four cells wide; some overlap) of expression that reveal molecular interactions between transcription factors and their target genes. (*Upper left*) In an early embryo, hunchback (red) and Krüppel (green) overlap (yellow); (*upper right*) Fushi tarazu (blue) and even-skipped (brown) are expressed in alternate cell stripes, marking the ∼14 segments that will eventually develop. (*Bottom*) Different colors show overlapping patterns of expression of various proteins that specify the ultimate developmental patterns of cells in different segments. (*C*) Individual *Hox* genes in both *Drosophila* and mammals are linearly arranged in chromosomal blocks. The linear genomic order corresponds to *Hox* gene expression, which is linear from anterior to posterior in the developing embryo. Remarkably, such linear order affecting body polarity from anterior to posterior is conserved in mammalian and *Drosophila* orthologs. (*A*, Reprinted, with permission, from Nüsslein-Volhard and Wieschaus 1980 [©Macmillan]; *B*, reprinted with permission, from Lodish et al. 2008; *C*, reprinted from Lodish et al. 2003.)

larval embryo into what later become body segments (Fig. 5-4) (St Johnston and Nüsslein-Volhard 1992).

A virtually universal central point of evolution was revealed by pursuing one of the families of genes uncovered by *Drosophila* genetics: the *Hox* genes. *Hox* is an abbreviation of *homeobox*, a DBD encoded by some homeotic genes that act as transcriptional activators during development. *Hox* genes exist in very similar form not only in *Drosophila* but in all segmented animals, including vertebrates. Remarkably enough, some of *Hox* genes serve the purpose of subdividing a developing mammalian embryo into segments just as *Drosophila Hox* genes do for larval segments (Fig. 5-4C). Thus, the "purpose" of *Hox* genes has been preserved through hundreds of millions of years of evolution. Perhaps the most astonishing aspect of this already amazing story is that many of the *Hox* genes are linearly arranged on a *Drosophila* chromosome, and they are "read out" (*transcribed*) to affect segments from front (head) to rear (see Fig. 5-14 on p. 270 for details). *And* the similar head-to-tail arrangement of *Hox* genes is maintained in blocks of corresponding mammalian genes in mammalian chromosomes (Fig. 5-4C). All of the developmentally important transcription factors and signaling pathways discovered in *Drosophila* have counterparts in all other animals, perhaps as convincing an argument as exists for Darwinian evolution.

Nüsslein-Volhard and Wieschaus shared the 1995 Nobel Prize with Edward B. Lewis, another distinguished *Drosophila* geneticist who preceded them in discovering the first genes that affect *Drosophila* segment differentiation.

Enhanceosomes

In mammals, many different transcriptional activators have been uncovered that are present mainly (but not exclusively) in particular types of cells. Liver cells, for example, have a half-dozen or more activators that are rarely present in any other cells (Xanthopoulos et al. 1991; Sheng et al. 2004). And liver cells are the sole or major source of dozens of proteins that are secreted into the serum to perform various tasks. Cell-specific expression of these liver-specific mRNAs (see Chapter 4, Fig. 4-42), however, is not brought about by any one of the liver-enriched transcriptional activators (Grayson et al. 1988). For a number of different genes, a varied assortment of the liver-enriched transcriptional activators, often binding to adjacent sites on DNA, act in concert during gene activation. It is clearly a *combinatorial* effort to activate the liver-specific genes (Lai and Darnell 1991).

This principle of several activators cooperating to achieve the maximal activation of a gene is probably the rule in mammalian cells. In one of the earliest and best-studied cases, Tom Maniatis and colleagues found that four different DNA-binding proteins bind within a 60-bp stretch of DNA to direct transcription of the interferon-β (*IFN*β) gene (Thanos and Maniatis 1995). An X-ray crystallographic picture of this cooperative arrangement shows, at the atomic level, the protein contacts among several of the cooperating activators (Fig. 5-5) (Panne et al. 2007).

Many clusters of DNA-binding sites for activators have been characterized in addition to those for IFN-β and liver-specific genes. These clusters of binding sites and their bound proteins are called *enhanceosomes*. Recent analyses show that clustered sites function in the regulation of many coordinated developmental events (Zinzen et al. 2009).

Transcriptional Activators in Stem Cells

In the late 1950s, John Gurdon showed that the nucleus of a frog egg (*Xenopus laevis*) could be removed and replaced by the nucleus of an embryonic cell and that the egg would almost always develop into a tadpole and sometimes into a frog (King and Briggs 1956; for review, see Gurdon and Melton 2008). This also worked with less efficiency with the nuclei of cells from differentiated tissue (intestinal or skin cells). These *nuclear transplantation experiments* settled the issue that at least all frog cells did, in fact, possess a full genetic complement and gave rise to the hope that vertebrate cells could be *reprogrammed*.

In recent years, reprogramming has actually been accomplished by introducing into growing undifferentiated human or mouse fibroblasts a set of gene expression constructs encoding four different transcription activators. These transfected cells revert to cells that resemble embryonic stem cells microscopically and in the panoply of expressed mRNAs (Takahashi and Yamanaka 2006). The mouse induced pluripotent stem (IPS) cells can actually contribute differentiated cells to a developing mouse. These remarkable cell biology experiments hold hope that human IPS cells can be induced to differentiate into a desired cell type that can be used in regenerative medicine (Yamanaka 2009). Whatever happens on the clinical front, these experiments dramatically emphasize the central role of a cascade of transcription factors in development.

Cell Signaling and Transcriptional Activators

In animals and plants, cell communication within an organism is critical for proper physiological responses. Much of this communication among cells (and organs) results in changing transcription patterns in a cell receiving a signal from outside (Evans 1988; Darnell 1997; Brivanlou and Darnell 2002). A brief discussion of two pervasively used classes of signaling pathways

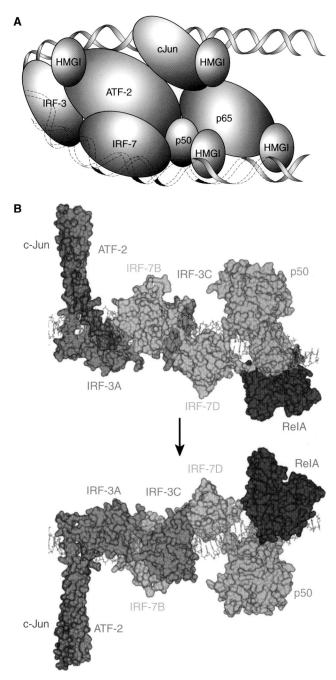

Figure 5-5. Enhanceosome on the interferon-β (*IFN*β) gene. (*A*) Drawing in 1995 of some of the proteins in the enhanceosome known to bind to an ~60-nucleotide region that controls rapid transcription of the IFN-β gene. (*B*) Space-filling model from X-ray crystallography of several of the regulatory proteins of the enhanceosome bound to a portion of IFN-β DNA. (*A*, Redrawn from Thanos and Maniatis 1995; *B*, reprinted, with permission, from Panne et al. 2007 [©Elsevier].)

important in invertebrates and vertebrates will illustrate the importance of signal pathways that result in transcriptional changes.

Steroid Receptor Superfamily

A large family of related proteins generically referred to as *nuclear receptors* (NRs) binds a wide range of signaling molecules that includes the steroids (e.g., cortisone and derivatives, estrogen, progesterone, testosterone) and other small molecules (e.g., thyroid hormone, retinoids, vitamin D). The signaling function of these circulating hormones depends on their capacity to dissolve in the lipid-rich cell-surface membrane and thus to enter the cell and bind a specific NR. In recent years, the importance of other NRs that bind metabolic products, particularly those of fat metabolism, has also been recognized (McKenna et al. 2009).

Some of the earliest discovered members of the steroid receptor super-family (estrogen, glucocorticoid, and testosterone receptors) bind DNA as homodimers. Others (vitamin D, retinoic acid, and thyroid receptors) form heterodimers with a common member of a related gene family called RXR (Naar et al. 1991; Umesono et al. 1991). These activators also emphasize the subtlety and specificity of homodimers and heterodimers for their specific binding sites in DNA (Fig. 5-6A). Homodimers (e.g., glucocorticoid and estrogen receptors) bind to inverted repeat sequences, as do many other homodimeric activators. In contrast, receptors that pair with RXR use two neighboring binding sites that are tandem repeats (Naar et al. 1991; Umesono et al. 1991).

In the absence of their activating hormone, some steroid receptors can act as repressors (by attracting protein complexes that act as corepressors) when bound to DNA. When the hormone for such NRs comes along, the corepressors are expelled and the NR is converted to activators that then recruit coactivators (Rosenfeld et al. 2006).

NRs were first detected in the 1960s by the binding of radioactively labeled estrogen to the estrogen receptor (for review, see Jensen and DeSombre 1972). On the basis of such specific hormone binding, the first NRs were purified (Payvar et al. 1983), and then steroid receptor genes were cloned in the mid 1980s (Hollenberg et al. 1985; for review, see Evans 1988). There are ~50 NR proteins in humans (Bookout et al. 2006; McKenna et al. 2009). The field of NRs has burgeoned into perhaps the most expansive of all of the studies on eukaryotic transcription factors, because of both the enormous importance of these ~50 proteins in vertebrate physiology and human disease (Fig. 5-6B) and their use in the ever-increasing numbers of basic science experiments that illuminate transcription factor and cofactor function.

A

GRE   5′ AGAACA(N)₃TGTTCT 3′
      3′ TCTTGT(N)₃ACAAGA 5′

ERE   5′ AGGTCA(N)₃TGACCT 3′
      3′ TCCAGT(N)₃ACTGGA 5′

VDRE 5′ AGGTCA(N)₃AGGTCA 3′
      3′ TCCAGT(N)₃TCCAGT 5′

TRE   5′ AGGTCA(N)₄AGGTCA 3′
      3′ TCCAGT(N)₄TCCAGT 5′

RARE 5′ AGGTCA(N)₅AGGTCA 3′
      3′ TCCAGT(N)₅TCCAGT 5′

B

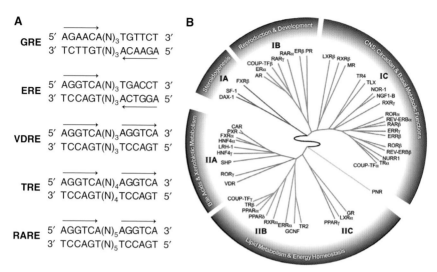

Figure 5-6. Nuclear receptor (NR) activity in mammalian physiology. (*A*) Composite of NR-binding sites (see Evans 1988; Umesono et al. 1991). The DNA-binding sites (or elements, E) for the glucocorticoid receptor (GRE) and the estrogen receptor (ERE) are inverted repeats, and these two receptors bind to GRE and ERE as homodimers. (Most homodimers do bind to inverted repeats.) The binding elements for vitamin D receptor (VDRE), thyroid receptor (TRE), and retinoic acid receptor (RARE) all require a monomer of the receptor to pair with the common retinoid X receptor (RXR). These heterodimers, composed of two different proteins, bind direct repeats. (*B*) A dendrogram (based on sequence similarity) divides the NRs according to contribution to various metabolic and physiological categories and proven biochemical functions in mice and human cells and tissues. (*B*, Reprinted, with permission, from Bookout et al. 2006 [©Elsevier].)

Transcription Factors That Interpret Signaling from Outside of the Cell

Circulating proteins or polypeptides that act as extracellular signaling molecules cannot enter cells but act as specific ligands for cell-surface receptors embedded within the membrane that send signals to transcription factors inside cells (Fig. 5-7). Many of these circulating extracellular ligands are small proteins (growth factors or cytokines ~100–150 amino acids long, or polypeptides ~25–40 amino acids long). Cell-surface receptor proteins have an extracellular ligand-binding domain and a largely hydrophobic domain 10–15 amino acids long that passes through the membrane (some longer receptors have several such domains that thread into and out of the membrane multiple times). These two domains are followed by an intracellular domain.

Following the arrival of an extracellular ligand, the intracellular domains of receptors convey a signal that eventually reaches a transcription factor (for review, see Levy and Darnell 2002; Schlessinger 2004; Papin et al. 2005). Some receptors are found on virtually all types of cells, and others are restricted to

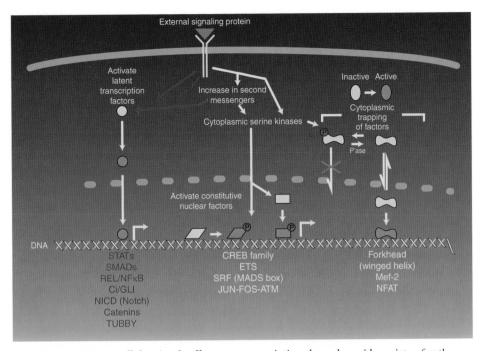

Figure 5-7. Extracellular signals affect gene transcription through a wide variety of pathways in all animal cells. (This figure shows mostly names of mammalian factors.) External signaling proteins bind in a highly specific fashion to their cognate cell-surface receptors. Depicted are several different intracellular events that subsequently affect transcription factors. The latent transcription factors (listed below the DNA helix) respond quickly in a pathway-specific fashion. STATs (signal transducers and activators of transcription) contain one special tyrosine that is phosphorylated (green arrow) at the cell surface by tyrosine kinases associated with the receptor; the STATs then accumulate in the nucleus as homodimers or heterodimers. SMADs undergo serine phosphorylation at a cell-surface receptor (green arrow) and accumulate in the nucleus as active heterodimers. The other latent transcription factors all require intracellular enzymatic events secondary to the cell-surface-receptor–ligand interaction that either promote or prevent an intracellular protease action. One member of the REL/NFκB family (RelA; p65) is released from a cytoplasmic inhibitor after the inhibitor is phosphorylated by a signal (short red arrows) from the cell surface that triggers proteolytic destruction of the inhibitor. The *cubitus interruptus* (*Drosophila* gene name; Ci/GLI) proteins require phosphorylation (red arrows) inside of the cell to be spared proteolytic cleavage, as do the catenins that depend on Wnt signaling at the cell surface. Activation of the Notch pathway involves proteolytic cleavage (purple arrows) to release the NICD, which has a TAD but no DBD. The NICD pairs with a protein (Let family) that binds DNA. In the least-well-studied pathway, Tubby is released from cell-surface binding by a phospholipase activated by phosphoinositide (a secondary messenger; blue arrow). All of the other transcription factors in the illustration are activated or inactivated by one or more of the hundreds of serine kinases or phosphatases triggered from a cell-surface-receptor–ligand interaction. For example, NFAT is converted to activity after a specific T-cell-receptor–ligand interaction leads to a rise in Ca^{2+} that triggers a release of a phosphatase (calcineurin; abbreviated "P'ase" in figure), the action of which frees NFAT to enter the nucleus. (*Legend continued on next page.*)

specific cell types. Each signaling complex at the cell surface signals a set of specific events inside of cells and affects a different set of genes. Currently, an enormous problem is trying to understand the multiple simultaneous signals that cells often receive.

Signal-Activated Latent Transcription Factors

There are a number of highly specific and well-understood extracellular signaling protein/receptor families, each of which activates a specific family of *latent transcription factors* (left pathways in Fig. 5-7). The intracellular biochemistry of these pathways differs, but all of the pathways increase transcription of a specific gene set after arrival of the now activated latent transcriptional activator from the cytoplasm. Some of these pathways are direct (unbranched) and result in quick delivery of a transcriptional activator to the nucleus.

There are many more cell-surface receptors that change the level of *second messengers* (e.g., cyclic AMP, diacylglycerol, phosphoinositides) controlling the activation/inactivation of the several hundred serine kinases and phosphatases in the cell. These metabolic changes also stimulate or inhibit specific transcription factors, but it is often difficult to define the precise pathway and which kinase has affected a particular transcription factor (Fig. 5-7) (Brivanlou and Darnell 2002).

After this abbreviated summary of some of the properties and physiological functions of a few types of the ~1000 transcriptional activators, we return to a necessary reality: Despite the central importance of activators in starting the chain of events that produce mRNA, they cannot by themselves activate maximal levels of transcription on in vitro templates or especially within cells. Further assistance is required.

COACTIVATORS AND EUKARYOTIC TRANSCRIPTIONAL INITIATION

How do regulatory proteins and the transcription machinery gain access to TSSs in chromatin to produce mRNA? The initial struggle to answer this

---

Figure 5-7. (*Continued*) The forkhead or winged-helix proteins (now known as Fox proteins) also enter the nucleus after dephosphorylation. Also depicted are a few of the dozens of other nuclear activators (at the center of the diagram, white) that are controlled by a required activating serine phosphorylation or inhibited by a serine phosphorylation, which must be removed, to allow release of the factor. It is in general the case that many different serine kinases/phosphatases can effect any particular example of the required change. (Redrawn from Brivanlou and Darnell 2002.)

question uncovered the existence of multiprotein complexes termed *coactivators* (for review, see Roeder 2005; Li et al. 2007) that cooperate with GTFs, activators, and Pol II on naked DNA during in vitro transcription. The first coactivators, TFIID (transcription factor for RNA Pol IID) and the Mediator complex, which contain between 12 and ∼30 proteins, were discovered because they interact directly with site-specific DNA-binding activators. Later with chromatin-covered templates, additional chromatin-modifying complexes generally containing from several to a dozen or so proteins were uncovered. These directly affect histones after binding to TFIID or Mediator complexes (or directly to activators) to enzymatically modify or physically relocate histones.

TFIID and TAFs: Bridging Coactivators

In the original separation by the Robert Roeder laboratory of factors required for in vitro Pol II initiation (Chapter 4, Fig. 4-39) (Matsui et al. 1980), one fraction was designated TFIID. When Nakajima et al. (1988) partially purified this factor, it was found to be a large protein complex that protected a stretch of ∼75 bp beginning upstream of and overlapping the major late adenovirus promoter. No single purified site-specific DNA-binding activator protein studied at this time protected more than ∼15–20 nucleotides at most. Roeder and colleagues began identifying individual proteins in TFIID, and Robert Tjian and colleagues joined in and purified the ∼10 proteins in the *Drosophila* TFIID complex (Fig. 5-8) (for review, see Gill and Tjian 1992; Takada et al. 1992).

The DNA templates originally used to detect TFIID all had required TATA boxes. One of the early proteins identified in the TFIID complex contacted the TATA box and was named TBP. Because the TFIID complex contained the TBP, the proteins of the complex were called TAFs (*TBP-associated factors*). By the mid 1990s, the proteins in *Drosophila* and human TFIID were also identified in yeast. By the early 1990s, it appeared that all eukaryotes had orthologs to virtually every one of the TFIID proteins (Fig. 5-8). The whole complex was the first identified *coactivator*, the designation for a protein complex that assisted the DNA site-specific binding activators to locate and bind to a TSS and position Pol II to start transcription correctly (Burley and Roeder 1996).

When knockouts of the genes for the TAF proteins were introduced into yeast or mice, variable results were found for the expression of different genes. For example, in yeast, many genes, but not all, required the individual TAF proteins for expression. In mice, knocking out most of the TAF genes individually proved to be lethal, but at different stages of development. It seemed that

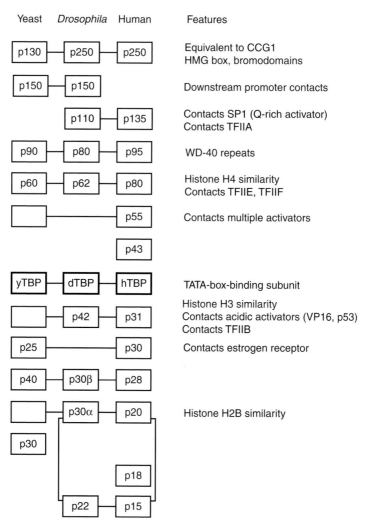

Figure 5-8. Proteins (TBP-associated factors, or TAFs) of the TFIID coactivator complex. Listed are the genes encoding each of the proteins of the TFIID complex from three eukaryotes; those connected by horizontal lines are orthologs, emphasizing that this complex is universally present in eukaryotes. "Features" column provides the 1996 version of how some of the proteins in the complex function. As research continued, it became clear that some differences in the number of proteins included in this complex occur in different animal cell types (Liu et al. 2009). TFIID was the first recognized coactivator. (Redrawn, with permission, from Burley and Roeder 1996 [©Annual Reviews].)

not all TAFs were required for transcription all of the time, but each of the proteins was important (Albright and Tjian 2000; Liu et al. 2009).

Mediator: Coactivator or Master GTF?

An independent stream of research on transcriptional initiation in eukaryotes came initially from the combination of genetics and biochemistry in yeast and resulted in the discovery of a second major coactivator that is universally important in Pol II initiation.

In 1987, Roger Kornberg and colleagues succeeded after lengthy trials in obtaining nuclear extracts from baker's yeast, *S. cerevisiae*, that would correctly start at known Pol II start sites (Lue and Kornberg 1987). Three yeast RNA polymerases—Pol I, Pol II, and Pol III—had already been purified. These had multiple subunits, the sequences of which closely resembled some sub-units of the three already known mammalian polymerases. Soon, yeast proteins were found that were very similar to mammalian GTF proteins. The yeast GTFs performed similarly in yeast Pol II transcription initiation to the earlier described functions of mammalian GTFs (Li et al. 1994).

Yeast cells were obviously a good general model for eukaryotic tran-scriptional functions. Continuing with cell-free transcription, Kornberg and colleagues then identified a large coactivator complex required for Pol II ini-tiation (Flanagan et al. 1991). When the complex was purified (Kim et al. 1994), it did not contain the polymerase, the previously recognized GTFs, or the TBP and associated TAFs (TFIID). But the complex was definitely required for efficient cell-free transcription.

Earlier genetic experiments with yeast in Richard Young's laboratory (Nonet and Young 1989; for review, see Hengartner et al. 1995) had identified a number of genes that affected transcription of mRNA, but the proteins encoded by these genes had not been purified or functionally characterized. When the purification in the Kornberg laboratory of the new yeast coactivator complex was complete, several of the constituent proteins were found to be encoded by the previously genetically identified genes required for mRNA synthesis (Kim et al. 1994). The new coactivator complex was christened *Mediator*.

Several laboratories using cell-free transcription with mammalian cell extracts later found large complexes, different from TFIID, that were required for steroid-receptor-driven transcription (Fondell et al. 1996, 1999; Rachez et al. 1999). When the proteins of the large mammalian coactivator complexes were identified, they were found to be extremely similar to the yeast Mediator proteins (for review, see Malik and Roeder 2000; Bourbon et al. 2004). It is now firmly established that every eukaryotic cell has a Mediator complex that increases in vitro transcription at all of the TSSs tested (Takagi and Kornberg 2006). Closely related proteins encoded in different genomes bring the total number of proteins found in different Mediator complexes to ~35. Small variations in the protein content of the Mediator in different cells make it likely that different activators regulating different genes will interact with different components in the Mediator complexes (Ebmeier and Taatjes 2010).

Although no definitive atomic structure of either the entire TFIID or the Mediator is yet available, computational sorting of images of single-particle electron micrographs show that the yeast Mediator changes conformation when bound to Pol II. These low-resolution structures also show that the human Mediator complex has some physical features that are similar to those of the yeast Mediator (Fig. 5-9) (Cai et al. 2009).

These two large complexes, the TFIID and the Mediator, form a *bridge* between transcriptional activators and the transcription machinery to promote pre-mRNA chain initiation, and they (one or the other or both) are very likely to be involved in every regulated pre-mRNA initiation event in eukaryotic cells. In fact, it has been suggested that because the Mediator complex binds directly to Pol II, it might be considered an obligatory factor—a master GTF—for every Pol II initiation event.

Crystallography of Transcriptional Initiation Proteins: TBP–DNA–TFIIB Structure

Throughout the late 1980s and 1990s, continuously improving techniques for gathering and computer processing X-ray-diffraction data occurred in parallel with the identification of the many proteins of the transcriptional machinery. Not surprisingly, the GTFs, the first group of proteins recognized to assist Pol II initiation of transcription, were studied crystallographically quite early. By the mid 1990s, all of the GTFs had been identified in yeast and human cells and their genes cloned (Table 5-2). At least 30 different proteins were involved in the GTFs. By this time, it had been proven that TBP contacted the TATA box but by itself bound weakly. However, TBP together with transcription factor TFIIB (one of the original GTFs that is not part of TFIID; both are named from the original separations; see Chapter 4, Fig. 4-39) could position purified polymerase to start transcription correctly in vitro.

In the early 1990s, the crystallographer Stephen Burley joined with the Roeder group to determine the structure of TBP and then TBP plus TFIIB bound to DNA. TBP itself was determined to be a bilobed structure reminiscent of a saddle complete with stirrups (Nikolov et al. 1992). The space on the underside of the saddle was exactly the size needed to accommodate DNA

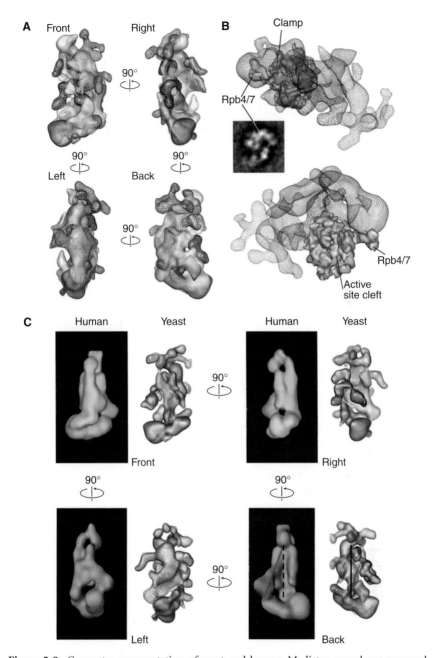

Figure 5-9. Computer representation of yeast and human Mediator complexes prepared from electron microscope images. (*A,B*) Colored images derived from computational conformational analysis of more than 1000 images of stained yeast Mediator particles both without (*A*) and with (*B*) RNA polymerase. RNA polymerase causes a conformational bend in Mediator; two polymerase subunits (Rpb4/7) have been proven to make contact with Mediator. (*C*) Images of the human Mediator can be compared with those of the yeast Mediator. (Reprinted, with permission, from Cai et al. 2009 [©Elsevier]; see also Taatjes et al. 2004.)

Table 5-2. Consensus of general transcription factors (GTFs) as of 1996

Factor		Number of subunits	MW (kDa)	Function
TFIID	TBP	1	38	Core promoter recognition (TATA); TFIIB recruitment
	TAFs	12	15–250	Core promoter recognition (non-TATA elements); positive and negative regulatory functions
TFIIA		3	12, 19, 35	Stabilization of TAP binding; stabilization of TAF-DNA interactions; antirepression functions
TFIIB		1	35	RNA Pol II–TFIIF recruitment; start-site selection by RNA Pol II
TFIIF		2	30, 74	Promoter targeting of Pol II; destabilization of nonspecific RNA Pol II–DNA interactions
RNA Pol II		12	10–220	Catalytic functions in RNA synthesis; recruitment of TFIIE
TFIIE		2	34, 57	TFIIH recruitment; modulation of TFIIH helicase, ATPase and kinase activities; direct enhancement of promoter melting (?)
TFIIH		9	35–89	Promoter melting using helicase activity; promoter clearance (?) by CTD kinase activity

The MW data are for human proteins. Exclusive of the Pol II 12 subunits, ∼30 proteins are involved in starting transcription at promoters containing a TATA box. (Reproduced, with permission, from Roeder 1996 [©Elsevier].)

(Fig. 5-10). When the TBP was cocrystallized with a TATA element (that of the adenovirus major late promoter), a startling structure was observed (Kim et al. 1993). The TBP saddle was filled with DNA, but the DNA was bent ∼110° as it passed under the saddle. The Burley-Roeder collaboration continued by solving a ternary complex of TBP, DNA, and TFIIB (Fig. 5-10D) (Nikolov et al. 1995). The downstream face of the ternary complex is formed by TFIIB where it is in a position to direct an engaged Pol II downstream from the appropriate start site.

Structure of Yeast Pol II and Mechanics of RNA Chain Elongation

As noted in the earlier discussion of the Mediator, Roger Kornberg and colleagues had achieved success in developing an in vitro system with yeast extracts (Lue and Kornberg 1987). Although there are 12 conserved subunits in yeast and mammalian Pol II, there is an easily purified 10-subunit Pol II complex in yeast that can transcribe DNA (Fig. 5-11). This complex was elected as a possible goal for crystallization and structure analysis. After 5

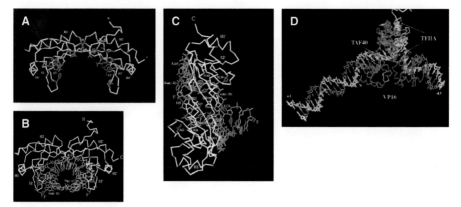

Figure 5-10. Structures of TBP, TBP bound to DNA, and TFIIB bound to the TBP–DNA complex. (*A*) The TBP structure is a symmetrical saddle-shaped structure with stirrups (the two loops projecting under the main structure). The molecule has an internal repeat that is responsible for the symmetry of the two halves. (*B*) TBP crystallized with a 40-bp sequence from adenovirus major late promoter that contains the TATA sequence. The DNA is seen to fill the underside of the saddle and engaged by the stirrups. (*C*) View of the DNA–TBP cocrystal from the top of the saddle, showing the extreme (~110°) bend introduced in the bound DNA (shown best by the orange strand). (*D*) Ternary complex of TBP (blue) with DNA ([green] coding strand, [yellow] noncoding strand) and the TFIIB (magenta and orange; an internal repeat in TFIIB gives rise to two globular halves of the molecule connected by a linker). Binding sites for various proteins (TAF40, TFIIA, VP-16), as determined by in vitro biochemistry, are indicated on TBP and TFIIB. (*A*, Reprinted, with permission, from Nikolov et al. 1992 [©Macmillan]; *B,C,* reprinted, with permission, from Kim et al. 1993 [©Macmillan]; *D,* reprinted, with permission, from Nikolov et al. 1995 [©Macmillan].)

to 6 years of strenuous effort and acute experimental insights (recounted by Kornberg in his Nobel Prize lecture [Kornberg 2007]), appropriate crystals were obtained and the enormous Pol II structure was solved (Cramer et al. 2001). The horseshoe-shaped Pol II complex has a cavity into which DNA can fit in order to be transcribed (Fig. 5-11A). This remarkable achievement was immediately followed up by success in obtaining crystals of a complex containing Pol II in the act of transcribing a constructed compound RNA:DNA oligonucleotide representing a region of DNA undergoing RNA transcription (Fig. 5-11C). When this structure was solved (Gnatt et al. 2001), it revealed a protein clamp that holds the template DNA in the cavity in the Pol II, confirming the active site for nucleotide addition. Ultimately, two new nucleotide binding sites were discovered: an entry site and a site where base-pair matching is determined to be correct. After the correct ribonucleotide triphosphate has entered the active site, a flexible "trigger loop" closes to ensure specific contacts and the addition of the correct nucleotide to the growing RNA chain (Wang et al. 2006).

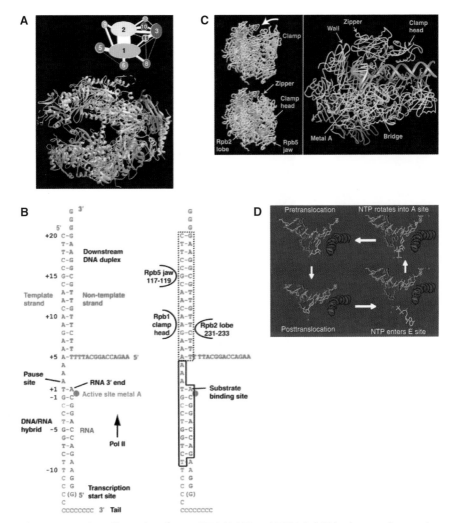

Figure 5-11. Crystallography of yeast RNA Pol II and RNA Pol II in the act of transcription. (A) Crystal structure of the 10-subunit yeast Pol II shows a cavity of the right dimensions to accommodate DNA. (*Upper right*) Coloring scheme shows the positions of some of the proteins of the polymerase. (B) Hybrid DNA–RNA oligonucleotide construct used to cocrystallize with Pol II shows (*left*) the site in which transcription occurs and (*right*) contact sites for specific Pol II proteins. (*C, left*) Template DNA (blue) entering a funnel that leads to the active site of RNA Pol II. (The coding strand is in green, and RNA is in red.) (*C, Right*) Template DNA held in place by a clamp and a bridging protein helix (green) extending between two arms of Pol II and Mg^{++} (metal A) at the active site. (D) Cycle of addition of one nucleotide. (*Top left*) Most recently added nucleotide (yellow) (pretranslocation). (*Bottom right*) Movement and rotation of the polymerase (part of which is shown in green) allow a correctly paired incoming NTP (yellow) to enter the E site of the polymerase and rotate to the A site to be joined to the growing chain. (*A*, Reprinted, with permission, from Cramer et al. 2001 [©AAAS]; *B,C*, reprinted, with permission, from Gnatt et al. 2001 [©AAAS]; *D*, reprinted, with permission, from Kornberg 2007 [©National Academy of Sciences].)

Recent crystallographic advances have brought an additional depth of understanding to the mechanics of start-site recognition by TBP and TFIIB and exquisite details of accurate chain elongation by Pol II (for review, see Cramer 2010).

HISTONES AND CHROMATIN FORMATION

In vitro transcription systems with purified proteins were central to all of the progress on transcription discussed so far. However, the DNA in eukaryotic cells is bound in *chromatin* with an equal weight of protein, largely histones. Beginning in 1974 (Kornberg 1974; Kornberg and Thomas 1974; Olins and Olins 1974), the regular distribution of histones on DNA was recognized, and by the 1990s, the structure of the histone octomer or *nucleosome* ($H2A_2$, $H2B_2$, $H3_2$, $H4_2$) around which DNA is wound had been solved (Fig. 5-12).

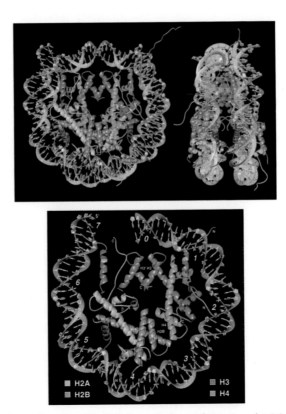

Figure 5-12. The nucleosome. (*Top*) Color-coded crystal structure (at 2.8 Å) of the octomeric histone core (a nucleosome) plus 147 bp of double-stranded DNA (*top* and *side* views). (*Bottom*) Top half of histone core, clearly showing the H3:H4 and H2A:H2B dimers. (Reprinted, with permission, from Luger et al. 1997 [©Macmillan].)

The Nucleosome, Histone Tails, and the Solenoid

The refined crystal structure of nucleosomes (Luger et al. 1997; Luger and Richmond 1998) shows an H3:H4 tetramer and an H2A:H2B tetramer that each occupies about half of the disc-shaped nucleosome (Fig. 5-12). Most important, the amino-terminal tails (30–40 amino acids), although not part of the crystal structure, project from the surface of the nucleosome and were not included in the crystal structure. These tails can be chemically modified (Fig. 5-13) and are discussed below.

Between each nucleosome in compacted chromatin is a single protein, histone H1 (or close relatives called by various names; e.g., H5 in birds), that is unrelated to the major histones and associates with its neighboring nucleosome during compaction. Chromatin containing H1 and nucleosomes at physiologic salt concentrations tends to condense into a 30-nm fiber referred to as a *solenoid*.

Nucleosomes Can Demonstrably Block Transcription

For decades, the reigning dogma was that compacted chromatin blocked transcription. It was widely and correctly anticipated that learning how eukaryotic RNA polymerase and accessory proteins circumvented this blockage would be required to understand regulated eukaryotic transcription.

A definitive negative effect of nucleosomes on transcription was clearly shown in vitro and in vivo in the 1980s. Robert Roeder and colleagues showed that in vitro transcription was blocked by covering the template with histones, but if a transcriptional activator was allowed to bind to DNA before the histones, nuclear extracts plus Pol II could still transcribe the chromatin template

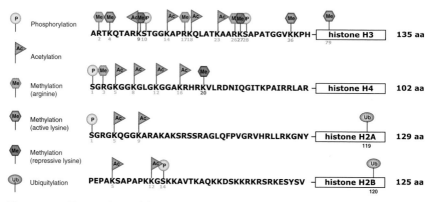

Figure 5-13. Enzymatic modifications of histone tails. (Reprinted, with permission, from Allis et al. 2007 [©CSHLP].)

(Workman and Roeder 1987), albeit more slowly than naked DNA. This result illustrated at once that chromatin can block transcription, but if an activator is bound to its site, the total proteins in a nuclear extract can furnish the additional proteins required at transcription start sites, and at least limited transcription of the chromatin-covered template can occur.

Michael Grunstein and colleagues (Han and Grunstein 1988) showed in yeast cells that removal of histones released the inhibition of transcription. They used an in vivo yeast system in which two of the four types of core histones were supplied from a plasmid that also carried selectable genes. Selecting against the plasmid stopped formation of these two types of histones and therefore of further nucleosome formation. "Unscheduled transcription" (i.e., from genes that were normally silent) resulted. Thus, it is surely true that histones can exert a negative transcriptional influence, both in cells and in vitro.

Early Discoveries of Histone Modification: Acetylation Is Linked to Transcription

Vincent Allfrey and others had detected chemical additions to histones—acetyl groups, methyl groups, and phosphates—in the 1960s and 1970s (Allfrey 1977). Already in 1972, Edman degradation from the amino terminus of histones had proven acetylation of the ε-amino groups of lysines near the amino terminus of both H3 and H4 (Candido and Dixon 1972).

Furthermore, a direct connection between histone acetylation and elevated transcription was established in 1988–1989. Tim Hebbes, Chris Turner, Alan Thorne, and Colyn Crane-Robinson described an antibody specific for acetylated lysine residues in histone H4 (Hebbes et al. 1989). In contrast to mammals, bird erythrocytes have nuclei that synthesize globin mRNA. Hebbes and colleagues treated chromatin from chicken erythrocytes with DNases to release individual nucleosomes and precipitated the nucleosomes with the antibody against acetylated H4. They recovered the antibody-bound nucleosomes, and globin DNA was enriched 30-fold compared to the DNA in the total nucleosome population (Hebbes et al. 1988). This clearly linked transcriptionally active chromatin to histone acetylation.

Yeast histones also were found in the late 1980s to have histone modifications. Michael Grunstein used genetic tricks (see above) to attack the problem of histone modification and chromatin transcription (Kim et al. 1988; Durrin et al. 1991). The H4 lysines at positions 5, 8, 12, and 16 were changed to arginine (still a basic amino acid but lacking the ε-NH_2 group of lysine). Acetylation of some lysine residues in the H4 amino terminus had already been documented. The Lys→Arg mutations caused a failure of induction of normal levels of specific mRNAs, implying a necessary positive role for histone acetylation in transcriptional control of specific genes.

DISCOVERING SPECIFIC HISTONE-MODIFYING ENZYMES AND COMPLEXES

Entering the 1990s, how transcription could initiate and proceed on chromatin either in cells or in vitro still remained an incompletely solved puzzle. A collection of coactivators—*chromatin-modifying* complexes—have been described since then (Li et al. 2007). These complexes interact with chromatin and perform two different tasks. Some complexes have enzymes that chemically modify histones in association with either increased or decreased transcription. Of equal importance, other ATP-dependent complexes can move (evict or slide) histones around. A discussion of how these capacities are integrated with transcriptional control follows.

By the mid 1990s, it was clear both that histones were enzymatically modified and that somehow transcription was not permanently blocked by nucleosomes. The stage was set for discovering the specifics of histone modification and any association of specific changes with regulated transcription of chromatin.

The breakthrough to the new and intense era of histone-modification studies came in 1996. David Allis and colleagues described the first specific enzyme that had the ability to add acetyl groups to lysine residues in histone tails and showed that this activity was required for mRNA synthesis. The first histone acetyltransferase (transacetylase) purified was from a protozoan—a single-cell eukaryote, *Tetrahymena thermophila* (Brownell and Allis 1996; Brownell et al. 1996). When the protein sequence of the *Tetrahymena* acetyltransferase was established, the acetyltransferase domain could be picked out by comparison with other acetyltransferase enzymes that were not specific for histones. Allis then compared the *Tetrahymena* histone transacetylase to proteins encoded in the genome of *S. cerevisiae*. The *Tetrahymena* transacetylase sequence closely resembled the sequence of part of a yeast protein called GCN5, whose enzymatic role was unsuspected. This sequence match was important because mutants in *GCN5* had already been established to block mRNA synthesis (Nonet and Young 1989). Some of these mutations in *GCN5* were later shown to prevent GCN5 acetyltransferase activity. Allis and colleagues switched to the genetically well-studied yeast system and showed that GCN5 could add acetyl groups to lysine in histone tails (Kuo et al. 1996). Finally, a link had been forged among specific histone modifications by specific enzymes that were required in the activation of transcription.

An intense period of research in this area has followed. (See *Epigenetics*, edited by C. David Allis, Thomas Jenuwein, and Marie-Laure Caparros [2007], a collection of essays covering a very wide range of topics on specific histone modifications and the enzymes involved. A list of histone modifications and references is in Appendix 2 on pp. 479–490.) Many enzymes that add and others that remove acetyl, methyl, and phosphate groups on specific sites on histone tails have now been discovered in all types of eukaryotic organisms. There are also enzymes (ligases) that attach short proteins such as ubiquitin and SUMO, a ubiquitin-related protein, to residues in H2A and H2B within the nucleosome cores. Mutations introduced into the genes of many of these enzymes cause difficulties that range from subtle to lethal. Thus, there is little doubt that many (perhaps *all*, if we knew more) of these enzymes perform specific critical tasks in the regulation of transcription of specific genes or some other important function of chromatin.

Virtually all of the enzymatic action on histones is performed not by single enzymes but by enzymes that are part of large protein complexes, dozens of which have been purified from various eukaryotic cells and analyzed as to exact protein content and biochemical roles. Details of interactions between various transcription regulatory protein complexes and modified histones have been described, and more are being uncovered through crystallographic studies of protein complexes (Ruthenburg et al. 2007; Taverna et al. 2007). Acetyl groups are recognized by *bromo* domains in various proteins and methyl groups by one of three domains (*chromo, tudor,* and *PHD*).

Histones and histone–DNA interactions are definitely important in many other aspects of DNA function in addition to transcription (repair, recombination, and replication). However, this discussion is limited to a few specific acetylations and methylations that seem clearly connected to transcription.

TECHNICAL NOTE: CHROMATIN IMMUNOPRECIPITATION (CHIP) ASSAYS

Two techniques have been chiefly responsible for the rapid progress in studying chromatin modifications. Mass spectrometric examination of histones has identified the amino acid sites and the chemical nature of amino acid modification of a library of such modifications (Fig. 5-13).

From the biochemical functional point of view, a workable technique for *ch*romatin *i*mmuno*p*recipitation—ChIP—has been invaluable not only for studies of modified histone distribution on chromosomal DNA but also for locating interaction sites for any DNA-binding proteins. Antibodies specific for many different histone modifications and for dozens of other nuclear proteins have been produced and used in the ChIP technique, which locates any protein associated with specific regions of DNA. Cultured *Drosophila* cells were first used (Solomon et al. 1988; Orando and Paro 1993) to develop a workable protocol. A covalent cross-link between the DNA and a bound protein is created by formaldehyde treatment of the *living* cell. Cell and chromatin breakage (usually by sonication) followed by precipation with epitope-specific antibody (e.g., an acetylated histone residue or an antigenic site on a transcription factor) allows collection of all DNA sites bound to the precipitated protein. Amplification of DNA sequences in the precipitate identifies those sequences (with a resolution of several hundred bases) bound to a specific protein in and around any specific gene(s) at the time of formaldehyde treatment. The analysis can be performed for an individual series of interest or by high-throughput sequencing of DNA (ChIP-Seq); the precipitated protein can be located at potentially all sites in the genome.

An extension of the ChIP technique locates specific protein–RNA interactions. Here, the most reliable technique involves cross-linking RNA with bound proteins by UV irradiation (cross-link immunoprecipitation, or CLIP), sonication, and epitope-specific antibody precipitation of the cross-linked RNA–protein. Conversion of precipitated RNA to DNA is then followed by high-throughput DNA sequencing (Licatalosi et al. 2008). (Note that the conversion of expressed RNA sequences to DNA followed by sufficient high-throughput sequencing identifies all expressed sequences in a cell.)

One caveat to readers not working with these techniques is appropriate. Antibody precipitation is seldom quantitative, and antibodies against different epitopes have different affinities, making quantitative comparisons of results with different antibodies unreliable. Furthermore, precipitated, cross-linked DNA (or RNA) is greatly amplified before sequencing. ChIP results are reliable for general conclusions regarding the distribution of particular proteins bound to particular nucleic acid sequences, *but* the results are not quantitative.

Furthermore, it is often desirable to know, but impossible to state accurately, how many specific genes in how many specific cells in any particular sample show a particular protein:DNA interaction or what fraction of a particular protein (to which an antibody is directed) has undergone a particular modification (e.g., on histones). The technique, although invaluable, does not answer these questions quantitatively.

HISTONE ACETYLATION REQUIRED FOR ACTIVE TRANSCRIPTION

Studies on histone modifications first focused on acetylation of lysine residues (Allis et al. 2007). For example, at least several different lysine residues in histones H3 and H4 can be acetylated, and others occur on histones H2A and H2B (Fig. 5-13).

Extensive studies both in vitro with purified histones used to construct chromatin (An and Roeder 2004) and in vivo with ChIP experiments have confirmed that histone acetylation (particularly of H3K9 and H3K14) occurs on histones near promoters during transcription of all genes examined. This acetylation is coincident with the arrival of Pol II and associated activators and coactivators. Acetylation occurs first around TSSs but most often spreads both upstream of and downstream from TSSs. (For an example, see Fig. 5-14A,B.) Overexpression of histone deacetylases reduces such acetylation and can block gene expression.

It seems to be universally agreed that histone acetylation is connected to active transcription *and* is the result of complexes containing acetyltransferases recruited by transcriptional activators and coactivators (Kouzarides and Berger 2006). It is widely (and reasonably) assumed that the negative charge on the acetyl group causes the chromatin to be decondensed, thus facilitating transcription. In contrast, hypoacetylation of histones enforced by histone deacetylases is the general rule in chromosome regions where transcription is not occurring. As discussed later, histone deacetylases are most often part of active repression complexes.

Histone-Remodeling Complexes and Nucleosome-Free TSSs

In addition to histone acetylation complexes, activation of transcription is associated with ATP-dependent histone-remodeling complexes. These complexes, characterized by the different ATPases that they contain, are extremely numerous—there are perhaps $\sim$25,000 complexes per human cell (Saha et al. 2006). In both yeast and mammalian cell-free systems, the ATP-dependent remodeling complexes are recruited by the Mediator complex and/or by transcriptional activators that then more efficiently recruit the transcriptional machinery. Chromatin-remodeling complexes can remove ("evict") histones from TSSs and can also cause histone displacement linearly along the DNA.

It is now agreed from whole-genome ChIP analyses of yeast, *Drosophila*, and mammalian chromatin that DNA at TSSs is often not bound by histones; these are nucleosome-free regions (NFRs). This is thought to be a consequence both of activity of the remodeling complexes and possibly of the DNA sequence characteristics around many TSSs to which histones bind with less efficiency (Cairns 2009; Milani et al. 2009).

Between the activity of acetylation complexes and the very large number of individual remodeling complexes, it is clear that protein-binding sites in DNA should not be considered permanently buried in chromatin. Furthermore,

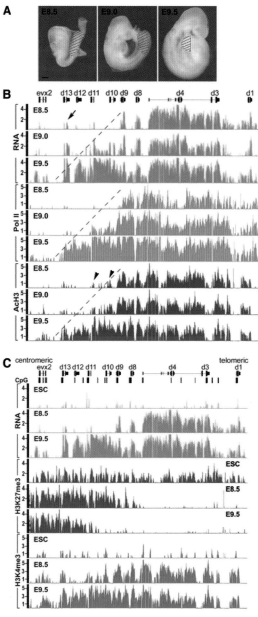

Figure 5-14. Association of acetylation and methylation of histone H3 with transcription of a set of *Hox* genes in the developing mouse embryo. (*A*) Portions of posterior tail buds (red stripes) of mouse embryos at embryonic day (E) 8.5, 9.0, and 9.5 were examined for expression of a series of *HoxD* genes by high-throughput sequence analysis. (*B*) The *Hoxd* genes (*d1–d13*) are diagrammed at top of *B* and *C*. mRNA production (green) for the *Hox* mRNAs was assayed, revealing increases in mRNA concentrations across the locus (right to left) from E8.5 to E9.5. Nucleosome modification and Pol II association were assayed by chromatin precipitation for Pol II (orange) or acetylated histone H3 (purple). The known progression of expression across this region is in the direction of d1 to d13. By E8.5, *HoxD* 1–9 are already active. (*Legend continued on facing page.*)

recent physical studies on model chromatin complexes make clear the important role of acetylation in relaxation of chromatin including the dissociation of histone H1 (Robinson et al. 2008). Even in a 30-nm fiber (a solenoid), there is probably sufficient flexibility to allow penetration by specific proteins (Kruithof et al. 2009).

Thus, site-specific DNA-binding activators can find binding sites and recruit both acetylation and ATP-dependent chromatin-remodeling complexes, leading to chromatin modifications (sliding, eviction, and acetylation). TSSs can thereby be efficiently located by activators and the recruited transcription machinery, allowing transcription initiation.

Methylation of Histones: Possible Roles in Transcription

As was the case with acetylation, methylation of proteins, including but not exclusive to histones (not to mention DNA and RNA), was documented in the 1970s (Allfrey 1977; Chukov et al. 2004). No role in transcription for the addition of methyl groups to histones was established before the current wave of interest in histone tail modifications.

Methylation of histone tails was discovered in 2000 (Rea et al. 2000), and soon thereafter, specific methylating enzymes were found. The cataloging of methylations on specific residues in the different histones was promptly performed (Fig. 5-13) (see Allis et al. 2007, Appendix 2, pp. 479–490). Both lysine and arginine residues can be methylated. Lysine residues can bear one, two, or three methylations (written as Kme1, Kme2, or Kme3, with K being the single-letter abbreviation for lysine). The meaning of the concurrence of active transcription and individual methylations is a matter of intense current research and is discussed below.

Histone Demethylases

Histone demethylases are among the latest enzymes added to the list of those engaged in histone modifications (for review, see Cloos et al. 2008). It was formerly believed that methyl groups, in contrast to acetyl groups, were

Figure 5-14. (*Continued*) Progression in the next day shows the appearance of Pol II and acetylated histone H3 and the production of mRNA from *HoxD* genes 10–13. (*C*) As in *A* and *B*, but precipitations with antibody against H3K4me3 and H3K27me3 are compared to ([ESCs] embryonic stem cells that are negative for *Hox* gene expression). The negative H3K27me3 mark is lost as the H3K4me3 is added. Note the low signal for H3K4me3 before induction and the presence all across the loci as transcription is increased. (Reprinted, with permission, from Soshnikova and Duboule 2009 [©AAAS].)

chemically stable, but clever biochemistry uncovered the existence and mechanism of both lysine and arginine demethylases (Shi et al. 2004; Wang et al. 2004).

The mammalian genome encodes dozens of demethylases, and mouse knockouts of single demethylase genes can be lethal (Shi 2007). Thus, demethylases have important roles. Specific demethylations that trigger transcription initiation of specific genes have not yet been documented; however, *Drosophila* homeotic genes that bear H3K9me3 and H3K27me3 marks early in development lose the K9 and K27 methylations and can be activated at later times in development (Ringrose and Paro 2007).

Transcriptional Activation: H3K4me3 and H3K36me3

Although a specific role in transcription is not yet clear for many of the histone methylations, certain important correlations are firmly established. Two histone methylations that correlate with active transcription have been most extensively studied. H3K36me3 is found in histones nearer to the 3′ regions of genes that are actively transcribed in both yeasts and mammalian cells (Kouzarides and Berger 2006). Recent reports suggest that this modification might occur in mammalian cell nucleosomes near sites at which splicing of pre-mRNAs occurs after the RNA transcript is formed (Spies et al. 2009). This can hardly be an exclusive function for H3K36me3 because relatively few yeast transcripts are spliced, yet H3K36me3 is prominent on the 3′ regions of active genes and is definitely increased after induction of transcription.

In contrast to H3K36me3, histone H3K4me3 is regularly associated with TSSs even in genes that are not demonstrably producing detectable levels of mRNA. This conclusion was reached by several groups who performed whole-genome analysis of H3K4me3 distribution (Fig. 5-15) (Barski et al. 2007; Guenther et al. 2007; Mikkelsen et al. 2007). In addition to the nearly ubiquitous H3K4me3 deposition near start sites, strong signals for H3K4me3 are present in many downstream nucleosomes in some actively transcribed genes (see Fig. 5-14C). Although H3K4me3 may *in a total cell population* be found to be present on many TSSs, it is not true that every gene in every cell in a population has the H3K4me3 mark. This important distinction is discussed on pp. 292–295.

Histone Modifications at Enhancers

The most recent clarification of how histone modifications may be related to transcription concerns enhancers. Several modifications were noted around previously identified enhancers before global genomic analyses by ChIP-Seq (precipitation of total fragmented, cross-linked chromatin by specific

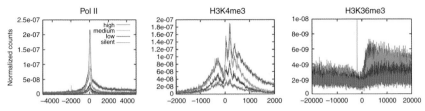

Figure 5-15. Methylated histone distribution around transcription start sites. Mononucleosomes from CD4+ lymphocytes were precipitated with antisera specific for Pol II, H3K4me3, or H3K36me3. DNA in each sample was subjected to high-throughput sequencing to determine the position of modified histones relative to transcription start sites on ~12,000 known genes. The number 0 on the ordinate is the transcription start site, and the number of nucleotides upstream and downstream is given. The expression level (mRNA concentration) of each gene was determined earlier by gene array analysis, and data are presented for genes expressed at three levels: high (red), medium (green), and low (blue) as well as for genes that are silent (purple; no detectable amount of mRNA in the gene array analysis). There is a definite correlation between amount of H3K4me3 and level of expression. There is also a definite presence of H3K4me3 and of Pol II on transcription start sites of genes not producing detectable amounts of mRNA (purple line). See text for discussion. (Reprinted, with permission, from Barski et al. 2007 [©Elsevier].)

antibodies, and massive sequencing of precipitated DNA) became widely used. The first of these enhancer modifications to be detected was H3K4me1. This was then detected by ChIP-Seq at hundreds of known enhancers, in some cases where the gene was not active (Heintzman et al. 2009; Bulger and Groudine 2011). Thousands of potential enhancers had the H3K4me1 and some of these were shown to actually be functional enhancers.

There is now general agreement (for review, see Zwaka 2010; Bulger and Groudine 2011) that H3K4me1 and H3K27ac are both found on enhancer regions of genes that are actively producing pre-mRNA and that H3K4me1 marks potential enhancers used at other times (Creyghton et al. 2010). It seems likely that site-specific DNA-binding proteins (activators) are responsible for bringing the methylase(s) to the H3K4me1 site before activation, and when activation occurs, the H3K4me1 is joined by a histone acetylase that marks H3K27 with an acetyl group.

Enhancers and Chromatin Loops

Ever since the discovery of enhancers that can lie hundreds to thousands of nucleotides away from TSSs of the genes that they regulate, it has been common to propose looping of DNA (or chromatin). Such loops could bring the enhancer and associated proteins close to the TSS and to the transcriptional machinery required for transcription. Strong evidence in favor of such loops has been obtained by showing that distant DNA sites can be captured near start sites by "chromosome confirmation capture." The technique involves in situ enzymatic DNA cleavage and rejoining, followed by sequencing to show linkage between distant chromosomal sites. Loops are particularly well demonstrated in the mammalian β-globin locus (for review, see Palstra et al. 2003; Dean 2005; Miele and Dekker 2008). The attraction between enhancer and TSS may be facilitated in some cases by the presence of the coactivator Mediator and a nuclear structural protein called cohesin, which may form rings around the two distant chromosomal sites (Kagey et al. 2011).

Methylations That Correlate with Repression: H3K9me3 and H3K27me3

Trimethylation of H3K9 and H3K27 is widely associated with chromatin compaction and gene silencing (Kouzarides and Berger 2006). The suggestion is frequently made that these negative-acting marks, which can be distributed widely around "repressed" genes or whole chromosome segments, are associated with keeping genes continually inactive from generation to generation. This possibility is also fully discussed below in the section Epigenetics and Transcriptional Control after repressors and corepressors are described (p. 288).

REGULATED TRANSCRIPTIONAL EVENTS AFTER INITIATION: PAUSED POLYMERASES AND COMPLETION OF PRE-mRNA

The last decade or so has witnessed numerous clarifying contributions to understanding molecular events required for RNA Pol II to successfully *complete* a pre-mRNA after *regulated* transcriptional initiation (for review, see Phatnani and Greenleaf 2006). Two streams of research have illuminated the importance of elongation control in eukaryotic transcription. First, proteins different from those required for in vitro initiation were found in the 1980s to be required for chain elongation. Second, it was discovered that the elongation step was regulatory in *Drosophila* genes whose transcription was triggered by heat. Recently, it has been proposed that elongation control may be quite widespread in vertebrates also and is likely to be an important step in gene regulation (Core and Lis 2008).

Phosphorylation of Pol II CTD Involved in Elongation

The largest subunit of eukaryotic Pol II contains on its carboxy-terminal domain (CTD) a repeated heptapeptide, $Y_1S_2P_3T_4S_5P_6S_7$. This CTD repeat was originally discovered in yeast Pol II, where there are 26 copies, whereas mammalian Pol II has 52 copies of the *same* heptapeptide (for review, see Young 1991). When Pol II is recruited to a start site, the CTD is not phosphorylated. The CTD repeat becomes phosphorylated on serine 5 by a kinase that is associated with TFIIH, one of the GTFs (Fig. 5-16). However, transcription then proceeds for only a short distance ($\sim$50 nucleotides or less), during which time m^7G caps are added to the nascent chain (Babich et al. 1980; for review, see Fuda et al. 2009). A series of much earlier experiments led to recognition of frequent premature termination and ultimately to proteins required for transcriptional elongation.

DRB and Transcriptional Elongation

Around 1980, experiments with ^{3}H-uridine or [^{3}H-methyl]methionine originally designed to study m^7G-capped heterogeneous nuclear RNA (hnRNA)

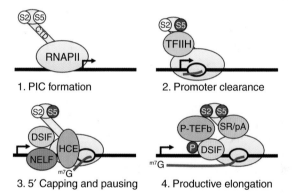

Figure 5-16. Early steps in Pol II transcription initiation and elongation. (1) RNA Pol II (RNAP II) is shown bound to a start site in the preinitiation complex (PIC). (2) Immediately, a kinase in TFIIH (one of the general transcription factors) phosphorylates the serine 5 residue of the CTD (red circle), allowing promoter clearance and synthesis of short ($\sim$50 nucleotide) RNA chains (red line). (3) During synthesis of the first 20–30 residues, capping of the nascent RNA occurs by capping enzymes ([HCE] human capping enzyme). Transcription is soon inhibited by two protein complexes, DSIF (contains the DRB sensitivity-inducing factor) and NELF (negative elongation factor), that act negatively to halt elongation. (4) Release from this restraint requires recruitment of yet another complex, P-TEFb, one subunit of which (a cyclin kinase) phosphorylates the serine 2 residue on the CTD, leading to release of NELF (also phosphorylated; not shown) and HCE, allowing RNA chain elongation. The DSIF complex becomes phosphorylated in one position and now acts positively to promote chain elongation. The SR/pA are RNA-processing-related proteins that join the now fully activated elongating Pol II machine. This summary is for human proteins; variations of this theme are performed in all eukaryotes by slightly different protein factors. (Reprinted, with permission, from Peterlin and Price 2006 [©Elsevier].)

revealed short (<200-nucleotide) RNA molecules with 5′ caps of two types (Salditt-Georgieff et al. 1980; Tamm et al. 1980). Most of the caps contained m^7G just as those on hnRNA and mRNA. A minority of molecules had $m^{2,2,7}G$ caps. The small molecules bearing $m^{2,2,7}G$ caps proved to be the RNAs found in the small nuclear ribonucleoprotein (snRNP) particles (U1, U2, U4, U5, and U6) that comprise spliceosomes (Busch et al. 1982). The small molecules containing the m^7G caps proved to be Pol II products, some, but not all, of which were eventually elongated to make pre-mRNA or other long hnRNA molecules. But many of the short RNAs were not elongated. The action of the antiviral drug DRB (5,6-dichloro-1-β-D-ribofuranosylbenzimidazole) proved to be decisive in defining different members of this group of short m^7G-capped molecules (Egyhazi 1974; Sehgal et al. 1976; Salditt-Georgieff et al. 1980; Tamm et al. 1980). Further study of these molecules ultimately led to the discovery of a complex containing the DRS-sensitive inhibitory factor (DSIF) that prevented elongation (Fig. 5-16).

When DRB was added 30 min before ^3H-uridine to HeLa cells, no label entered either long hnRNA or polyribosomal poly(A)-containing mRNA, but labeling (with [^3H-methyl]methionine) of short m^7G-capped molecules continued. If 45-sec label times with ^3H-uridine were allowed before DRB was added, longer labeled molecules (~500 bases and longer) continued to elongate (Sehgal et al. 1976), and a limited amount of labeled poly(A)-terminated mRNA was formed, suggesting that if synthesis of an RNA had passed a critical step, it could be completed, processed, and enter the cytoplasm. The number of [^3H-methyl]methionine-labeled, m^7G-capped short RNAs were increased twofold to 3.5-fold in cells treated with DRB compared to untreated cells. This was interpreted to mean that most capped nuclear RNA molecules were normally destined to escape the DRB-imposed stop site but that one-third to one-half of all m^7G-capped molecules were short transcripts not affected by DRB (Salditt-Georgieff et al. 1980; Tamm et al. 1980).

These early findings led biochemists in the late 1980s and 1990s to search for and ultimately find a DRB-sensitive protein (Wada et al. 1998; for review, see Kohoutek 2009 and Core and Lis 2008). The DSIF protein complex contains the DRB-inhibited factor, a serine kinase, together with another small protein subunit. Further experiments found that DSIF (also referred to as Spt4/5, the yeast protein name) associates with a second complex, NELF (negative elongation factor). Together, these complexes restrict Pol II to the first ~50 or fewer nucleotides of a pre-mRNA (Fig. 5-16). For elongation to proceed, another positive-acting complex, pTEFb, is then recruited. The pTEFb complex contains a serine kinase (CDK9, a cyclin kinase), the action of which dispels NELF and phosphorylates the large subunit of DSIF, which now acts as a positive elongation factor. The pTEFb-associated kinase also phosphorylates serine 2 of the Pol II CTD. These events allow continued elongation (Peterlin and Price 2006; Kohoutek 2009).

Continued elongation to complete the pre-mRNA requires yet other sets of factors that displace histones, most likely by a partial dissociation (the H2A/H2B dimers are thought to be displaced as dimers) followed by a necessary restitution of nucleosomes in the wake of Pol II passage to prevent promiscuous RNA starts on nucleosome-free DNA (Fig. 5-17) (Workman 2006).

Paused Polymerases and Regulation of Transcription at the Elongation Step

With the great interest in transcriptional initiation and its control, John Lis and colleagues in the late 1980s turned their attention to the details of the transcription of *Drosophila* "heat shock" genes before and after heat shock, both in larvae and in cell culture (Gilmour and Lis 1986; Rougvie and Lis 1988; for review, see Rasmussen and Lis 1995; Core and Lis 2008). These genes encode mRNAs that are quickly increased above a low background level

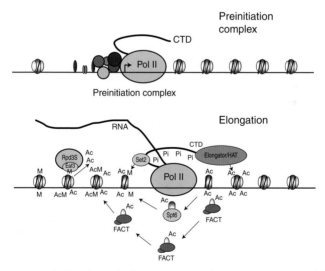

Figure 5-17. Transcription through chromatin-covered genes. (*Top*) A simplified version of a preinitiation complex shows that nucleosomes lie ahead of an engaged Pol II. Events of promoter departure are not depicted (see Fig. 5-16). (*Bottom*) Pol II in "mid-stream." Histone acetylation (by histone acetyltransferase elongation [HAT], orange) precedes (or accompanies) Pol II. Methylation of histones (e.g., at H3K36) also occurs as transcription proceeds. (Set2, in yellow, is a methyltransferase.) FACT (red) is a multiprotein complex that allows histone removal and restitution as the transcribing Pol II moves through a region of chromatin. (Modified, with permission, from Workman 2006 [©CSHLP].)

after *Drosophila* cells experience elevated temperature (above 30°C). Through an extensive and careful set of experiments, these investigators proved in cells at a normal temperature (25°C or below) that a Pol II molecule, phosphorylated on serine 5 but not on serine 2, was paused at the beginning of several heat shock genes (Fig. 5-18). When the ChIP technique was perfected, a transcriptional activator (a GAGA factor) that presumably assisted in getting the paused Pol II to the TSS was also found to be present on the paused heat shock genes. This is a most important point relevent to the primary event in regulation of these genes (discussed below).

The paused polymerases on several different heat shock genes had made ~30 nucleotides of the pre-mRNA and then stopped. After heat shock, the Pol II is phosphorylated on serine 2 of the CTD (Park et al. 2001). The heat shock factor (HSF), a site-specific DNA-binding activator, is also part

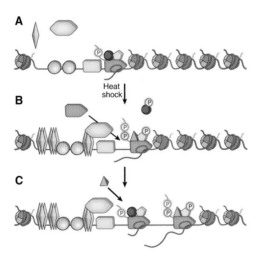

Figure 5-18. Pol II is paused for action on *Drosophila* heat shock genes. (*A*) Nucleosome octomers (green spheres) are interrupted at the TSS of a heat shock gene where Pol II (red bullet), phosphorylated on serine 5 of its CTD (red encircled P), is paused, having made short RNA (purple line projecting from the bottom of Pol II). Also present is a site-specific DNA-binding factor, GAGA (yellow circles), and GTFs (blue rectangular box). The pausing is effected by DSIF/NELF complexes (DSIF, pink pentagon; NELF, purple circle). (*B*) Heat shock sets in motion several events: The heat-shock factor trimerizes (yellow diamonds), forming a high-affinity, positive-acting DNA-binding factor that recruits coactivators (green hexagon), including Mediator. (*B,C*) pTEFb (Blue pyramid/triangle) then phosphorylates both NELF (purple circle in *B*), releasing it from DSIF, and serine 2 on the Pol II CTD (blue "Ps" in circles). (*C*) DSIF (pink pentagon) becomes phosphorylated (blue P in circle) on one subunit, now acts as a positive elongation factor, and RNA chain elongation proceeds. (Modified, with permission, from Fuda et al. 2009 [©Macmillan].)

of the paused complex. The HSF changes conformation (trimerizes) after heat shock and now binds tightly to the promoter to recruit coactivators including Mediator and pTEFb (Fig. 5-18). Now, the paused polymerase elongates pre-mRNAs. Exactly what proteins are left behind bound to the promoter of heat-shock-activated genes is not clear. Very likely, reinitiation of transcription requires the HSF to continue to recruit all of the necessary components for a successful high level of transcription (Fuda et al. 2009). An important point is that the paused polymerase is not "permamently" paused. Rather, the production of mRNA from genes activated by heat shock rises by >20–40-fold (Ardehali and Lis 2009). Because during maximal heat shock gene expression the residence time of Pol II at a TSS is no more than 10–15 sec, the paused polymerase remains paused for 10 min or less (J. Lis, pers. comm.).

With the initial *Drosophila* studies as a background, researchers using Pol II antibodies to Ser5 on the CTD of Pol II in both *Drosophila* and mammalian cells have now identified paused polymerases at a large fraction (50%–60%) of all Pol II TSSs ("genes") in animal cells. (These "global" experiments [Barski et al. 2007; Guenther et al. 2007] rely on ChIP and HTS; see Technical Note: ChIP Assays, p. 267.) Some but not all of the paused polymerases have the ∼30–60-nucleotide paused transcript. In *Drosophila*, the protein complex NELF and a transcription factor, GAGA, that are known to be present on all heat shock genes with paused Pol II are also present in a large proportion (60%–80%) of all genes with paused polymerases (Lee et al. 2008). Thus, there appears to be a frequent and perhaps regular series of events that establish paused Pol II complexes at many genes even if they are not producing (or are producing little) full-length pre-mRNA and mRNA (Fig. 5-18). Obviously, the presence of methylation of histones (H3K4me3) at the transcription start sites of such genes could result from recruitment of methylases during establishment of the paused polymerases, rather than H3K4me3 being deposited independent of transcriptional initiation.

A distinct physiological effect of paused Pol II with or without a short nascent chain was reported in cells of the developing fly embryo (Boettiger and Levine 2009). During early embryogenesis, before division of the fly larva into segments, many genes are known to be activated in a cascade (see p. 246). By ChIP analysis, RNA polymerase was detected on the TSSs of many but not all of these early genes. As the fertilized fly egg develops, the genes with the paused polymerases become synchronously activated (within ∼3 min) in most of the appropriate cells of the embryo, whereas the genes that lack a paused polymerase become activated but in a stochastic fashion during an ∼15-min window (Fig. 5-19). The paused polymerases clearly allow a quicker transcriptional response.

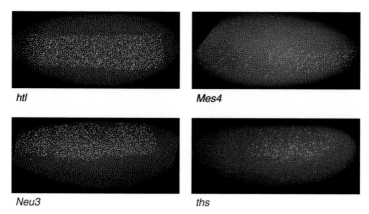

Figure 5-19. Paused polymerases contribute to synchronous gene expression. Early *Drosophila* embryos stained with fluorescent probes (small yellow dots) to specific RNAs (*htl*, *Mes4*, *Neu3*, *ths*) that are known to be expressed in an early phase (~30 min) after fertilization. (Probes are to long intronic regions that stain pre-mRNAs in nuclei.) Two of the genes do not have paused polymerases (*Mes4*, *ths*), and two do (*htl*, *Neu3*). The expressions of *htl* and *Neu3* (*left*) occur synchronously within an ~3-min window; the expression of *Mes4* and *ths* (*right*) occurs during a longer time, within an ~15-min window. (Reprinted, with permission, from Boettiger and Levine 2009 [©AAAS].)

A final point regarding the relationship between paused polymerases and transcriptional regulation is essential. Although the output of genes with paused polymerases is regulated by the frequency of release of paused polymerases, it remains true that the assembly of an active polymerase *before* pausing is the initial regulated event. Transcriptional *initiation* remains the primary step in regulation. Elongation seems to be an important *second* regulatory step.

Short Promoter-Proximal Transcripts

The pioneering work on short RNAs at paused sites by Lis and colleagues focused attention on considerable additional work on short RNAs both in the Lis laboratory and in other laboratories (summarized in Core et al. 2008; Seila et al. 2008, 2009).

With the availability of high-throughput DNA sequencing, a number of large laboratories and consortia found that short RNAs are produced from a large fraction of all recognized loci from which pre-mRNA initiation is documented. These short RNAs are transcribed in both directions around TSSs as well as from other sites in the mammalian genome, such as repeated elements throughout the genome, and even around poly(A) sites (for review, see Kapranov 2009). It is clear that the majority of the short RNAs do start at a previously recognized TSS and that the molecules (from 20 to 200 in length) bear an m^{7}G cap (see Carninci et al. 2005, 2006; Taft et al. 2009). The existence of the short RNAs is clearly proven, but the frequency compared to molecules that are eventually elongated is unclear in mammalian cells. The presence of an m^{7}G cap on short RNA is reminiscent of the short-capped DRB-resistant RNAs described much earlier (Salditt-Georgieff et al. 1980; Tamm et al. 1980). In these earlier experiments using cultured animal cells, the short-capped DRB-resistant short RNAs represented about one-half of all capped molecules. Analysis of all nascent transcripts in yeasts shows, however, that the number of short RNAs is small compared to normal longer Pol II transcripts (Churchman and Weissman 2011). The physiological meaning of these undeniably present small promoter-proximal RNAs is not clear. Their existence could have important implications for the understanding of H3K4me3 modifications on "genes" that produce no detectable mRNA.

Concluding Remarks Regarding Locating Transcription Start Sites

How do all of this feverish transcriptional activity at TSSs and the idea of so many paused polymerases accord with the earlier attention to carefully orchestrated, purposeful escorting of Pol II to start sites by transcription activators and coactivators to effect regulated transcriptional control?

Let us step back and examine what is clearly proven: The *rate* of synthesis of different pre-mRNAs resulting in finished mRNAs clearly changes from cell to cell and from time to time (see Chapter 4). There is a huge body of information that this regulated mRNA synthesis both in cultured cells and in specialized cells in animals depends on the presence and action of many hundreds of specific transcriptional activators. For example, the *rate* of synthesis—proven by run-on transcription—of pre-mRNA for globin mRNA in an erythroblast, or liver-specific mRNA in hepatocytes as they mature, or steroid- or cytokine-induced mRNAs in cell culture and in animals is strictly under transcriptional control (see Chapter 4). In these cases, it is not just one initial transcriptional event but continued transcription for a sustained period that controls the final amount of a particular mRNA in a cell. Virtually nothing is known in vivo regarding the actual frequency of initiation and what if any difference exists in the galaxy of proteins required for the first specifically induced transcription event in comparison to each succeeding round of transcription.

With respect to the widely distributed, paused polymerases, it seems, at least in the case of heat shock genes and certain *Drosophila* development genes, that the paused polymerase allows a hair-trigger response to increase transcriptional rates. But the proteins required for continued rounds of transcription are not known. Moreover, an active transcription complex has to be assembled as the first step in production of an RNA from a "paused" gene.

What all of the new short, divergent, promoter-proximal RNAs referred to in the previous section indicates first of all is that start sites for Pol II are not "hidden," even in cells that will never produce particular RNAs. Global analysis of histone distribution shows that in yeast, *Drosophila*, and human chromosomes, many (perhaps most) TSSs lie between two neighboring nucleosomes (see Fig. 5-15) (Guenther et al. 2007; Lee et al. 2007; Mavrich et al. 2008). Thus, the TSS itself is often "free." Given the opportunity, Pol II plus some of the plentiful coactivators and GTFs as well as potentially even some widely distributed transcriptional activators with only moderately fastidious DNA-binding specificity may recruit Pol II to TSSs in an "unregulated" fashion. These potentially unregulated Pol II engagements lacking the full collection of coactivators might then produce abortive short RNAs or become paused and remain unable to complete a pre-mRNA transcript until some further event. However, *purposeful* physiologically programmed *rapid* transcription rates are most likely regulated in the stepwise "canonical" fashion—activators recruit a full complement of coactivators that recruit GTFs and Pol II to make full-length pre-mRNAs (for discussion, see the introduction to Fuda et al. 2009).

BLOCKING TRANSCRIPTION IN EUKARYOTIC CELLS: "REPRESSION"

Much of the negative control of transcription in eukaryotes depends on site-specific DNA-binding proteins that, instead of attracting coactivators, attract corepressors that make initiation sites in chromatin less accessible (for review, see Rosenfeld et al. 2006). Another broad category of proteins that act negatively to inhibit transcription acts more simply. This category acts directly on either transcriptional activators or directly on the proteins of the transcriptional machinery itself. Only after activator proteins and the various types of positive-acting coactivators were identified could the existence of and mode of action of the many negative-acting factors be uncovered.

Direct-Acting Repressor Proteins

Although there are many variations on the theme of proteins acting negatively on eukaryotic transcription without involving chromatin, a few examples will illustrate the principle.

Site-specific DNA-binding proteins that activate transcription are often members of a sequence family that act as homodimers. Negative-acting members of such families may form dimers but lack DNA-binding capacity or TADs. Such proteins can obviously greatly reduce transcription from a gene.

HLH and Id Proteins

An excellent example of a family that has some members with transcriptional-activating capacity and some without is the helix-loop-helix (HLH) family of mammalian DNA-binding proteins. This family gets its name from characteristic structural elements in the proteins, namely, two helical protein sequences with an unstructured loop in between (Lamb and McKnight 1991; Sikder et al. 2003). Through interaction of the helical regions, members of the family of HLH activator proteins dimerize and bind DNA. HLH proteins can bind either as homodimers or heterodimers; these can be positive acting by binding to an enhancer. However, within the HLH family, there are versions of HLH proteins, called *Id proteins*, that lack DNA-binding capacity. When an Id protein interacts with a positive-acting HLH partner, it can block DNA binding and thus abort gene activation (Fig. 5-20, I) (Benezra et al. 1990; Sikder et al. 2003). It is the relative amounts of such positive- and negative-acting HLH proteins that govern transcription from a usually positive-acting DNA-binding site.

Hedgehog Pathway and Ci Proteins

Another theme of direct negative-acting proteins is illustrated by the *Drosophila* transcription factor called *cubitus interruptus* (Ci), the transcriptional regulatory protein in the Hedgehog pathway (Figs. 5-7, 5-20II) (Méthot and Basler 1999, 2001). This pathway participates in the correct development of many different cell types in *Drosophila*. It exists in all animals and in an expanded form in mammals. The *Ci* gene encodes a large DNA-binding protein. When this pathway is activated by the Hedgehog ligand binding at the cell surface, the large protein survives to activate transcription by binding to enhancers. However, when the pathway is not activated, the large protein is cut almost in the middle; the DNA-binding half survives, but the portion containing the TAD is destroyed. Thus, the DNA-binding portion now sits on enhancers in target genes, and it acts negatively because further steps in transcription initiation do not occur (Méthot and Basler 2001). The several different mammalian activators in this pathway are called GLI proteins because they were first discovered in brain tumors called gliomas that originate in glial cells.

Inhibiting Action of a DNA-Bound Activator

Another class of direct transcriptional-inhibitory proteins can bind to and block the required interaction of an activator or coactivator. Many examples

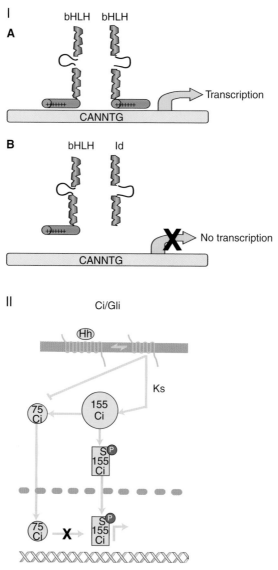

Figure 5-20. Direct inhibitors of activator function. (*I*) Id proteins. Helix-loop-helix (HLH, or bHLH for basic HLH) proteins are a very large family of transcriptional activators that participate in (among other activities) muscle development, lymphocyte maturation, and cell cycle progression. (*A*) Two positive-acting bHLH proteins dimerize (homodimers and heterodimers are known) and bind DNA using a DBD (green). (*B*) Id proteins lack the DBD and vitiate the activation potential of their partners. (*II*) *Cubitus interruptus* (*Ci*) gene product can be a transcriptional activator or inhibitor. Full-length Ci protein (155 kDa) is proteolytically cleaved unless it receives a signal from Hedgehog (Hh) at the cell surface, resulting in a serine phosphorylation. Full-length molecule activates transcription; the 75-kDa cleaved product lacks a TAD and blocks transcription. (*I*, Redrawn and modified, with permission, from Sikder et al. 2003 [©Elsevier]; *II*, redrawn, with permission, from Brivanlou and Darnell 2002 [©AAAS].)

of such interruptions have been studied in both yeast and mammalian cells (Dyson 1998; Zhang et al. 1998; Carman and Henry 2007; Jiang et al. 2009). The first such arrangement in eukaryotic cells to be clearly described involves the regulation of galactose metabolism in baker's yeast. The GAL4 protein is a positive transcriptional activator for genes encoding enzymes that metabolize galactose. But in cells grown on glucose (or carbon sources other than galactose), the GAL4 protein does not act. This is because another protein, GAL80, binds to the GAL4 protein and blocks its ability to attract coactivators, even though GAL4 may still be bound to enhancer sites in DNA. When galactose is furnished as the only carbon source, galactose binds to a protein, GAL3, that, in turn, interacts with the GAL80 protein, disrupting the GAL80–GAL4 interaction (Diep et al. 2008; Jiang et al. 2009). GAL4 is then free to act as a transcriptional activator, so that the enzymes required for galactose metabolism are formed (Wightman et al. 2008). As the galactose concentration declines, the process is reversed, and GAL80 returns to block transcription.

In none of these negative-acting systems does inhibition of transcription result from a negative-acting protein binding directly to—occluding—a TSS, as repressors in bacteria do.

Transcriptional Inhibition by Chromatin-Modifying Factors

Perhaps the most pervasive, and certainly the most extensively studied, *negative*-acting factors are those that affect chromatin. There are large inhibitory complexes that either already contain DNA-binding proteins or can bind to proteins already bound to specific sites in DNA. Some site-specific DNA-binding proteins that would otherwise be site-specific DNA-binding activators can bind corepressors and thus act negatively (Rosenfeld et al. 2006). The inhibitory complexes that bind to the already DNA-bound proteins often bring with them deacetylases and/or methylases (or perhaps demethylases) that enforce gene silencing. Moreover, some of these complexes can, by interacting with one another in a chain-like fashion, shut down great swaths of a chromosome.

Groucho/TLE Repression

A widely used repressive mechanism involves proteins that were originally discovered genetically in *Drosophila* and in yeast but that exist in mammals also. The operative repressor proteins are known as Groucho (abbreviated Gro) in *Drosophila* and in mammals and as TLE (transducin-like enhancer of split) proteins in yeast. A Gro protein can effect repression in several ways but always in the company of a site-specific DNA-binding protein that, on its own, may or may not show positive gene activation potential. By binding to such site-specific DNA-binding proteins, the Gro repression mechanism operates at specific sites on a chromosome (Fig. 5-21). Histone deacetylases or histone methylases that target H3K9 (a transcriptionally active mark when acetylated or a repressive mask when trimethylated) are included in the Groucho/TLE complex to perform repression. Recently, it has been recognized that serine phosphorylation of the Groucho/TLE protein can relieve the repression caused by the Gro-organized complex (Cinnamon and Paroush 2008).

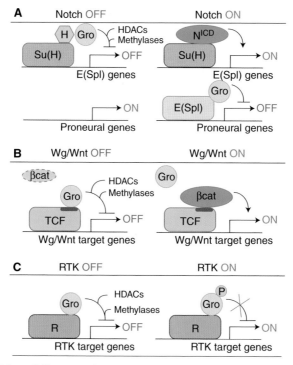

Figure 5-21. Three different mechanisms of Groucho (Gro) gene repression in *Drosophila* and mammals. (*A*) In *Drosophila*, the suppressor of Hairless (Su[H]) protein binds its cognate DNA site but has no transactivation domain. When cells are signaled to activate the Notch pathway, the NICD is released (see Fig. 5-7) and displaces Gro, and the repression is relieved. By itself, the NICD is not an activator because it cannot bind DNA. (*B*) TCF is a mammalian homolog of Su(H) that also lacks a necessary component of a transcriptional activator although it can bind DNA. Gro protein bound to TCF ensures repression at this site until the Wnt (wingless) extracellular protein binds to the cell surface, leading to the preservation of β-catenin. β-Catenin can displace Gro and reverse the repression. (*C*) Gro proteins bound to some genes can be phosphorylated on serine (P) (by growth factor activation of receptor tyrosine kinases [RTKs], which then activate serine kinases). This removes Gro, to derepress genes. (Modified and redrawn, with permission, from Cinnamon and Paroush 2008 [©Elsevier].)

Polycomb Repression

A very important, pervasively used, collection of inhibitory proteins in animals are referred to as *Polycomb* (Pc) complexes (for review, see Grossniklaus and Paro 2006; Ringrose and Paro 2007). Proteins in the Pc complexes and chromosomal sites of their action were discovered in *Drosophila* through mutations in several different genes that apparently released repressed abnormal phenotypes.

The Polycomb group of genes in *Drosophila* initially gained attention because they affected morphological changes in structures that are produced by particular imaginal discs. Cells in individual imaginal discs are set aside early to eventually form parts of the adult fly. The cells in imaginal discs grow and divide as the larva develops through three larval stages. Imaginal disc cells then form legs, wings, head parts, eyes, etc., at the end of larval development. Thus, the early determination of these groups of cells lasts for many cell generations as the larva develops until a particular adult structure is finally constructed during the late third instar and pupation. Polycomb mutations changed the cell fates of imaginal disc cells—creating so-called *homeotic* transformations. For example, leg discs produce legs with a known number and position of sex combs. The Polycomb mutation *extra sex comb* causes legs with more than the normal number of sex combs. The implication is that an "epigenetic" state of repression is maintained for generations until the now-determined cells in the imaginal disc perform the final stage of growth and morphogenesis. The basis for the negative activity of Polycomb complex proteins through many generations has been of great interest (Grossniklaus and Paro 2006; Ringrose and Paro 2007).

Two types of Polycomb protein complexes, Pc1 and Pc2, have been described; these are thought to act consecutively (Pc2 and then Pc1). Proteins within Pc2 either bind to ill-defined DNA regions or to identified site-specific DNA-binding proteins, the genes for which were originally detected as *repressors*. The Pc2 complex contains histone methylases that target H3K9 and H3K27 for trimethylation. The Pc1 complex contains several other proteins including the Polycomb (*PC*) gene itself, whose mutation gave this repression phenomenon its name, as well as other proteins including deacetylases. Because the Polycomb complexes can interact with one another, the silent chromatin state can spread over a large chromosomal region, which may contribute to maintaining repression from generation to generation (Bonasio et al. 2010; Guenther and Young 2010; Margueron and Reinberg 2011).

HP1 Repression

Another important negative-acting protein that appears to facilitate continuous chromatin condensation throughout successive cell cycles is called HP1, for heterochromatin protein 1 (for review, see Maison and Almouzni 2004). The centromeric and telomeric regions of eukaryotic chromosomes, which contain few if any genes, are heterochromatic (condensed). From yeast to human cells, this condensation is orchestrated by HP1. HP1 protein binds to a histone methyltransferase and also self-associates; thus, when a centromeric region is duplicated, there is ample opportunity for new nucleosomes to be covered with HP1 and to be trimethylated on H3K9 (see the models of Probst et al. 2009 and the next section). In addition to methyltransferases, histone deacetylases are also recruited to centromeric heterochromatin. Heterochromatic regions are also found in islands along the lengths of chromosomes, and HP1 is often found in these regions as well, where it attracts histone deacetylase and histone methylase complexes.

When it was first discovered in *S. pombe* in Shiv Grewal's laboratory, an unusual and unexpected participant in centromeric heterochromatin formation was noted. *S. pombe* has an HP1-like protein that serves to help form heterochromatin just as HP1 does in other organisms. In *S. pombe*, the machinery for making short inhibitory RNAs (siRNAs) is discussed on p. 317. This machinery is required for centromeric heterochromatin to form in *S. pombe*. The short RNAs are probably generated from RNA transcripts of both strands of the repetitive centromeric DNA. These can self-anneal to form double-strand RNA (dsRNA). siRNAs ($\sim$25 nucleotides long) are carved out of the dsRNAs. A large protein complex, including both deacetylases and methylases, is recuited by binding to the siRNAs in the *S. pombe* centromeres (Fig. 5-22) (Sugiyama et al. 2005; for review, see Bonasio et al. 2010). These experiments clearly show "epigenetic" propagation of heterochromatin as cells divide; they also illustrate that even centromeric heterochromatin can be invaded by an RNA polymerase that makes the siRNA precursor (Guenther and Young 2010). Recent experiments now show that siRNAs are formed in Polycomb-repressed chromatin just as is the case in heterochromatin (Kanhere et al. 2010). Affinity for the short RNAs by proteins in corepressors helps to recruit the repressive complexes. Thus, production of these short RNAs within repressed DNA regions is now thought to be a necessary part of an overall repression of genes within repressed regions (Fig. 5-22) (Bonasio et al. 2010; Guenther and Young 2010; Kanhere et al. 2010).

"EPIGENETICS" AND TRANSCRIPTIONAL CONTROL

We have discussed certain histone modifications and some chromatin-associated proteins that are definitely associated with transcriptional regulation. These findings have been linked to the term *epigenetics*, producing some confusion regarding transcriptionally related histone modifications

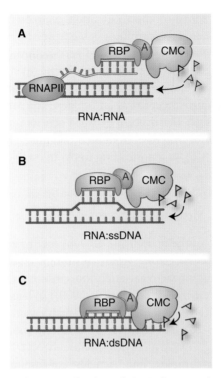

Figure 5-22. Possible mechanisms of action of siRNA in repression of heterochromatin. Three possible modes of action (*A*, *B*, and *C*) are shown. (Green) The RNA-binding protein (RBP), A (pink) is an adaptor protein that recruits the chromatin-modifying complex (CMC) siRNA is the short, straight yellow line. (*Panel A*) siRNA hybridized with a nascent RNA (longer wavy yellow line), (*panel B*) siRNA hybridized with one strand of an affected DNA region, and (*panel C*) siRNA forming a triplex with an affected DNA region. In each case, the complex includes the CMC, which could include a histone methylase and/or histone deacetylase. (Reprinted, with permission, from Bonasio et al. 2010 [©AAAS].)

during the cell cycle. Dan Gottschling, in reviewing the promiscuous use of *epigenetics*, gives a masterful review of what this term has meant to investigators through the years since being coined by C.H. Waddington, the eminent developmental biologist (Gottschling 2006; also, for review, see Bonasio et al. 2010). Adrian Bird, long attentive to chromatin studies, discusses advantages of the current permissive use of the term (Bird 2007).

There are certainly events that are not, strictly speaking, "genetic" but when in force can change cell behavior over many generations. Gottschling cites the temperate bacteriophage λ, where an RNA-polymerase-inhibiting protein—the repressor—continues to be formed in each generation, thus establishing the lysogenic state that persists for many generations. Production of a long

RNA (the Xist RNA) somehow inactivates most genes on one of the two female X chromosomes of mammals, forming a stable repressive effect on gene expression that begins early in embryogenesis and persists throughout life.

These are clear instances of carryover during each cell division of an abundant repressive substance that perpetuate a state not strictly determined genetically. This is widely accepted as an "epigenetic" state, although the precise means of ensuring X inactivation between generations (immediately after DNA replication) is not proven although considerable progress has occurred in solving interactions of long noncoding RNAs in establishing X inactivation (Tian et al. 2010).

DNA Methylation: Accepted Epigenetic State

A brief consideration of the major class of DNA methylation (the 5-carbon of cytosines in CpG islands) clarifies what constitutes a case of epigenesis that ensures a continuous biochemical effect on chromatin from generation to generation (for review, see Probst et al. 2009). When a methylated CpG site in DNA is replicated, one of the two daughter strands is momentarily unmethylated (Fig. 5-23). The hemimethylated state is recognized by specific DNA methylases (e.g., DNA methyltransferase 1, or DNMT-1) that *accompany* the replication complex, and virtually concomitant with replication, the new DNA (on both new chromosomes) becomes methylated (Schermelleh et al. 2005). These events occur rapidly because fork movement is between 30 and 60 nucleotides/sec and no nonnucleosomal stretches in the wake of growth fork passage have ever been detected. This well-integrated DNA methylation program linked to DNA replication fulfills the "strict" definition of *epigenetics.*

However, even here it has recently become clear that DNA methylations during interphase are not immutable. They can and do change in the history of a particular cell lineage. For example, extensive demethylation of DNA takes place just after mammalian cell fertilization; this is followed by programmed methylation that is then largely stable thereafter. The basis for the programmed remethylation could easily involve site-specific DNA-binding proteins. Furthermore, genome-wide definition of methylated DNA bases recently showed site-specific methylation changes in the "methylome" (the total methylated base profile) as a stem cell differentiates or as an embryonic cell differentiates, and these changes almost certainly involve site-specific DNA-binding proteins (Lister et al. 2009). (These experiments were performed by sodium bisulfite reactions with DNA [bisulfite coverts C to U but does not affect 5-methylcytosine] followed by high-throughput sequencing that theoretically locates every single 5-methylcytosine.)

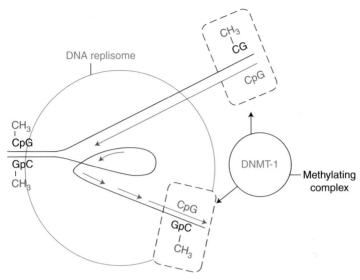

Figure 5-23. Epigenetic repair of hemimethylated DNA follows immediately after replication. When the DNA replication fork passes sites in DNA where cytosine is methylated (green CH₃), the two new duplexes have hemimethylated sites. Accompanying the replisome (the entire collection of proteins required to replicate DNA) is a complex of proteins including those that recognize the hemimethylated state. A protein complex, DNA methyltransferase 1 (DNMT-1), methylates the appropriate cytosine opposite the methylated CG residue in the complementary strand. (Long red arrow) Leading strand in DNA replication, (short red arrows) lagging strands in DNA replication. (Redrawn and modified, with permission, from Probst et al. 2009 [©Macmillan].)

Finally, DNA demethylases have recently been identified (Okada et al. 2010). Thus, even "permanent" DNA modifications quite possibly change as a *result of* changing *transcription factor binding* or binding of other DNA-binding proteins in coordination with demethylases (Lister et al. 2009). It remains true, however, that generally speaking, m⁵CpG methylations are maintained as a cell goes through division because there is a replication-connected process of epigenesis to ensure the continuation of methylated DNA sites.

Thus, DNA methylation is an established epigenetic state. But are histone methylations epigenetic?

Polycomb Repression and Heterochromatin Repression are Epigenetic Modifications

Several characteristics of repression of genes by Polycomb complexes and of heterchromatin repression make it likely that the term *epigenetics* does correctly apply to these cases. These are the best examples of epigenetic propagation of a transcription state, namely, continuous repression over many generations. The reasons are as follows: (1) The biology is established: Active repression of selected genes or for accumulated transcripts from heterochromatin over many successive generations is proven. (2) Polycomb and HP1 complexes are found over extensive blocks of chromatin and remain in place for generations. (3) From the molecular point of view, a plausible scheme is established for ensuring postreplicative continuation of repression.

In both cases, the continuous methylation of H3K9me3 and H3K27me3 through many generations may involve continuous production of short RNAs in the repressed regions to ensure repressive complex binding (Figs. 5-22 and 5-24). In addition, the plethora of H3K9me3 and H3K27me3 nucleosomes surrounding TSSs ensures that on both daughter chromatids a nucleosome with modified histones will be close to another modified nucleosome (Fig. 5-24). At least one of the protein subunits of the histone-modifying methylase (the EED subunit in Fig. 5-24) can recognize a modified H3K9me3 and/or a similarly modified K27 (Fig. 5-25) (Margueron et al. 2009). By bringing the methylase into the area, this recognition could facilitate methylation of a neighboring new histone at K9 or K27. No direct association of a histone methylase with the DNA replication machinery would be required to ensure continued repression. Repression by HP1 and Polycomb complexes where methylation of nucleosomes is so dense seems to qualify as epigenetic. Note, however, that a definite experiment has not been done showing that the majority of methylation of H3K9 and H3K27 in repressed regions actually occurs during S phase.

H3K4me3: What Directs It to TSSs?

Whereas the case for epigenetic continuation of repression is quite strong, the opposite is true for the single-most-prominent histone modification, H3K4me3, associated with *transcriptional initiation*.

The discovery in global distribution studies of mammalian cells of H3K4me3 at most TSSs, whether a gene was "active" or not (Fig. 5-15) (Barski et al. 2007; Guenther et al. 2007), initially posed a threat to the interpretation that transcriptional activators and coactivators lead the transcriptional machinery to a TSS to initiate purposeful transcription. The question was raised: Could it be that some entirely unanticipated mechanism delivers an H3K4me3 "mark" to the TSS that then serves as a signpost to begin all transcriptional initiation by Pol II? If so, how was a TSS so precisely recognized?

Reflection on this question, in view of all of the recent evidence, suggests that H3K4me3 most likely is deposited near TSSs by transcriptional activators

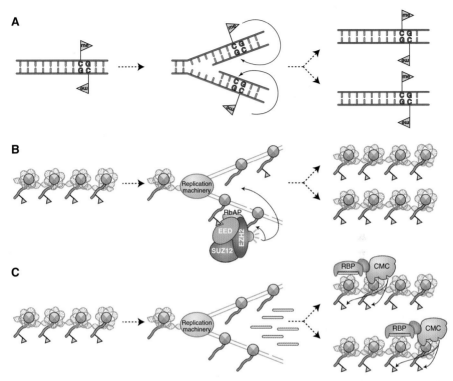

Figure 5-24. Transmission of epigenetic states. Three mechanisms are shown (*A*, *B*, and *C*). (*A*) In a simplified version of CpG methylation, the long-accepted epigenetic methylation of DNA is diagrammed (see Fig. 5-23 for involvement of DNMT, the methylating enzyme that accompanies the replication fork to recognize hemimethylated DNA). (*B*) Hypothetical model of epigenetic passage of H3K27me3 to new histones just after replication. H3K27me3-decorated histone tails (blue lines with yellow flags) attract Polycomb protein complexes (blue discs) that include EED, which itself can recognize H3K27me3, and EZH2, a histone methylase. Key points are as follows: (1) Nucleosomes just after the growing fork present one nucleosome with and one without H3K27me3 (this assumes no histone octamer splitting, for which no evidence exists) and (2) the recognition of an H3K27me3 histone modification by EED allows the EZH2 methylase to add —CH₃ groups to the H3K27 of any nearby nucleosome (arrow). (*C*) Model of epigenetic transfer of methylation and repression in heterochromatin. The repetitive DNA in heterochromatin is constantly transcribed to produce short interfering RNAs (siRNAs, yellow bars; see pp. 319–320, Figs. 5-37 and 5-38). These siRNAs are capable of recruiting histone methylases through RBPs associating with adaptor proteins (A) and CMCs that contain H3K27me3 methylases. (The addition of methylations in heterochromatin is discussed on p. 288.) (Reprinted, with permission, from Bonasio et al. 2010 [©AAAS].)

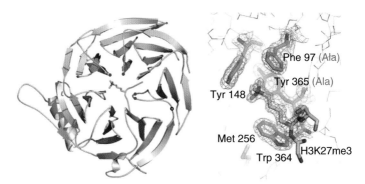

Figure 5-25. Trimethyl-lysine peptide (like the peptide in histone tails) binding to EED, a domain of one of the proteins in the PC2 complex (one of the two major Polycomb complexes). (*Left*) A peptide (yellow) from the tail of trimethylated histone H3K27 crystallized with the EED. (The H3K27me3 in histones is associated with gene repression.) The ribbons (gray) represent the β-propeller structure of the EED into which the peptide (yellow) binds. (*Right*) Amino acid contacts in the bindings. Physical measurements showed very strong binding between the two components in the figure and in addition between EED and an H3K9 trimethylated peptide (also a repressive mark) as well as very weak or no binding to H3K4me3 and H3K36me3 peptides (positive histone marks). (Reprinted, with permission, from Margueron et al. 2009 [©Macmillan].)

and/or coactivators that recruit an active methylase and not by a continuous epigenetic mechanism.

1. Some genes have H3K4me3-modified nucleosomes scattered over the entire body of the gene *after* but not before the gene is induced (Fig. 5-14C). These modifications are clearly the result of induced methylation by recruited methylases.

2. Many genes—whether transcribed at a low or high rate as well as genes that apppear to be transcriptionally silent—have *only* one or two H3K4me3 nucleosomes on either side of the actual TSS (see Fig. 5-15). This highly restricted H3K4me3 distribution is the same in silent and highly active genes (Barski et al. 2007; Guenther et al. 2007; Mikkelsen et al. 2007). However, the signal in ChIP-Seq analyses for H3K4me3 is at least five to 10 times higher in rapidly transcribed genes for these *same two nucleosomes*. If every highly transcribed gene in every cell had the H3K4me3 modification on the nucleosomes neighboring the TSS, then no more than 10%–20% of the silent genes in the total collection of cells could have the H3K4me3 modification. By this argument, the modification cannot be a *universal mark* of all TSSs in all cells on all genes that would serve as a constant signpost for Pol II to start transcription.

3. As discussed earlier, many, if not all, TSSs in human cells are active sites for the synthesis of short aborted RNAs or short paused RNAs. Many of these sites do not score as producing mRNAs. What brings Pol II to these sites to begin synthesis is not clear at this point, but it could well be that even an incomplete premature initiation complex would contain histone methylases, accounting for H3K4me3 signals at "inactive" TSSs.

4. If the marking of TSSs with H3K4me3 were an obligatory property of all genes in all eukaryotic cells, this presumably would occur at replication, and there is presently no evidence that it does.

In summary, it seems likely that H3K4me3 at TSSs results from recruitment of methylase-containing enzymes to the TSS and that it is not epigenetic. This does not dismiss the possible importance of H3K4me3 because repeated transcription initiation is likely important to achieve regulated higher levels of mRNA. Several proteins that are important in transcriptional initiation (e.g., the TAF3 protein in the TFIID coactivator) (Vermeulen et al. 2007) have protein domains that could and probably do recognize H3K4me3 during repeated initiations. But the notion that H3K4me3 is a *preexisting chromosomal instructional mark* for transcriptional initiation seems to be incorrect.

DIFFERENTIAL PROCESSING OF PRE-mRNA

From all of the foregoing discussion in this chapter, it is clear that a large fraction of the genetic and synthetic capacity of cells is devoted to choosing particular genes for expression. The choice of particular TSSs, remodeling and chemically modifying chromatin, and pausing to decide to complete a pre-mRNA are all regulated choices. But decisions on the path to final mRNA production are hardly finished at that stage. Elaborate selective processing of the pre-mRNA definitely occurs but is comparatively poorly understood.

During the 1980s, mRNA sequences and sequences of the genomic loci from which the mRNAs are derived were completed in large numbers, revealing many different choices of poly(A) sites and of splice sites in pre-mRNA. High-throughput sequencing of whole genomes and cDNA sequencing of "expressed sequence tags" has now made clear that nearly all (>90%) human pre-mRNA transcripts can be differentially spliced (Wang et al. 2008), either in a single cell or in different cell types in various tissues. Moreover, >50% of pre-mRNAs have more than one poly(A) site that is used in making mRNA.

Some pre-mRNAs can be spliced to yield many different proteins. One famous *Drosophila* locus, *Dscam*, has 95 alternative exons and can be spliced to yield more than *30,000* different protein products, many thousands of which have actually been proven to exist (Zhan et al. 2004). The variety of resulting proteins furnishes a code for neurite (dendrite) associations, a major purpose of which is to prevent a cell from connecting to itself (Fig. 5-26) (Hattori et al. 2009).

The extensive differential pre-mRNA processing belies the grossly misleading statement made repeatedly that "the human genome contains 25,000 genes." Because the large majority of the splicing and poly(A) choices result in different proteins, it is clear that our genomes actually code for many more than 25,000 different proteins, possibly several hundred thousand. A given pre-mRNA transcript encodes as many different proteins as there are differently spliced mRNAs derived from it. Because each protein could possibly be specifically damaged by a mutation, as many "genes" as there are proteins derive from one transcription unit (for review, see Nilsen and Graveley 2010).

Molecular Recognition Mechanisms in mRNA Processing

In Chapter 4 (pp. 201–207), we briefly discussed the very important discoveries in the early 1980s that quickly followed the discovery of splicing—the finding that consensus sequences border introns, the discovery of transesterification, and the detection and early characterization of the spliceosome. In addition, 3'-end processing—cleavage followed by polyadenylation—was recognized as necessary to make an mRNA. In this section, information regarding these events is updated before discussing some specific cases of differential pre-mRNA processing.

Cleavage and addition of poly(A) at the 3' end of the cut pre-mRNA is often the first step in pre-mRNA processing (Weber et al. 1980; Hofer and Darnell 1981; Amara et al. 1984; Galli et al. 1988). This decision settles which

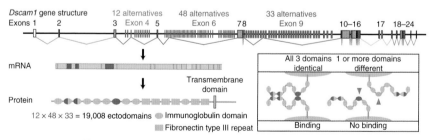

Figure 5-26. Differential splicing in the *Dscam* gene of *Drosophila*. Members of the extensive family of Dscam proteins are expressed in dendrites of arborization neurons. Dendrite binding or nonbinding results from splicing choices among different classes of exons. Deletion mutations of exons of the *Dscam* gene, thus reducing the possible combinations, leads to aberrant dendrite overlapping and self-connections. (Reprinted, with permission, from Hattori et al. 2009 [©Macmillan].)

sequences are going to appear at the 3′ end of the mRNA (both in the terminal exon and in the 3′-untranslated region). Selecting from the array of possible exons near the cap site (and, of course, the first exon along with the cap) at the 5′ end of the pre-mRNA can certainly occur before reaching the poly(A) site. This is especially true in extremely long pre-mRNA molecules (Singh and Padgett 2009). In the processing of long pre-mRNAs (some are 100,000 nucleotides long or even longer), splicing factors certainly bind to the still-growing pre-mRNA before the poly(A) site is reached, and there is clear evidence of some intron removal before transcription is complete (Beyer and Osheim 1988; Lemaire and Thummel 1990; Singh and Padgett 2009; Wada et al. 2009). However, the removal of introns requires some time (Singh and Padgett [2009] estimate that this can take 5–10 min), and Pol II synthesizes RNA at 3–4 kb/min (i.e., >50 nucleotides per second) (Sehgal et al. 1976; Singh and Padgett 2009; for review, see Ardehali and Lis 2009). This means that a Pol II has traversed ~20,000 nucleotides by the time a splicing event occurs (5 min = 300 sec; 60 nucleotides × 300 sec = 18,000 nucleotides). Thus, splicing and transcription can hardly be considered to be concomitant events. Finally, it is clear that poly(A)$^+$ nuclear RNA is much longer than mRNA, meaning that the poly(A) has been added before splicing is complete (Chapter 4, Fig. 4-22) (Salditt-Georgieff et al. 1976).

Finally, any pre-mRNA that has two (or more) poly(A) sites must necessarily await the most distal poly(A) choice to choose the exons nearest to that distal site. This was illustrated clearly in the primary late adenovirus transcript, which has five (or six) poly(A) choices (see Chapter 4, Fig. 4-31). The most distant poly(A) site in the major late transcript is 25 kb (synthesis time 6–7 min) from the start site, but splicing selection of the body of mRNAs in this multiply spliced transcript with five poly(A) stations awaits poly(A) addition (Weber et al. 1977; for review, see Nevins and Chen-Kiang 1981). Aside from the necessity of splicing at the last exon–intron gene junction awaiting the poly(A) choice, there is no evidence of coordination between splicing and poly(A) addition.

Poly(A) Addition Complexes and Choice of Poly(A) Sites

The formation of the complex required for recognition of a poly(A) site, cleavage of the pre-mRNA, and the addition of poly(A) is intimately tied up with starting a pre-mRNA transcript. The first processing event, adding the 5′-methylated cap to the nascent transcript, occurs by the time that the RNA chain is ~20 nucleotides long, that is, before the decision is made regarding elongation of the nascent transcript (see Fig. 5-16) (Babich et al. 1980). As soon as the antitermination factor (pTEFb) acts and serine 2 phosphorylation of the CTD of Pol II occurs, the first of the series of factors necessary for poly(A) addition is recruited (Richard and Manley 2009). In vitro biochemistry has proved that there is an enormous 3'-end processing complex containing ~85 proteins (Fig. 5-27A,B). This complex is required first to recognize poly(A) sites, then to cleave the pre-mRNA, and finally to lead the poly(A) polymerase subcomplex to add the poly(A) (Shi et al. 2009).

Experiments with yeast and mammalian cells have identified three sequence motifs that likely contribute to choosing a poly(A) site (Fig. 5-27C). The first of these to be recognized in mammalian cells is an A(A/U)UAAA motif, which is present in almost half of mammalian pre-mRNAs. This A-rich motif is positioned ~20 nucleotides upstream of the cleavage site. Two other regions—a UGUAN(A) sequence further upstream and a U-rich region downstream—are present not only in yeast and mammalian cells,

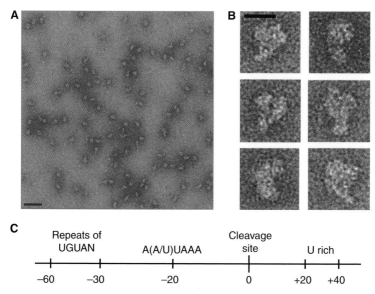

Figure 5-27. 3'-End pre-mRNA processing complex. (*A*) Electron micrographs of the purified negatively stained complex reveal its large size and complexity. Bar, 100 nm. Purified complex contains ~85 separate proteins including the original ~15 proteins known for some time to be able to cleave and add poly(A) to a substrate RNA that contains a signal to add poly(A). Complex contains many proteins of unknown function that may interact with regulatory proteins that choose a poly(A) cleavage site. (*B*) Gallery of typical images. Bar, 20 nm. (*C*) Sequences that participate in signaling poly(A) addition. (*A,B,* Reprinted, with permission, from Shi et al. 2009 [©Elsevier]; see also Venkataraman et al. 2005; Richard and Manley 2009.)

but also in a similar form in plants. Moreover, all of the many proteins that participate in creating the 3′ end of the mRNAs share similarities in protein sequence among all of the eukaryotes. (Excellent reviews of all of these events are given in Venkataraman et al. 2005; Moore and Proudfoot 2009; Richard and Manley 2009.)

Termination of transcription always occurs downstream from the poly(A) cleavage site, sometimes by hundreds of bases (Hofer and Darnell 1981; Richard and Manley 2009). If the poly(A) site is not recognized and followed by cleavage, termination does not occur. If 3′ cleavage and poly(A) addition are executed properly, the downstream sequences are rapidly degraded. Termination then occurs, perhaps because a 5′-to-3′ exonuclease recognizes the cut end of the nascent transcript and catches up with the polymerase. The molecular elements of termination are current topics of ongoing research (Richard and Manley 2009).

Recognition of Splice Sites and Splicing Machinery

Soon after the discovery of splicing, consensus sequences at the borders of the vertebrate introns were recognized to be GU and AG, the first two (5′) and last two (3′) intronic nucleotides, respectively (Chapter 4, Fig. 4-32) (Breathnach and Chambon 1981). Of interest, in yeasts a longer 5′ consensus, GUAAGU, is similar but longer and rarely varies, although the 3′ AG is not longer. As more genomic sequences from animals and plants accumulated, it became clear that the consensus at the intron borders was not absolute but was favored.

Major and Minor Splice Site Sequences and Spliceosomes

One additional important fact about splice site recognition came from experiments from Joan Steitz and colleagues. In continuing to look for small RNAs that were similar to U1, U2, U4, U5, and U6, which are part of the major spliceosome, they discovered in 1998 two additional low-abundance snRNAs, U11 and U12 in human cells (Montzka and Steitz 1988). These molecules had the trimethyl guanosine cap ($m^{2,2,7}G$) characteristic of the major U-rich spliceosomal RNAs. Her laboratory then discovered (Tarn and Steitz 1996) that a minor spliceosome removed introns bordered by AT–AC instead of those of the major class bordered by GT–AG. In addition to U11 and U12, these spliceosomes also contain U4 and U6 analogs designated as U4atac and U6atac. Only a few percent at most of introns are in this minor class, but they are present in both animals and plants and must be removed by the minor spliceosome (Tarn and Steitz 1997).

Sequences That Enhance and Silence Splicing

After definition of intron boundary sequences, investigators searched for other sequences that were important in splicing. These experiments took the form of mutagenizing intronic sequences and border sequences in exons that promoted or compromised efficient splicing. Enhancing and silencing sequences were both found. Thus, exon splice enhancers (ESEs) and intron splice enhancers (ISEs), and exon silence sequences (ESSs) and intron silence sequences (ISSs), exist that likely bind effector proteins (Maniatis and Tasic 2002; Black 2003). Many such effector proteins are now known.

Two specific sequence regularities in introns were discovered not by mutagenesis but by sequence comparisons in one case and by biochemistry that examined the released intron in another. First, in the great majority of vertebrate introns, there is a pyrimidine-rich region ~15 nucleotides upstream of the 3′-splice site (Fig. 5-28A) (Maniatis and Tasic 2002; Black 2003). Second, there is always an A residue ~30 nucleotides upstream of the 3′-splice site. This residue gained attention when the excised intron was found to be not linear but a branched structure, with the conserved A residue at the branch (Padgett et al. 1984; Ruskin et al. 1984). (Thus, both the 2′-OH and 3′-OH residues of the ribose in this A residue are linked by phosphates to another nucleotide in the excised lariat structure.) The many steps in the required transesterifications that are now known are depicted in Figure 5-28B,C.

Most experiments have shown that the silencing and enhancing proteins for splicing are present in all cells. As might be expected, a number of these proteins are obligatory for survival (as shown when these genes are knocked out in mice) because each is probably involved in splicing hundreds to thousands of different pre-mRNAs. In a few instances (see the next section), *cell-specific RNA-binding proteins* govern the choices of which exons are included in particular pre-mRNA processing events.

Current Picture of the Spliceosome and Its Action(s)

The spliceosome contains five small nuclear RNAs (snRNAs)—U1, U2, U4, U5, and U6—that are similar in all eukaryotes, a remarkable evolutionary fact (Wahl et al. 2009). Each snRNA forms a ribonucleoprotein particle. There are 100–150 such proteins in all species (more than in ribosomes) and somewhat more proteins in vertebrates than in yeasts. Each U1, U2, U4, and U6 snRNP contains a set of seven shared proteins plus many other unshared proteins. The active spliceosome (not the individual snRNPs) is apparently assembled each time a splicing event occurs (Fig. 5-28B,C)

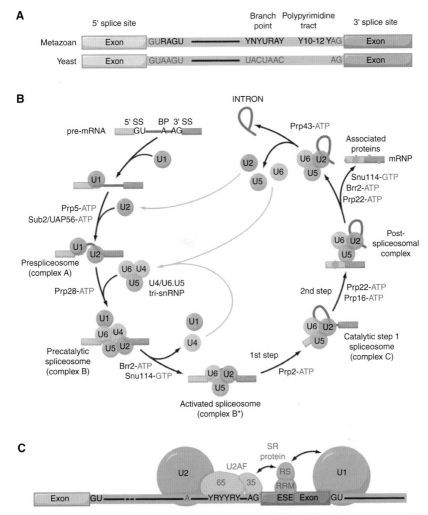

Figure 5-28. Eukaryotic splicing reaction: Stepwise assembly of the spliceosome. (*A*) Diagram of conserved sequences in mammalian and yeast pre-mRNAs. (*B*) Initial spliceosome assembly requires the ATP-dependent binding of U1 small nuclear RNA (snRNA) at the 5′ exon–intron junction. The U1 RNA base-pairs with sequences in both the exon and the intron. In mammals, the U1 snRNP binding is stabilized by accessory proteins (the so-called SR proteins, which are serine and arginine rich). Also early in the assembly, proteins bind the branchpoint around the A residue near the 3′-splice junction. These *branch-point binding proteins* include a factor that recognizes the U2 snRNP and recruits it to the branchpoint complex, ensuring that the nucleotides at or near both ends of the intron have been recognized. The dual U4/U6 snRNP, held together by base-pairing, along with the U5 snRNP, now join the complex. All of the snRNPs are present, but the spliceosome is still not activated. The unwinding of the base-paired U4/U6 and discharge of U1 leaves the U6 free to form a base-pair arrangement with the RNA in the U2 snRNP. (*Legend continued on next page.*)

(Wahl et al. 2009). This assembly pathway and its biochemical/biophysical consequences have been learned mainly from biochemical experiments with pure RNA substrate plus purified accessory proteins and partially pure spliceosomal components isolated from cells. Some of the key points have been reinforced by mutagenesis of snRNAs and proteins with the introduction of recombinant DNA constructs into cells (both yeasts and mammalian cells). In addition, X-ray structural studies are beginning to yield impressive information regarding events at the center of the spliceosome (Zhang et al. 2009). Because the spliceosome is so huge (larger than a whole ribosome), crystallography of the whole structure has not been achieved. Figure 5-28B provides details known or thought to occur during each step in a pre-mRNA splicing event.

The events depicted in Figure 5-28B have been determined with relatively short introns, mainly through biochemistry. It is widely assumed that splicing out very long introns involves the facilitating splice sequences (ESEs, ISEs, etc.), with bound proteins bringing upstream exon termini close to downstream intron termini so that precise splicing out of the intron can occur (Fig. 5-28C) (Maniatis and Tasic 2002; Black 2003). In cases where regulated cell-specific splicing factors have been uncovered, the generally distributed RNA recognition proteins also may be involved, but splicing decisions are ultimately determined by cell (or tissue)-specific splicing factors.

Although it is an important subject on its own, tRNA splicing, which is performed by specific protein enzymes, is not discussed here but is thoroughly covered in reviews (Westaway and Abelson 1995). These enzymes can recognize exon–intron borders, cleave the pre-tRNA, remove introns, and reunite exons.

Figure 5-28. (*Continued*) The splicesome containing U2, U5, and U6 is now activated. The two ends of the 5′ and 3′ exons plus the branchpoint are brought into close proximity. A very large protein called Prp8 (not shown), first identified in yeast but present in all eukaryotic cells, is central to assembling these various elements into the same physical location. Two transesterification reactions occur: The 2′ OH of the branchpoint adenosine attacks the 5′–3′ phosphate linkage at the upstream exon–intron boundary and attaches to the 5′ phosphate. The now-free 3′ OH on the first nucleotide of the upstream exon attacks the phosphate linking the last nucleotide of the intron to the first nucleotide in the downstream exon, and a second transesterification completes the splicing of the exons and releases the intron in lariat form. (The helicases and ATPases that are known to function at various steps in the cycle are highlighted with red ATP and GTP.) (*C*) Because transcription of very long pre-mRNA can take many minutes (a 100-kb pre-mRNA takes 30–60 min to be synthesized), it is thought that splicing assemblies form on nascent transcripts but await binding an exon splice enhancer (ESE) to complete splicing. U1 and U2, with the U2 ancillary factor (U2AF), lead SR proteins plus an RNA recognition protein (RRM) to the complex. (Reprinted, with permission, from Wahl et al. 2009 [©Elsevier].)

Cell-Specific Splicing Factors

Several clear instances of cell-specific differential splicing or poly(A) choice are illustrative of both specially dedicated splicing factors and of the physiological importance of differential splicing.

Sex Determination in Drosophila *Is a Lesson in Differential Splicing*

Male and female embryos of the fruit fly *Drosophila melanogaster* produce two different forms of the Dsx protein (encoded by the gene *double sex*, or *dsx*) that determines sexual differentiation (Fig. 5-29). The *dsx* gene was discovered because mutants produce flies that have some characteristics of each sex (Baker 1989; Cline and Meyer 1996; Schutt and Nothiger 2000).

During the development of sex-determining cells of the fly embryo, the *dsx* gene produces two different proteins as a result of a cascade of differential splicing and poly(A) choices. The action of the two Dsx proteins is to repress genes required for sexual development—female Dsx represses genes required for male development (and vice versa for the male Dsx protein).

This remarkable set of protein:pre-mRNA interactions represents perhaps the most fully understood case of cell-specific differential poly(A) and splicing choices, from both molecular and physiological perspectives. (Note in Fig. 5-29 that the *sxl* and *tra* mRNAs vary in males and females as a result of differential splicing and share the same poly[A] site, whereas the *dsx* mRNAs differ in the exons used in males and females as a result of choice of poly[A] sites.) These clear results emphasize the power of both kinds of differential processing choices in producing different mRNAs.

Calcitonin/CGRP: *Same Primary Transcript, Two Different Proteins in Brain and Parathyroid*

Some years ago, the human parathyroid was found to produce and secrete a hormone, calcitonin, that is required for Ca^{2+} retention. When the gene for this protein was cloned, it was recognized that another short protein, termed CGRP (calcitonin gene-related peptide), was also encoded by this gene (Amara et al. 1984). The differential formation of the CGRP and calcitonin mRNAs was one of the earliest physiologically important processing choices to be recognized. Both mRNAs arise from the same pre-mRNA, which has two poly(A) sites. Choice of the first poly(A) site excludes downstream exons and results in the production of calcitonin (Fig. 5-30). Choice of the second poly(A) site results in the production of CGRP, which excludes one exon found in calcitonin. Thus, based on different poly(A) choices, different 3'-terminal exons are chosen. Thus, in two different cell types in two different

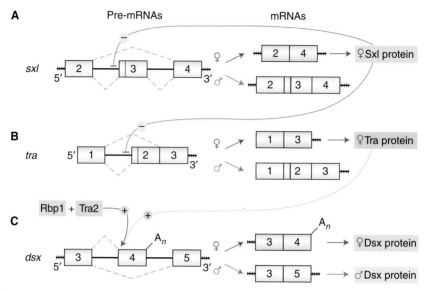

Figure 5-29. Pre-mRNA processing controls sex determination in *Drosophila*. The formation of mRNA for each of the Dsx proteins that determine sex in the fly embryo depends first on differential processing of a pre-mRNA for a protein called sex lethal (Sxl), itself a negative-splicing factor. (Proteins and pathways in pink are female; those in blue are male.) The *sxl* gene encodes several exons, in three of which (2, 3, 4 in the diagram) alternative splicing occurs. A protein stop codon (heavy red bar), when included, prevents formation of a full-length protein molecule. An Sxl protein produced without exon 3 (enforced by the female version of Sxl itself) allows translation of exon 4. The female Sxl protein also binds to the pre-mRNA for a protein called Tra (for "transformation"), again resulting in female-specific splicing, producing *tra* mRNAs with two exons and no stop codons. The functional Tra protein in females, in combination with two constitutively expressed proteins (Rbp1 and Tra2), directs the choice of the first poly(A) site in the *dsx* pre-mRNA, thereby including exon 4 in the mRNA. This choice distinguishes female Dsx proteins from male Dsx mRNA (and proteins), where the second poly(A) site is chosen and exon 4 of the *dsx* pre-mRNA is omitted, but exon 5 is included. This cascade of splice and poly(A) choices thus accounts for two different Dsx proteins. (Reproduced, with permission, from Moore et al. 1993 [©CSHLP]; for additional details, see Maniatis and Tasic 2002.)

organs form two different mRNAs encoding two different proteins from the same pre-mRNA are formed.

Nova Proteins of Mice and Brain-Specific Splicing

Directed by catastrophic medical illnesses that cause a rapid onset of debilitating neurological illness, Robert Darnell (a neurologist and neuroscientist) and colleagues have uncovered a series of mammalian RNA-binding proteins that serve in differential splicing in nerve cells. Nova proteins (similar proteins

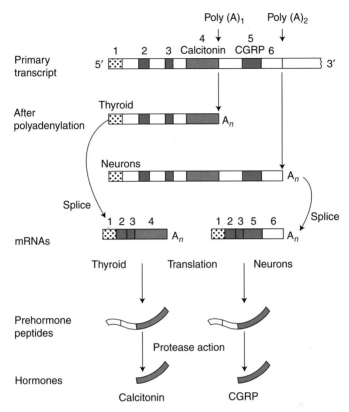

Figure 5-30. Differential processing of a primary transcript from the calcitonin gene results in two products: calcitonin in the thyroid gland and calcitonin gene-related product (CGRP) in certain neurons. The poly(A) site choice and subsequent splicing choices are coordinated. (Solid colored shapes) Coding exons, (stippled boxes) upstream untranslated exons. (Redrawn from Darnell et al. 1990 [©Scientific American Books, used with permission of WH Freeman]; see Rosenfeld et al. 1983.)

from two genes: *Nova1* and *Nova2*) are produced only in the brain, although in many different types of nerve cells in the brain, and are located mainly but not exclusively in the cell nucleus (Ule et al. 2005).

Human cancers sometimes produce Nova proteins aberrantly. Rare individuals respond to the appearance of these usually brain-specific proteins by an overexuberant autoimmune response that kills nerve cells in the brain and leads to severe debilitation, including paralysis and/or mental degeneration and eventually death.

The antiserum of these unfortunate patients allowed identification of the Nova protein and the eventual cloning of the *Nova* gene(s). Domains in the Nova proteins bind strongly to RNA. The favored RNA-binding sequence for the Nova proteins appears in many pre-mRNAs produced exclusively in brain cells.

Early evidence using reporter constructs in cultured cells suggested a role for Nova proteins in differential splicing. However, much more dramatic results came from mice that had undergone a gene knockout of *Nova1*, *Nova2*, or both (Ule et al. 2003). On the functional side, the knockout mice are paralyzed in their hind limbs, mimicking symptoms seen in some patients with the autoimmune human disease. Specific splicing patterns of particular brain-specific mRNAs that occur in normal (wild-type) mice are upset in the *Nova* knockout mice (Fig. 5-31). These experiments clinched the role of Nova in directing at least some specific RNA splicing patterns in nerve cells.

Recent experiments on pre-mRNA sequences cross-linked to Nova protein by UV irradiation and precipitated by Nova antibody (CLIP sites) have then used high-throughput DNA sequencing (HTS) of the cross-linked RNA (converted to DNA) to provide deeper insight into how *Nova* affects splice choices (Ule et al. 2005; Licatalosi et al. 2008). These experiments identify *Nova*-binding sites in the RNA of brain tissue that occur on any genomic site. Depending on whether the binding site occurs before or after a splice junction, a particular exon is included or omitted in the processing of the mRNA. Because there are so many *Nova*-binding sites, this suggests a widespread role for *Nova* in pre-mRNA splicing choices in the brain.

Of particular interest in these experiments is that many of the mRNAs in whose splicing the Nova proteins are involved encode proteins that are found in nerve synapses (Ule et al. 2005). These structures contain the excitatory or inhibitory chemicals required to pass along or suppress signals between nerve cells (Fig. 5-32). Thus, the Nova protein has an important role in synaptic establishment and function.

As more and more specific cases of differential splicing are understood at the molecular level, the physiological importance of differential splicing is brought home ever more forcibly (Licatalosi and Darnell 2010).

RNA-Binding Proteins Do Not Just Aid Splicing

RNA splicing focused attention on the specificity of RNA–protein interactions, but the completion of splicing is only the beginning of a manifold series of mRNA–RNA–protein interactions. The overall function of another set of mRNA-directed protein complexes is *quality control*, i.e., detection of errors in the newly produced mRNA, on the one hand, and the initiation of the processes of destruction if errors are detected, on the other. Following completion of splicing, exon junction complexes (EJCs) bind ∼20–24 nucleotides upstream of an exon–exon junction (for review, see Nott et al. 2004).

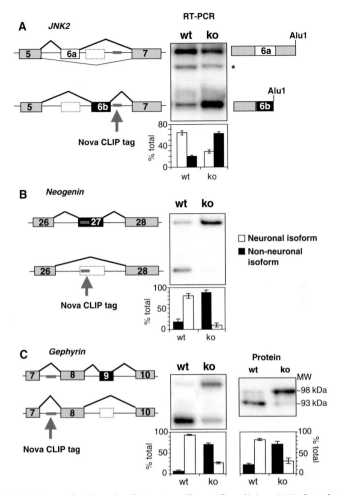

Figure 5-31. Evidence for the role of Nova in cell-specific splicing. RNA from brain tissue from early developing mice (~7 d after fertilization) was examined for position of bound Nova protein to total cell RNA (including nuclear RNA, where pre-mRNA would be found). Tissue from wild-type and knockout animals was irradiated with UV to cross-link protein and RNA. The RNA–protein complexes were precipitated with Nova antibody, and the bound RNA was converted to DNA and sequenced to locate CLIP sites. These CLIP sites were largely in introns (99 of 121), although some were in pre-mRNAs that were suspected to have functions controlled by Nova through differential splicing. (Red "tags" show Nova binding to specific sequences within the predicted Nova-controlled pre-mRNA splices; no RNA with these sequences was precipitated in the knockout animals.) Three selected genes—(A) *JNK2*, (B) *neogenin*, and (C) *gephyrin*—were tested for effects on differential splicing comparing wild-type and knockout mice. The intron–exon arrangement in the three genes tested is shown, as is the difference in splicing patterns determined by copying the brain poly(A)-terminated mRNAs by reverse transcriptase–polmerase chain reaction (RT-PCR). Large differences in differential splicing are obvious between RNA for wild-type (wt) mice and Nova knockout (ko) mice in the neuronal iso-forms of each mRNA, confirming the importance of Nova in splicing of brain-specific mRNAs. (Reprinted, with permission, from Ule et al. 2003 [©AAAS].)

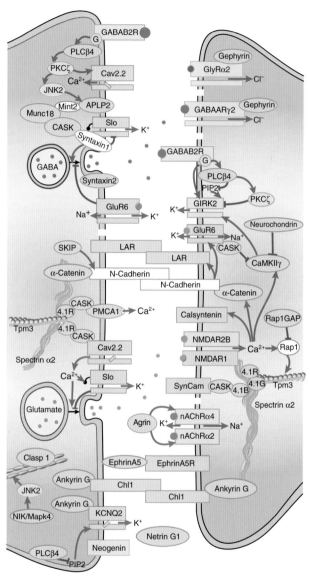

Figure 5-32. Nova proteins act in differential splicing in many proteins found in synaptic vesicles. Figure shows a synaptosome (junction of two synaptic vesicles), highlighting proteins known to have a role in synaptic transmission whose mRNAs show Nova-regulated differential splicing. In an HTS-CLIP analysis using wild-type/knockout mouse brain tissue comparisons, many new Nova targets were identified (see Fig. 5-31). These validated targets are in yellow (Ule et al. 2003) and orange (Ule et al. 2005). RNAs with the highest number of cross-links in the HTS-CLIP analysis are in blue. The putative connections between the various proteins (blue and red arrows and lines) are taken from the literature. (Reprinted, with permission, from Ule et al. 2005 [©Macmillan].)

At least 10 proteins are involved in this action, and in model systems, the complexes demonstrably speed up nucleocytoplasmic export. A very detailed pathway called nonsense-mediated decay (NMD) is integrated with EJCs by protein: protein associations. The action of NMD occurs in the cytoplasm, where nonsense codons within ~50 nucleotides upstream of an exon–exon junction are recognized, targeting the potentially errant mRNA for destruction.

The field of eukaryotic mRNA destruction in the cell cytoplasm is now so well studied that two whole volumes on the subject have recently been published (Maquat and Kiledjian 2008a,b). Some of this work covers newly discovered small RNA–mediated mRNA decay, but much of it concerns other RNA–protein interactions throughout the cytoplasm, including those of NMD.

Finally, recognition of sequence features of an RNA, often in stem loops, serves novel roles in positioning mRNAs at specific sites in the cell cytoplasm (Holt and Bullock 2009). A growing list of mammalian mRNAs is now known to be conducted to specific cellular destinations, where the protein encoded by a specific mRNA will be used when translation occurs.

Each of the subjects mentioned briefly here has proven to be a fertile area of cell biological exploration and emphasizes the enormous biochemical importance of the existence of recognizable physical structures in mRNA beyond the "simple" protein-coding function of mRNA (Cruz and Westhof 2009).

MRNAS ARE UNSTABLE MOLECULES

Bacterial mRNA Turnover

From the first studies of bacterial mRNA, it was recognized that mRNA molecules had an evanescent lifetime as the basis of prompt changes in enzyme synthesis. Some mRNAs lasted perhaps as long as 10 min, but others were gone within a minute or two after their synthesis ceased.

Recent work has described the enzyme complexes that participate in destroying bacterial mRNAs (Deutscher 2006). For example, in *E. coli*, RNase E is an essential protein. The degradative enzyme polynucleotide phosphorylase—the enzyme capable of forming polyribonucleotides when given high concentrations of nucleoside diphosphates—is also important in mRNA turnover. It is established that endonucleolytic cleavage is likely the first step of turnover, and a redundant battery of nucleases probably can all take part in returning the mRNA to 5′ ribonucleotides that can then be reused. Many studies have also revealed that culture conditions can change the half-lives of mRNA (Kuechenmeister et al. 2009).

Eukaryotic mRNA Turnover

The considerable difference between bacterial mRNA—chemically unadorned and ready for immediate transcription—and eukaryotic mRNA—with its 5′-m^{7}GpppN cap and 3′-poly(A) tail—leads, not surprisingly, to more complicated pathways of turnover in eukaryotes.

The instability of eukaryotic mRNA was hinted at in the first experiments that detected polysomal mRNA (Penman et al. 1963). When HeLa cells were treated with actinomycin, which stops all RNA synthesis, protein synthesis on polyribosomes declined with an average half-life of 2–3 h (Chapter 4, Fig. 4-11), presumably because of mRNA instability.

Because the use of an inhibitor such as actinomycin risked not reflecting an accurate decay rate, later studies followed radioactivity into mRNA and the loss of radioactivity in time with no drug treatment. The complicating factor in these experiments was the inability to completely and quickly "chase" the labeled RNA precursor from the large intracellular ribonucleotide pools (Puckett et al. 1975; Puckett and Darnell 1977). Nevertheless, estimates of half-life under the best chase conditions suggested that HeLa cell mRNAs exist with half-lives of from ∼1 h or less to many hours (12 h or more). When cDNAs allowed individual mRNAs to be measured, the rate of accumulation to a maximum level of radioactivity in individual mRNAs (another method of measuring turnover rate) proved that both short- and long-lived eukaryotic mRNAs existed (Harpold et al. 1981).

Most recently, the enzymatic mechanisms of mRNA turnover have been thoroughly explored. Several dozen proteins participate in mRNA turnover (Wilusz et al. 2001; Parker and Song 2004). It suffices here to say that a general pathway is central: poly(A) shortening occurs as mRNA ages and, probably at a critical short length, the mRNA enters the turnover pathway. Decapping (removal of the m^{7}GpppN cap) then occurs, and any one of several nucleases (5′ to 3′ or 3′ to 5′) completes the task. It is interesting, and in keeping with what has been discovered regarding RNA synthesis and processing, that yeast (*S. cerevisiae*) and mammalian cells have similar basic turnover machinery, with mammalian systems having added additional proteins. Almost 20% (∼4000) of all mammalian mRNAs have sequences referred to as AU-rich elements (AREs) in their 3′ untranslated regions (UTRs). These elements are particularly common in mRNAs controlled by cytokines and growth factors (as well as these factors themselves) and in transcription factors, all of which have quite short half-lives (Bakheet et al. 2006). Proteins (complexes) that recognize these AREs are well known and are responsible for ushering mRNAs, leading these sequences into a turnover pathway.

Excellent thorough reviews of the myriad details of mRNA turnover and how differential turnover times can be imposed on mRNAs are available (Garneau et al. 2007; Khabar 2010).

It is obvious that regulated mRNA turnover is a crucial element in overall gene control.

SMALL RNAs SUPPRESS CYTOPLASMIC mRNA FUNCTION

One of the most explosive areas of research in eukaryotic gene regulation in the 1990s and after 2000 deals with so-called microRNAs (miRNAs) and short interfering RNAs (siRNAs). These molecules, ubiquitous in plants and animals but not present in at least some single-cell eukaryotes (e.g., the choanoflagellate *Monosiga brevis* and baker's yeast), were a surprise discovery of the 1990s and early 2000s. Three different concomitant pathways of discovery, two involving the nematode *Caenorhabditis elegans* and one involving a variety of plants, uncovered the existence and regulatory functions of miRNAs and siRNAs.

C. elegans: A Genetically Tractable Developmental Target

The story leading to the discovery of miRNAs and siRNAs begins with the great geneticist Sydney Brenner. After contributing so much to bacterial genetics from the late 1950s through the early 1970s, Brenner started a school of research in the mid 1970s to study the nematode *C. elegans* (Brenner 1974) to provide a "simple" model organism with which to study development, especially neurobiology. A comprehensive genetic map of this organism's six chromosomes was obtained (as was the entire genomic sequence much later). Perhaps the most amazing initial discovery came from careful microscopic examination of this little transparent creature by John Sulston along with Brenner. (Using Nomarski optics and light microscopy, the nucleus of each cell can be identified by the trained observer.) The adult worm contained an exact number of cells (959) in exactly the same place in each of the animals. Furthermore, the 959 adult cells came from a precise number of precursor cells. The tissues of the animal are formed by cell division plus specific *programmed cell death* of 131 of the cells during the development of the worm. For example, each of the 302 nerve cells in the *C. elegans* adult comes from a precise division schedule of precursor cells, with specific cells continuing and others dying. These remarkable properties made this organism a favorite for studying genes that are responsible for specific cell decisions that lead to proper tissue differentiation (summarized in Riddle et al. 1997). Knowing the biological details then led to a set of remarkable and surprising molecular findings with *C. elegans*.

One of Brenner's younger colleagues, Robert Horvitz, established at The Massachusetts Institute of Technology (MIT) one of the (now dozens of) leading United States outposts for *C. elegans* work. Horvitz and colleagues identified the genes responsible for the programmed cell death pathway (Horvitz 2003).

A similar cell death pathway occurs in the cells of all animals and is important for normal development. For example, the fingers and toes of mammals originate with webbing between them, like those of a duck's feet. The tissue in this webbing is then reabsorbed by regulated programmed cell death as the fetus develops.

Programmed cell death occurs by *apoptosis*, a stepwise activation of pro- teases that digest key proteins in the targeted cells. Normal cells that suffer genomic damage (e.g., from X rays) enter this cell death pathway and are removed. Cancer cells somehow avoid apoptosis subsequent to genomic dam- age and aberrant chromosome arrangements, and they continue to grow. Brenner, Sulston, and Horvitz shared the 2002 Nobel Prize in Physiology or Medicine for their discoveries (Brenner 2003; Horvitz 2003; Sulston 2003).

lin4 *and* lin14: *Cytoplasmic RNA–RNA Regulation Discovered*

In the late 1980s, two of Horvitz's younger colleagues, Victor Ambros and Gary Ruvkun, in their own independent laboratories, were studying hetero- chronic genes. These genes are involved in the correct *timing* of the four larval stages of *C. elegans* (L1–L4) on their way to becoming adult animals. Genetic experiments with the heterochronic mutants (Ambros 1989) indicated that, as the animal went through larval development, a gene called *lin4* repressed a gene called *lin14*. When molecular genetics made it possible to clone any desired gene from *C. elegans*, it was found by the Ruvkun laboratory that *lin14* encoded a protein—no surprise (Ruvkun and Giusto 1989). Using anti- bodies to score lin14 protein, they found the protein to be very prominent in L1 but virtually gone in L2. They then found that the sequences in the *lin14* gene required for *lin4* to repress *lin14* lay in the 3' UTR of the *lin14* mRNA, not in the amino-acid-coding portion of the mRNA (Wightman et al. 1991).

When *lin4* was cloned by the Ambros laboratory, the sequence did not encode a protein at all (Lee et al. 1993). No long sequence uninterrupted by stop codons existed. However, they knew that they had the right piece of DNA in their clone. The wild-type DNA, 693 nucleotides long and from the genetically mapped *lin4* region, was introduced into the fertilized egg of *lin4* mutant animals. It cured the developmental defect caused by the *lin4* mutation. In the same paper, Ambros and his students showed that the *lin4* gene caused the production of two short RNAs, ∼25 and ∼60 nucleotides long (Lee et al. 1993). The 60-nucleotide product contained the sequence of the ~25-nucleotide product and could be folded into a hypothetical stem–loop structure.

Ambros and Ruvkun were in close contact, and they discovered, by comparing the sequence of short RNAs from the *lin4* gene with the 3'-UTR sequence of the *lin14* gene, that the two regions had complementary segments between the short RNA of *lin4* and the 3' UTR of *lin14* (Fig. 5-33) (Wightman et al. 1993). Deletion of these complementary sequences from *lin14* blocked *lin4* repression of *lin14*. Deletion of the two stretches with the closest base-pair match had the strongest effect. As a harbinger of what was to come, the *lin4/lin14* complementarity was interrupted in the middle by an ~10–20-nucleotide stretch. The 5' end of the small RNA had the strongest base-pair match with the several matching sequences in the *lin14* 3' UTR. This turned out to be a rule in the regulatory short RNAs discovered later. The Ambros and Ruvkun laboratories had clearly discovered an RNA:RNA control

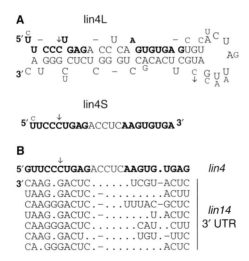

Figure 5-33. Short RNAs encoded by *C. elegans lin4* have complementary regions to *lin14* mRNA. A 693-nucleotide DNA sequence suffices to rescue *lin4* mutants, and this was sequenced. (*A*) RNase protection assays in wild type but not in *lin4* mutants (total RNA) identified a less common ~60-nucleotide (*lin4L*) and a quite common ~20-nucleotide molecule (*lin4S*), the sequence of which resided within the 693-nucleotide *lin4*-containing DNA. The *lin4L* molecule is shown as a possible folded "hairpin." The sequence near the 5' end of *lin4L* (boldface characters) is preserved in *lin4S*. (The faint nucleotides written above various loci are sequence changes between *C. elegans* and *Caenorhabditis briggsae*, another nematode.) (*B*) 3'-UTR region of the *lin14* mRNA (Wightman et al. 1991) contains seven regions (aligned) with at least partial complementarity to *lin4* (boldface characters). (Reprinted, with permission, from Lee et al. 1993 [©Elsevier].)

of a particular *C. elegans* mRNA. But perhaps this was just a molecular curiosity (Ambros 2008).

let7: *Widely Shared Small RNA*

It was several years later that research in the Ruvkun laboratory, in collaboration with the Horvitz laboratory, found a second small RNA encoded by another heterochronic regulatory gene called *let7* that negatively regulated the *C. elegans lin41* gene (Reinhart et al. 2000). The sequence of the short *let7* RNA broke open the field of small RNAs in animals. Ruvkun and colleagues (Pasquinelli et al. 2000) found, by comparing whole-genome sequences, that the *Drosophila* and human genomes had sequences almost exactly like that of the *let7* gene of *C. elegans*, which could, as transcribed RNA molecules, fold into stem−loop structures (Fig. 5-34A). Moreover, by RNA analysis (northern blot), the *let7* sequence was present early in development across a dozen or more animal phyla. When the comparisons to available genome data were complete, it was obvious that all animals above cnidarians, but no single-cell eukaryotes, had a gene similar to *let7* (Fig. 5-34B).

The floodgates were about to open on small RNA inhibition of mRNA function. Two other near-simultaneous approaches to RNA gene silencing were proceeding, and these need to be explained before describing how RNA inhibition of mRNA function works.

Injected dsRNA Triggers mRNA Suppression

In 1998, two laboratories, those of Andrew Fire and Craig Mello, also working with *C. elegans*, reported a remarkable conclusion to work begun several years before (summarized in Fire et al. 1998). During his postdoctoral time in Sydney Brenner's laboratory in the mid 1980s, Fire had tried to stop mRNA function by injecting various nucleic acid constructs into early embryos. Specifically, he attempted to prevent function of the wild-type *unc22* mRNA, which encodes a muscle protein (Fire 2007). Worms with an *unc22* mutation show a peculiar twitching behavior, and this phenotype allowed a test for whether an injected nucleotide stretch had affected the *unc22* gene. In an attempt to disrupt function, recombinant DNA expression constructs producing *antisense* RNA fragments to the *unc22* mRNA were injected into larvae, occasionally resulting in twitching worms. However, similar results were achieved using DNA constructs that produced only *sense* RNA.

The puzzling results with *unc22* DNA constructs (Fire et al. 1991) were later extended by illuminating experiments of Su Guo and Ken Kemphues (1995). These investigators injected antisense RNA to *PAR1* mRNA in an attempt to phenocopy mutations of the gene, and they succeeded. However, again sense

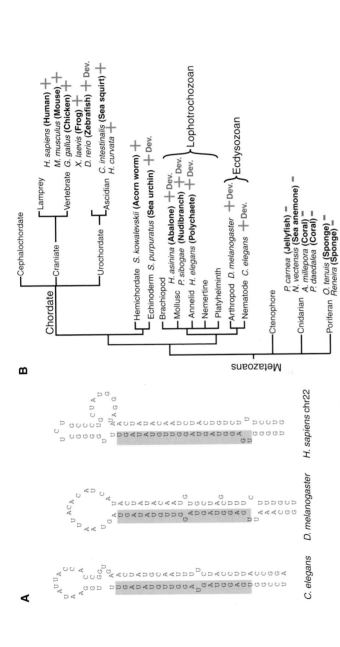

Figure 5-34. Sequences similar to *C. elegans let7* RNA: phylogenetic comparison. (*A*) Shown are *let7* DNA sequences in *C. elegans*, *D. melanogaster*, and *Homo sapiens*, drawn as expressed RNAs folded into stem loops. The gray area corresponds to the short RNA expressed in cells from each species. (*B*) Phylogenetic distribution of *let7* sequences (+, green) shows *let7* expression. "Dev." (blue) indicates expression in adults. Species lacking *let7* are marked in red (−). (Modified, with permission, from Pasquinelli et al. 2000 [©Macmillan].)

RNA worked almost as well. Craig Mello also used direct injection, again of sense and antisense RNA, to inhibit expression of yet another gene (Rocheleau et al. 1997). Because the phenomenon was *not* based exclusively on antisense RNA, it was referred to by Mello as *RNAi* (for "RNA inhibition").

Mello and Fire then teamed up to try to explain this phenomenon. The two laboratories systematically tested a variety of *unc22* sequences by injecting separately single-stranded sense and single-stranded antisense RNA as well as *dsRNA* molecules. Double-stranded *unc22* sequence-specific RNA proved to be more than 100 times more effective in producing twitching animals than either the sense or antisense single strands alone (Fig. 5-35).

This experiment was a bombshell. Laboratories all over the world working on animal cells of all kinds rushed to make dsRNAs of sequences of many different mRNAs to reduce the concentration of particular proteins. It was as if the research community was suddenly gifted with the ability to mutate any particular gene at will to study the functional role of its encoded protein. These so-called *knockdown experiments* became among the most frequent experiments in the new century. How dsRNA accomplished this near miracle is described following the next section.

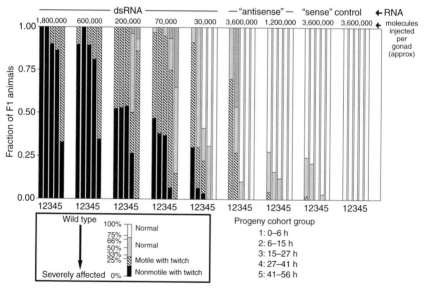

Figure 5-35. Double-stranded RNA inhibits UNC22 function in *C. elegans*. Andy Fire, Craig Mello, and colleagues injected groups of *C. elegans* adults with nucleic acid constructs representing the coding regions of the *Unc22* gene in various doses. They scored for a "twitching" phenotype in the progeny (fractions of animals) at various times after injection. Purified dsRNA was vastly more effective than either sense or antisense RNA in producing twitching. (Modified, with permission, from Fire et al. 1998 [©Macmillan].)

Inhibitory Small RNAs (siRNAs) in Plants

The earliest entries in the story of natural inhibitory RNA actually came from plant biology. It was observed many decades ago that when tobacco ring spot virus infected the lower leaves of plants, the virus did not enter the meristem to infect the upper leaves. Moreover, recovery from plant virus infection was followed by resistance to reinfection. This phenomenon was not explicable by immunity involving antibody-forming cells, as in animals (for review, see Herr and Baulcombe 2004).

When the introduction of transgenes into plants was perfected, transgenes encoding plant mRNAs or viral sequences were installed into recipient plants. Expression of any mRNA from a transgene eventually was accompanied by repression of the homologous host gene or virus. It seemed that any foreign RNA that the plant encountered—viral RNA or RNA products of transgenes—was recognized, and subsequently any such sequence of foreign or host origin was destroyed (Lindbo et al. 1993; Longstaff et al. 1993).

To explain the silencing phenomenon, David Baulcombe proposed that a short inhibitory RNA molecule (or siRNA) was involved—the virus genome that was first recognized to be effective in instigating the phenomenon was, after all, RNA. Andrew Hamilton in the Baulcombe laboratory examined plants in which a transgene had silenced a host gene and found short RNAs ($\sim$25 nucleotides long) in the suppressed plant cells; the sequences of these RNAs were complementary to the mRNAs of the genes whose activity was suppressed (Fig. 5-36) (Hamilton and Baulcombe 1999; for review, see Baulcombe 2006). In these experiments, both sense and antisense short RNAs were present, and it was conjectured that perhaps RNA-dependent RNA polymerase was involved. But neither the mode of action of the short RNAs nor their route of synthesis was clear.

GENOMES ARE FULL OF SEQUENCES THAT ENCODE miRNAs

The genie was out of the bottle—small RNAs were found in abundance in plants, worms, flies, and humans. Copying all RNAs less than $\sim$30 nucleotides long into DNA (cDNA) using reverse transcriptase and then cloning and sequencing the resulting cDNA identified literally hundreds of small RNAs. Sequences were matched against total genomic sequences to locate the genes that the cell might copy to make the small RNAs (for review, see Ambros 2008).

The genomic sequences encoding what came to be called miRNAs can be located within introns of protein-coding genes, where they are likely to be transcribed by Pol II, as well as in separate genes. Some are found in clusters that are all transcribed together, probably also by Pol II. Some that are

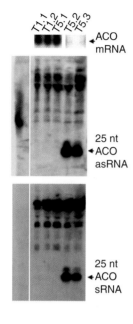

25 nt
◄ACO
asRNA

25 nt
◄ACO
sRNA

Figure 5-36. Plants contain short inhibitory RNAs (siRNAs) targeted against transgenes. RNA from five transgenic tomato lines was examined by electrophoresis. The sequence of *ACO* (1-aminocyclopropane-1-carboxylate oxidase) was installed behind a virus promoter to perform transgenesis. In three of the transgenic lines, the host *ACO* was not suppressed, and in two, the *ACO* mRNA was suppressed. Short *ACO*-specific sense and antisense RNA sequences (~25-nucleotide *ACO* sRNA and asRNA) were present in the two lines in which the endogenous gene was suppressed. (Reprinted, with permission, from Hamilton and Baulcombe 1999 [©AAAS].)

scattered among repetitive DNA regions such as centromeric or telomeric DNA are probably transcribed by Pol III (Bartel 2004; Borchert et al. 2006). There are more than 800 identified miRNAs in mammals (Aravin and Hannon 2008; Bartel 2009), more than 100 each in *Drosophila* and *C. elegans*, and perhaps many thousands in plants (Bartel 2009). Many of the mammalian miRNAs are expressed in a tissue-specific fashion, and this is likely true in all animals. By performing PCR elongation with antisense miRNA strands and total nuclear RNA, it became clear that longer (150–200-nucleotide) precursor molecules, pri-miRNAs, are the primary transcripts from which the miRNAs arise. These primary products, pri-miRNAs, can be capped, polyadenylated molecules that form stem–loop structures (see Fig. 5-37).

Pri-miRNA and Pre-mRNA Processing: RISC and Argonaute

Because the miRNAs derive from longer nuclear RNA transcription products, RNA processing is clearly required to produce the miRNAs. Early *C. elegans* and *Drosophila* genetics uncovered several genes that did not encode miRNAs but were thought to be involved in producing miRNAs (e.g., *rde1-4*) (for review, see Fire 2007). Biochemistry tracked down the proteins encoded at these sites. Two of these were processing enzymes (Drosha, Dicer), and others were found in cytoplasmic complexes containing the finished miRNA (Fig. 5-37 and 5-38).

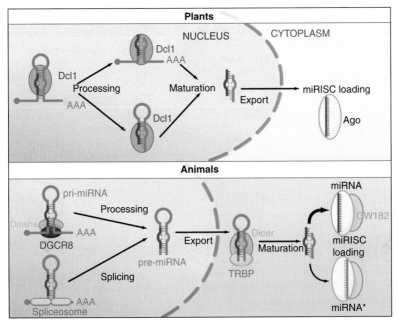

Figure 5-37. Nuclear processing of pri-miRNAs in plants (*top*) and animals (*bottom*) and loading of siRNAs into cytoplasmic RNPs. Primary transcripts—pri-miRNAs—can be capped and polyadenylated. (Also shown in animals is a possible derivation by splicing of the pri-miRNA from an intron of a coding gene.) The Drosha nuclease (or Dcl1 in plants) reduces the pri-miRNA to a pre-miRNA that is exported to the cytoplasm. In animals, Dicer-directed events lead to production of active RNA:protein complexes that suppress mRNA. (Reprinted, with permission, from Carthew and Sontheimer 2009 [©Elsevier].)

The original cleavage of the pri-mRNAs is performed in animals by the nuclear enzyme *Drosha* and in plants by a similar enzyme, Dcl1 (Fig. 5-38) (Lee et al. 2003; Carthew and Sontheimer 2009). The cleavage leaves a 60–70-nucleotide-long pre-miRNA that is further cleaved in the cytoplasm by the enzymes in the Dicer complex (Bernstein et al. 2001), which have both cleavage and helicase domains (for review, see Bartel 2004). Dicer activity leaves a hybridized product containing both the functional 22-nucleotide miRNA and its complement, denoted miRNA* (Fig. 5-38).

When mechanistic studies began to determine how small RNAs functioned, miRNA and siRNA were thought to be separate pathways. miRNA was thought to be "endogenous" (i.e., the genome contained information for its production), and siRNA was thought to be "exogenous" (i.e., resulting from dsRNA of invading viruses or from injected dsRNA). However, overlapping transcription of both strands of a DNA region in the cell nucleus gives

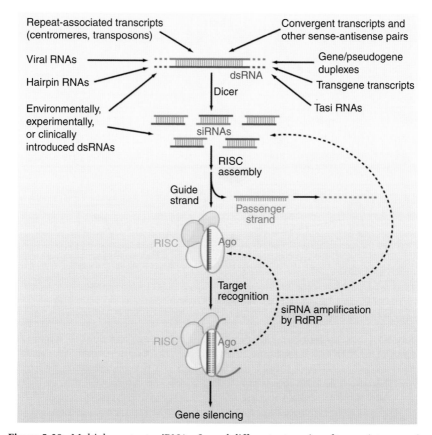

Figure 5-38. Multiple routes to siRNAs. Several different categories of transcripts can exist as dsRNA structures that can be processed by Dicer into siRNAs. These can derive from molecules transcribed in the cell or dsRNA introduced from outside the cell. The RNA duplexes can be intramolecular or intermolecular, and although most are perfectly base paired, some or not. The dicer product (here, referred to as siRNA) [in red]) consists of a guide strand (red) that assembles into a functional RNA-induced silencing complex (RISC) ribonucleoprotein and a passenger strand (blue), which is ejected and degraded. All forms of RISC contain the functional strand of siRNA bound to an Ago (Argonaute) protein, and many, if not most, forms of RISC contain additional protein factors. Target mRNAs are then recognized by base pairing, and silencing ensues by either cleavage or translation repression. (Reprinted, with permission, from Carthew and Sontheimer 2009 [©Elsevier].)

rise to dsRNA and thus furnishes an "endogenous" double-stranded precursor. Dicer activity processes whatever double-stranded substrate it is offered and delivers the immediate trimmed double-strand product to the RISC complex (Fig. 5-38), where the effective, siRNA single strand is selected. The strand selection from the double-stranded precursor (not understood) is performed in the RISC protein complex (Fig. 5-37). One RNA strand, referred to as the *passenger*, is ejected/destroyed; the *guide* strand survives. This removal is performed by the "slicer" enzyme activity of the RISC also (Liu et al. 2004). The RISC also contains proteins from the Argonaute (Ago) family. There are multiple Argonaute family members in plants and animals. Each of the small RNAs, regardless of origin, associates with an Argonaute protein in an RNP complex. Suppression or cleavage of the target mRNA is directed by base-pairing of the siRNA or the miRNA to the mRNA, to effect either translational repression or cleavage of the target mRNA.

There is, however, one important difference in siRNAs of some organisms. Surpisingly, siRNAs are amplified in both plants and *C. elegans*. There exist RNA-dependent RNA polymerases (RdRPs) that can copy single-stranded RNAs. This amplification results in huge numbers of copies of siR-NAs per cell (as many as 50,000 copies per *C. elegans* cell) (for review, see Lim et al. 2003; Guidiyal and Zamore 2009).

Sites in mRNA as Targets for Small RNAs

One of the most daunting challenges for understanding miRNAs is to discover their in vivo targets. This problem was already suggested after the discovery of *lin4* and *lin14* (Fig. 5-33), because *lin4* has only short and imperfect base-pair matches with the purported target *lin14* mRNA.

Using computational methods, Bartel and colleagues (Bartel 2004, 2009) have been instrumental in pointing out the most important *potential* complementary target sequences—so-called *seed sequences*—in mRNAs and miRNAs. Seed sequences can be located in the 3′ UTRs or in the body of mRNAs. The regions of approximate complementarity between miRNA and the target site in an mRNA are only 6–8 nucleotides long and occur by chance quite often in the human genome of 3×10^9 bp (8 mers occur every ~65,000 bp).

A recent report provides a substantial improvement on locating sites in mRNAs to which the RISC complexes are actually bound (Chi et al. 2009). These experiments used UV irradiation of thin brain slices to cause in situ cross-linking of Argonautes in the RISC with the presumed interacting miRNA–mRNA oligonucleotides. From the crystal structure of an archaeal Argonaute protein–RNA complex (Wang et al. 2008), it seemed to be likely that the molecular contacts were sufficiently close to allow UV cross-linking of Argonaute protein to each of the RNA oligonucleotides expected within the RNA–protein complex, one in the miRNA and one in the mRNA. In fact, when the Argonaute-precipitated RNA from the UV-cross-linked complexes was sequenced, both interacting RNA sites were recovered.

Approximately 40% of the apparent RISC-bound sites were recovered in the 3′ UTRs of the mRNAs and 25% were in coding regions. A large fraction of the cross-linked regions were in computationally predicted sites. However, many computationally predicted sites did not show up as being cross-linked.

Piwi-Interacting RNA

Recently recognized small RNAs that deserve status as a separate grouping are called Piwi-interacting RNAs (piRNAs). The name comes from the *Drosophila* gene *piwi*, which encodes a protein with a domain common to Argonaute proteins that is necessary for proper germ cell formation. The Piwi-type proteins (Piwi, Aubergine) are expressed in male and female germ cells in many animals (for review, see Aravin and Hannon 2008) and have been studied most thoroughly in *Drosophila* and mice.

The small RNAs associated with Piwi proteins in *Drosophila* (Vagin et al. 2006; Brennecke et al. 2007) are slightly larger than miRNAs (up to 30 nucleotides compared to 22 nucleotides) and less homogeneous in size than the previously recognized siRNAs and miRNAs. Most of these RNAs are complementary (antisense) to many different coding regions of *retroposons*, the repeated sequences scattered throughout the *Drosophila* and mammalian genomes. There is another striking difference. They do not depend on Drosha or Dicer for their manufacture.

At present, RNAs associated with the Piwi protein are thought to arise from cleavage of dsRNA due to transcription of both strands of a retroposon. Logically, a small RNA that binds to a retroposon transcript, leading to its destruction, would protect the genome by limiting the spread of the retroposon, providing at least a teleological reason for the existence of piRNAs (for review, see Hannon 2008; Malone and Hannon 2009).

Specific Actions of Small RNAs in Mammals

The totally unanticipated discovery of the miRNA and siRNA molecules has had profound effects on gene regulation research as well as on a practical level, promising the use of such RNAs to control levels of particular mRNAs, e.g., in disease. The precise roles(s), if any, of all of these molecules are not yet clear. For example, it was noted in the Robert Horvitz laboratory that individual removal of the great majority of the miRNAs in *C. elegans* produced no discernible effects on the animals, at least under laboratory conditions (Miska et al. 2007). However, from the original *C. elegans* work on *lin4* and *let7*, it is clear that individual miRNAs do, indeed, have specific biological roles.

In mice, evidence is rapidly gathering that different cell types, including stem cells, *do* produce different arrays of miRNAs (for review, see Hannon 2008; Bartel 2009). Solid evidence for the function of specific miRNAs on specific mRNAs in different mammalian cell types is accumulating rapidly. A well-studied case occurs in mammalian cardiac and skeletal muscle (Williams et al. 2009; Small and Olson 2011). The synthesis of muscle-specific miRNAs (the miR-133 group and miR-206) is controlled by transcription factors (MyoD and myogenin) that were known for years to control transcription of pre-mRNAs for a variety of muscle-specific proteins (Fig. 5-39). Dramatic evidence of the physiological importance of the miRNAs comes from specific deletions in mice. When miR-133a1 and miR-133a2 were deleted, the hearts of the resulting animals were dilated as they aged, and this resulted in sudden death in the mice (for review, see Williams et al. 2009). It is notable, however, that the miRNA deletion is not lethal but results in a late phenotype in adult mice.

Another example of a single miRNA of demonstrable physiological importance is miR150, which is crucial in late-stage B-cell differentiation and function (Xiao et al. 2007). miR150 has an apparent role in balancing the concentration of a transcription factor, cMyb, within a narrow ~twofold range. Another molecule, miR155, is crucial in the development of specialized regulatory T cells (Tregs). The absence of miR155 results in a several-fold rise in SOCS1 (suppressor of cytokine signaling 1). SOCS1 has the function of suppressing the action of the transcription factor STAT5. The increase in the suppressor resulted in a decreased number of Treg cells not in their total

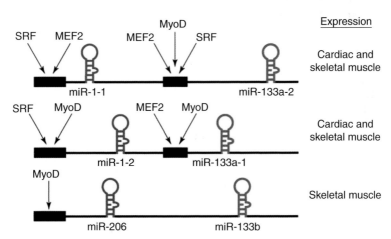

Figure 5-39. Muscle-specific miRNA genes in mice. Shown are three sets of tandemly clustered miRNA genes (abbreviated miR) and binding sites for muscle cell regulatory activator proteins (MyoD, MEF2; black rectangles). (SRF) General transcription activator in many cells. (Reprinted, with permission, from Williams et al. 2009 [©Elsevier].)

absence. Klaus Rajewsky and colleagues suggested that perhaps the physiological role for miRNAs in general may be to help balance the concentrations of particularly critical proteins such as transcription factors (Lu et al. 2009).

It is even clear that, in specific instances, miRNAs, by virtue of having their target sequence embedded in a crucial site, can directly affect transcription. The miR-320 sequence is in the promoter of the cell cycle gene *POLR3D*, in the antisense direction. This miRNA can recruit the Pc2 Polycomb complex as part of an Argonaute-containing complex in the nucleus and repress this gene (Kim et al. 2008).

It certainly appears to be true that in many instances, the reported effect of an individual small RNA is to control mRNA or protein content by only a factor of 2–3, nevertheless resulting in a distinct phenotypic change. One obvious lesson to draw from such results is that many specialized cells depend on a delicate balance of proteins to function normally.

LONG NONCODING RNAs

Studies on mammalian RNA in the 1960s and 1970s simply involved incorporating radioactive nucleosides into cellular RNA, extracting the RNA, and using the methods available at the time examining the RNA for size, determination, base composition, detection of chemically modified bases, and specific measurement of the relative synthesis rate using cDNAs in run-on analyses. These explorations naturally led to the recognition of ribosomal precursor RNA and extranucleolar long RNA, referred to as hnRNA (Darnell 1968). Later studies (Darnell 1975; Salditt-Georgieff et al. 1981; Salditt-Georgieff and Darnell 1982) showed that not all of the hnRNA appeared to be pre-mRNA. Nine individual polysomal mRNA sequences were present mainly in the ~40% of long nuclear RNA that was polyadenylated. The long nonpolyadenylated nuclear RNA was greatly depleted in these sequences. Apparently, hnRNA contained many large capped molecules other than pre-mRNA (Salditt-Georgieff and Darnell 1982).

Until recently, only one long noncoding nuclear RNA, discovered in 1991, has claimed much attention. This RNA, the *Xist* RNA, is involved in X-chromosome repression in mammals and is discussed below. With the advent of high-throughput DNA sequencing, thousands of specific long noncoding RNAs have been discovered and clearly seem to be of great importance in regulating chromosomal events.

Xist, the X-Inactivation RNA: Noncoding RNA with a Function

Mary Lyon showed in 1961 (Lyon 1961, 1998) that in adult female somatic cells of mammals, one of the two X chromosomes was inactive. The inactivation in somatic cells was random but occurred in all cell types. The basis for this near-chromosome-wide repression has been the object of intensive research ever since.

The genomic region of the inactivated chromosome that is responsible for this repression is the X-inactivation center (Xic). The Xic is near the center of the X chromosome and is rich in repetitive, noncoding sequences, whereas most of the relatively few protein-coding genes of the X chromosome are located in the terminal one-fourth of the chromosome. In 1991, Carolyn Brown and colleagues in Huntington Willard's laboratory (Brown et al. 1991) reported a long (17–19 kb) noncoding RNA transcript named *Xist* that is copied from a site within the Xic. The *Xist* gene region also encodes a 5–6-kb noncoding RNA, termed *Tsix*, and a shorter antisense transcript called *Xite*. Both *Tsix* and *Xite* RNAs are antisense to *Xist* and are probably effective in *preventing* X inactivation. In mammals between days 3.5 and 5.5 postfertilization, uterine implantation of the blastocyst signals the onset of differentiation and triggers the random choice to inactivate one of the two X chromosomes in the cells of the inner cell mass, which becomes the embryo proper. At this time, *Xist* RNA becomes the dominant transcript compared to *Tsix* and *Xite* (Chow and Heard 2009). (In the remaining unsuppresseed X chromosome of female cells, *Tsix* remains the dominant RNA product.) The *Xist* RNA accumulates on the chromosome from which it is transcribed and is found associated with virtually the entire suppressed X chromosome. All but a few genes residing on the X chromosome are inactivated (Wutz 2007).

Recent studies with histone antibodies have shown that extensive *Xist*-covered regions of the inactive X lose H3K9 acetylation and H3K4 methylation, histone marks associated with transcription. Subsequently, Polycomb complexes are recruited, and H3K27me3, the hallmark of repressed chromatin, accumulates on the X chromosome as it undergoes inactivation. The recruitment of Polycomb complexes and the establishment of repression have been linked to the transcription of repetitious DNA elements (long interspersed elements, or LINES), and these are abundant in the X chromosome. Proteins in the Polycomb complexes (Pc2) are thought to recognize and bind to new short dsRNA regions that are derived from transcription of both strands of the repetitive DNA elements (Chow et al. 2010; Kanhere et al. 2010).

RNA Profiling in the High-Throughput DNA Sequencing Era

By the middle years of the first decade of the 21st century, complete "annotated" genome sequence databases became available for many species. *Annotated* within this context meant "recognizable" gene products—by then the "standard" RNAs including miRNA genes. High-throughput DNA sequencing and complete genome tiling arrays (overlapping single-stranded DNA segments covering an entire genome, coupled to a substrate to allow hybridization to the arrays) were used to study the total transcriptional output of cells. In these analyses, total cell RNA (sometimes just nuclear RNA) was enzymatically converted to DNA and sequenced or hybridized to tiling arrays whose sequences were known. The recovered sequences could then be matched to a fully sequenced genome. These techniques are very sensitive and uncovered extensive and unexpected RNA formation from "noncoding" regions (i.e., regions outside of the previously annotated genomic coding regions).

Although seldom stated, a word of caution regarding the existence of these RNAs is appropriate. The techniques were capable of picking up sequences that were extremely rare, i.e., present in far less than one copy per cell (for review, see Mercer et al. 2009).

The current problems with the bioinformatics and molecular biological sorting of the immense amount of these transcribed sequences are nicely brought into focus by quoting the review by John Mattick and colleagues (Mercer et al. 2009), whose laboratory has contributed significantly to developing this field:

> Coding and noncoding RNAs (ncRNAs) can be difficult to distinguish. In eukaryotes, a protein-coding transcript is commonly defined by the presence of an ORF [*open reading frame*] greater than 100 amino acids. However, a long ncRNA might contain such an ORF by chance alone, and many well-characterized long ncRNAs do indeed contain long ORFs. Reciprocally, proteins smaller than 100 amino acids might also be translated, with functional peptides as small as 11 amino acids being reported in *Drosophila* species (Kondo et al. 2007). [Discussed later.] The observation that selection favours synonymous over nonsynonymous mutations to preserve coding usage has been exploited to help distinguish between transcripts with true rather than spurious ORFs (Lin et al. 2007). Nevertheless, despite such improvements in the annotation of transcripts in recent years, we still lack a satisfactory definition of ncRNAs and there remain many ambiguous transcripts that exhibit both coding and noncoding traits.

Long Noncoding (lnc) RNAs from Defined Transcription Units

A recent global search for long noncoding RNAs (lncRNAs) in human cells used the logic that chromatin marks surrounding the transcription units for such RNAs might show the same chromatin modifications as those found in pre-mRNA transcription units (Guttman et al. 2009).

The entire human genome was searched for potential transcription units that contained no protein-coding regions but were marked by the presence of histone modification H3K4me3 at a presumed 5′ end (a TSS) and histones modified by H3K36me3 downstream. More than 5000 such genomic areas that do not encode protein but do produce lncRNAs were identified (Guttman et al. 2009; Khalil et al. 2009). These RNAs were not antisense to known coding regions.

In most cases, it is not established how or if the lncRNAs function (Goñi et al. 2004; Ponting et al. 2009). However, active research on lncRNAs is rapidly revealing how at least some of these RNAs function in regulation of gene expression possibly through effects on chromosome structure. Several examples in which progress has been made illustrate how these molecules are being studied.

Regulatory lncRNA from a Hox *Gene Locus: HOTAIR*

As mentioned in Chapter 4 and in this chapter (e.g., Fig. 5-4), *Hox* genes are crucial in organizing the anterioposterior and dorsoventral body plan of bilaterians. In addition to the known protein-coding *Hox* genes, a number of antisense lncRNAs emanating from the *Hox* loci have been found (Rinn et al. 2007). One of them, named *HOTAIR* (for Hox antisense inhibitory RNA), was found to have a definite, if as yet incompletely understood, molecular function.

The choice of this region for examination rested on an earlier remarkable finding regarding *Hox* gene expression in cultured human (and mouse) fibroblasts. Cells described as fibroblasts will grow out from skin biopsies of different parts of the adult—scalp, toes, arms, back, abdomen, etc. It was also conventional to think of cultured cells from disaggregated embryos as just "fibroblasts." All of these cells—from embryos or adults—looked alike microscopically, regardless of the part of the embryo or skin from which they came.

Howard Chang and colleagues (Chang et al. 2002) challenged the notion of similarity of all of these "fibroblasts" (Rinn et al. 2007). The human skin "fibroblasts" from different locations on the body showed distinct patterns of gene expression (by microarray analysis of mRNA content), suggesting that they retained at least some memory of a transcriptional pattern established in embryogenesis (see Fig. 5-40).

Among the genes that retained distinct expression patterns in adult fibroblasts was the *HoxC* gene cluster, which is differently expressed in different regions (from anterior to posterior) of mammalian embryos. In addition to the protein-coding *Hox* genes, several lncRNA genes were identified,

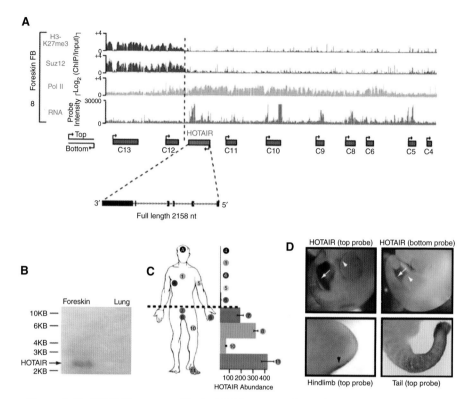

Figure 5-40. HOTAIR represses *HoxD*. (*A*) The top two lines show repression of the C13 and C12 loci based on the ChIP-identified presence of a repression methylase (Suz12, a Polycomb complex protein; purple) and the presence of H3K27me3 histones (blue). Pol II is present in the neighboring chromosomal region (line 3, orange), where *HOTAIR* and other *HoxC* RNAs were detected through sequencing of nuclear RNA (line 4, red). The lncRNA *HOTAIR* (purple box and gene diagram below line 4) is transcribed from the human *HoxC* locus, in opposite direction to neighboring *HoxC* protein-coding genes (brown). (*B*) *HOTAIR* is expressed in foreskin fibroblasts but not in lung/fibroblasts (northern blot). (*C*) Different amounts of *HOTAIR* mRNA are found in fibroblasts from different parts of the human body (all sites are skin fibroblasts except for number 4, which is lung). Abundance was measured by real-time PCR. (*D*) Expression of mouse *HOTAIR* by in situ hybridization (purple) in a developing mouse embryo. The genomic sequence of mouse *HOTAIR* is ~90% identical to that of human *HOTAIR*, and the *Hox* gene arrangements are conserved; thus, function may be also. (Adapted, with permission, from Rinn et al. 2007 [©Elsevier].)

including the *HOTAIR* gene, from the *HoxC* region on chromosome 2 (Fig. 5-40).

In fibroblasts, the *HoxD* region on chromosome 2 is associated with the repressive Polycomb (PC2) complex. An siRNA knockdown of *HOTAIR* lncRNA transcribed from chromosome 2 resulted in removal of the PC2 and derepression of both coding and noncoding regions in the *HoxD* locus on chromosome 12 (Rinn et al. 2007). Although the mechanisms underlying the repression by *HOTAIR* of a locus on another chromosome are still being explored (Tsai et al. 2010), there seems little doubt that removal of *HOTAIR* lncRNA functions to somehow relieve repression of the *HoxD* loci (Fig. 5-41).

Gene Repression by a p53-Dependent lncRNA

It has been established for well over two decades that p53 is a crucial protein in halting growth and instituting apoptosis in damaged cells. A large fraction of human tumors have lost p53 function and resist apoptosis. Genes that are positively regulated by p53 were studied throughout the 1980s and 1990s. Examination of total mRNA in cells where p53 was activated also revealed decreases in some mRNAs that proved to be due to p53-dependent repression. The basis for at least some of this repression is apparently due to an lncRNA (denoted by the authors as a large intergenic noncoding RNA, lincRNA-p21), the transcription of which is positively regulated by p53. This remarkable finding may have at least two general implications. First, a positive-acting transcription factor, p53, can produce a repressor. Second, the repressor can be a lncRNA (Huarte et al. 2010).

lncRNAs Can Also Increase Transcription

By computational analysis, Ulf Ørom and colleagues (2010) in Philadelphia and Barcelona, Spain, located several thousand lncRNAs that were expressed differently in different tissues and that were near but did not overlap neighboring protein-coding genes. In experiments designed to test the function of the lncRNAs, they treated skin cells (keratinocytes) with a known stimulant, a phorbol ester (TPA, 12-O-tetradecanoylphorbol 13-acetate). A number of the lncRNAs increased, as did some of the mRNAs encoded in genes neighboring these lncRNAs. Furthermore, they reduced some of the lncRNAs with siRNA treatment and found a corresponding decrease in mRNAs encoded in a neighboring gene. Thus, it seems clear that these lncRNAs have a positive influence on gene expression, presumably at the level of transcription.

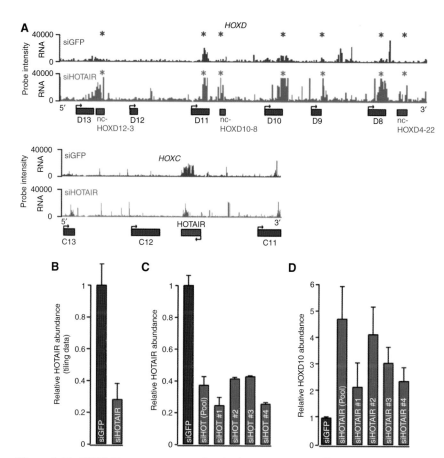

Figure 5-41. HOTAIR represses expression in the *HoxD* region of human fibroblasts. (*A*) Expression data in the *HoxD* region obtained by sequencing DNA copied from entire fibroblast cell RNA. Although *HOTAIR* is transcribed from the *HoxC* locus on chromosome 12 (see Fig. 5-40), when it is removed by siRNA, transcription increases on specific sites on the *HoxD* locus that is on chromosome 2. The top two lines in *A* show the relative amounts of transcribed RNA across the *HoxD* locus when cells are transfected with an irrelevant siRNA (blue) and after *HOTAIR* siRNA (red). Lines 3 and 4 (blue and red) are transcript abundance in the *HoxC* locus to show that the *HOTAIR* siRNA greatly decreased *HOTAIR* transcripts (line 4, red). (*B*) Plot of the bottom two lines, 3 and 4, of *A*. (*C*) Measurement of abundance of *HOTAIR* RNA by quantitative RT-PCR (qRT-PCR) after siRNA treatment as in *A*. (*D*) Measurement of increase of *HoxD*-encoded specific mRNAs by qRT-PCR after siRNA treatment as in *A*. (Reprinted, with permission, from Rinn et al. 2007.)

Flo11 and Transcriptional Regulation by Two Competing lncRNAs

A particularly well-understood example of control via transcription of two negative-acting lncRNAs was recently described (Bumgarner et al. 2009). The *Flo11* gene in *S. cerevisiae* encodes a protein required in pseudohyphae formation. The histone deacetylase complex (Rpd3L) is present in ChIP arrays bound to a region close to the TSS for *Flo11*, under conditions in which *Flo11* is rapidly transcribed. This was puzzling because deacetylase complexes are usually negative acting. Genetic and molecular genetic experiments solved the puzzle. There are two lncRNAs, *PWR1* and *ICR1*, that are transcribed in regions just upstream of the *Flo11* promoter (Fig. 5-42). The formation of these two lncRNAs controls *Flo11* transcription. Two DNA-binding proteins, Sfl1 and Flo8, in turn, are responsible for controlling the lncRNAs by competing for closely spaced binding sites (Fig. 5-42; blue site binds Flo8, red site binds Sfl1). If Flo8 dominates, it joins with Rpd36, the deacetylase complex, to outcompete Sfl, and activates formation of PWR1, which, in turn, represses the transcription of *ICR1*. General transcription factors then produce *Flo11* mRNA, and pseudohyphae then form. If Sfl dominates (bottom half of Fig. 5-42), *PWR1* transcription is blocked. *ICR1* is then transcribed, which blocks *Flo11* formation and pseudohyphae formation. Thus, control of *Flo11* is in the hands of two lncRNAs.

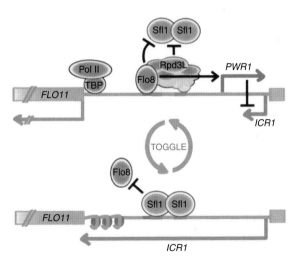

Figure 5-42. Control of transcription by relief of promoter occlusion. In *S. cerevisiae*, the *FLO11* gene is controlled by *promoter occlusion*, the term used to describe inaccessibility to a promoter because of upstream transcription across the occluded promoter. (See text for details.) (Reprinted, with permission, from Bumgarner et al. 2009.)

Some "Noncoding RNAs" Are Misnamed

As the quote from Mercer et al. (p. 326) emphasizes, the designation of "long noncoding RNAs" has difficulty in that some of these molecules may (or do) have buried within them regions that encode short polypeptides that, if translated, may account for their physiological function. Biochemists have known for decades that short polypeptides can have important physiological roles. The two SH-linked chains of insulin, e.g., are only 21 and 30 amino acids long, and the 241-amino-acid proopiomelanocortin protein is split into six functional pieces, with each short peptide retaining specific properties. Cleavage of longer "polypeptides" to yield functional extracellular circulating molecules is well established. Short peptides that act intracellularly are at the moment thought to be much rarer.

Two groups, one in Japan (Kondo et al. 2007, 2010) and one in England (Galindo et al. 2007), have recently completed examination of the underlying cause of mutations in a *Drosophila* gene called *pri* (for "polished rice"). Mutations in this gene were originally recognized as causing a defect in bristle formation on the fly head and in formation of a hook-like projection on the leg. *pri* mutations were later found to cause deformities in spiracles, the actin-fiber-supported "tracheal" tubes through which O_2 exchange occurs. The *pri* RNA transcript, which is 1549 nucleotides long, was at first thought not to encode proteins (i.e., to be a lncRNA). The recently reported and startling finding, however, is that *pri* does encode amino acids *but only short peptides* (Galindo et al. 2007; Kondo et al. 2007). There are at least five scattered coding regions for peptides, the sequence for each of which begins with a translation start codon and ends with a translation stop codon. These peptide-coding regions are conserved in a wide variety of insects. If in a transgene the coding sequence for a marker protein, GFP (green fluorescent protein), is inserted before the stop codon of the short peptides, green fluorescence is detected in cells that contribute to bristles and in the trachea, evidence that the peptides are produced. Thus, *pri* is a long RNA, *but* it is definitely not noncoding. And the peptides it encodes function in two different ways. The peptides regulate the fate of a transcription factor in bristle formation and assist in correctly bundling actin fibers in the trachea. A burning question at the moment is how many such "long noncoding RNAs" may, in fact, harbor information for short peptides.

Conclusion: Long Noncoding RNAs

Several things are clear at this early date in the study of long noncoding RNAs. These plentiful RNAs can have important physiological functions and are not transcriptional noise, as was suggested earlier. Some of these RNAs are doubtless antisense regulators. Some, like *Xist* and *HOTAIR*, have negative functions distant from their chromosomal origin, whereas others seem to promote transcription of neighboring protein-encoding genes. What seems safe to predict is that learning the roles of these many new lncRNAs will be a growth industry in the immediate future.

RNA ON DISPLAY: RECENT PROGRESS IN EUKARYOTIC GENE EXPRESSION

After the summary in this chapter of roughly three decades of very rapid advances in knowledge regarding the regulation of mRNA formation in eukaryotes, of the roles of short and long RNAs in transcription, and of mRNA lifetime and transcriptional effectiveness, perhaps the best way for you to catch your breath is to return to glance at Table 5-1 (p. 234). The intricate involvement of regulatory RNAs in the production and destruction of mRNA is dizzying. Many steps to using this knowledge fully lie ahead, but it will *largely* be a case of continuing to study *RNA molecules* and *RNA:protein complexes*. This is the challenge of the present century.

The depth of our present understanding of how our cells each performs its specialized gene control tasks has advanced amazingly from the middle of the 20th century. A cynic (only a cynic, in my opinion) might hold to the belief that after Crick and Watson, Jacob and Monod, and the solution of the genetic code, the pathway to where we are today was "completely obvious." Perhaps in decades—centuries—to come, that will be the inevitable and even correct conclusion. After all, once it was realized by 1966 that a long string of nucleic acid sequences was *all* that the DNA in chromosomes consisted of, that the genetic code was universal, and that there was sufficient molecular specificity in a nucleic acid sequence to be recognized by the endless possibilities of folds in a protein, it *only* remained to match protein:protein and protein:nucleic acid interfaces to explain all of regulation. And such matches may be handled entirely by machines after genomic sequencing and advanced bioinformatics completely take over in the new century.

I cannot refute such a hard-bitten look at the last 50 years of molecular biological achievement, but it leaves out the great thrill of hundreds of lesser and greater discoveries along the way, an overwhelming number of which center on eukaryotic RNA processing and mRNA regulation, both transcriptionally and posttranscriptionally. Virtually none of these functions was expected before research on eukaryotic gene expression that began in studies on cultured animal cells and cells infected with animal viruses in the 1960s and have been swept along particularly since the 1970s by studies in *Drosophila*, *C. elegans*, and yeast.

Perhaps most of all, it should be easy to grant that the immense practical promise of molecular biology required the accumulation of all the science of the last 50 years, not to mention that of the next 50. These detailed advances have brought, and continuing advances will continue to bring, great progress to agriculture, medicine, and even perhaps to fields such as clean energy technology, as we learn how better to engineer single-cell eukaryotes. Thus, those among the thousands of us who did not discover the structure of DNA, the universal genetic code, or the first proteins that can and do control genes can still justifiably enjoy our achievements and be secure in the knowledge that we took part in a great and continuing enterprise.

REFERENCES

Albright SR, Tjian R. 2000. TAFs revisited: more data reveal new twists and confirm old ideas. *Gene* **242**: 1–13.

Allfrey VG. 1977. Post-synthetic modifications of histone structure: A mechanism for the control of chromosome structure by the modulation of histone–DNA interactions. In *Chromatin and chromosome structure* (ed. HJ Li, R Eckhardt), pp. 167–191. Academic Press, New York.

Allis CD, Jenuwein T, Reinberg D, Caparros M-L, eds. 2007. *Epigenetics*. Cold Spring Harbor Laboratory Press, Cold Spring Harbor, NY.

Amara S, Evans RM, Rosenfeld MG. 1984. Calcitonin/calcitonin gene related peptide transcription: Tissue-specific expression involves selective use of alternative polyadenylation sites. *Mol Cell Biol* **4**: 2151–2160.

Ambros V. 1989. A hierarchy of regulatory genes controls larva-to-adult developmental switch in *C. elegans*. *Cell* **57**: 49–57.

Ambros V. 2008. The evolution of our thinking about microRNAs. *Nat Med* **14**: 1036–1040.

An W, Roeder RG. 2004. Reconstitution and transcriptional analysis of chromatin in vitro. *Methods Enzymol* **377**: 460–474.

Aravin AA, Hannon GJ. 2008. Small RNA silencing pathways in germ and stem cells. *Cold Spring Harbor Symp Quant Biol* **73**: 283–290.

Ardehali MB, Lis JT. 2009. Tracking rates of transcription and splicing in vivo. *Nat Struct Mol Biol* **16**: 1123–1124.

Babich A, Nevins JR, Darnell JE Jr. 1980. Early capping of transcripts from the adenovirus major late transcription unit. *Nature* **287**: 246–248.

Baker BS. 1989. Sex in flies: The splice of life. *Nature* **340**: 521–524.

Bakheet T, Williams BRG, Khabar KSA. 2006. ARED 3.0: The large and diverse AU-rich transcriptome. *Nucleic Acids Res* **34**: D111–D114.

Banerji J, Rusconi S, Schaffner W. 1981. Expression of a β-globin gene is enhanced by remote SV40 SNA sequences. *Cell* **27**: 299–308.

Barski A, Cuddapah S, Cui K, Roh T-Y, Schones DE, Wang Z, Wei G, Chepelev I, Zhao K. 2007. High-resolution profiling of histone methylations in the human genome. *Cell* **129**: 823–837.

Bartel DP. 2004. MicroRNAs: Genomics, biogenesis, mechanism and function. *Cell* **116**: 11–29.

Bartel DP. 2009. MicroRNAs: Target recognition and regulatory functions. *Cell* **136**: 215–233.

Baulcombe D. 2006. Of maize and men, or peas and people: Case histories to justify plants and other model systems. *Nat Med* **14:** 1046–1049.

Benezra R, Davis RL, Lockshon D, Turner DL, Weintraub H. 1990. The protein ID: A negative regulator of helix–loop–helix DNA binding proteins. *Cell* **61:** 49–59.

Bernstein E, Caudy AA, Hammond SM, Hannon GJ. 2001. Role for a bidentate ribonuclease in the initiation step of RNA interference. *Nature* **409:** 363–366.

Beyer AL, Osheim YN. 1988. Splice site selection, rate of splicing and alternative splicing on nascent transcripts. *Genes Dev* **2:** 754–765.

Bird A. 2007. Perceptions of epigenetics. *Nature* **447:** 396–398.

Black DL. 2003. Mechanisms in alternative pre-messenger RNA splicing. *Annu Rev Biochem* **72:** 291–336.

Boettiger AN, Levine M. 2009. Synchronous and stochastic patterns of gene activation in the *Drosophila* embryo. *Science* **325:** 471–473.

Bonasio R, Tu S, Reinberg D. 2010. Molecular signals of epigenetic states. *Science* **330:** 612–616.

Bookout AL, Jeong Y, Downes M, Yu RT, Evans RM. 2006. Anatomical profiling of nuclear receptor expression reveals a hierarchical transcriptional network. *Cell* **126:** 789–799.

Borchert GM, Lanier W, Davidson BL. 2006. RNA polymerase III transcribes human micro-RNAs. *Nat Struct Mol Biol* **13:** 1097–1101.

Bourbon HM, Aquilera A, Ansari AZ, Asturias FJ, Berk AJ, Bjorklind S, Blackwell TK, Borgarefe T, Carey M, Carlson M, et al. 2004. A unified nomenclature for protein subunits of mediator complexes linking transcriptional regulators to RNA polymerase II. *Mol Cell* **14:** 553–557.

Breathnach R, Chambon P. 1981. Organization and expression of eukaryotic split genes coding for proteins. *Annu Rev Biochem* **10:** 349–383.

Brennecke J, Aravin AA, Stark A, Dus M, Kellis M, Sachidanandam R, Hannon GJ. 2007. Discrete small RNA-generating loci as master regulators of transposon activity in *Drosophila*. *Cell* **128:** 1089–1103.

Brenner S. 1974. The genetics of *C. elegans*. *Genetics* **77:** 71–94.

Brenner S. 2003. Nature's gift to science (Nobel lecture). *Chembiochem* **4:** 683–687. http://nobelprize.org/nobel_prizes/medicine/laureates/2002/brenner-lecture.htm.

Brivanlou AH, Darnell JE Jr. 2002. Signal transduction and the control of gene expression. *Science* **295:** 813–818.

Brown CJ, Ballabio A, Rupert JL, Lafreniere RG, Tonlorenzi MGR, Willard HF. 1991. A gene from the region of the human X inactivation centre is expressed exclusively from the inactive X chromosome. *Nature* **349:** 38–44.

Brownell JE, Allis CD. 1996. Special HATs for special occasions: Linking histone acetylation to chromatin assembly and gene activation. *Curr Opin Genet Dev* **6:** 176–184.

Brownell JE, Zhou J, Ranalli T, Kobayashi R, Edmondson DG, Roth SY, Allis CD. 1996. *Tetrahymena* histone acetyltransferase A: A homolog to yeast Gcn5p linking histone acetylation to gene activation. *Cell* **84:** 843–851.

Bulger M, Groudine M. 2011. Functional and mechanistic diversity of distal transcription enhancers. *Cell* **144:** 327–339.

Bumgarner SL, Dowell RD, Grisafi P, Gifford DK, Fink GR. 2009. Toggle involving *cis*-interfering noncoding RNAs controls variegated gene expression in yeast. *Proc Natl Acad Sci* **106:** 18049–18050.

Burley SK, Roeder RG. 1996. Biochemistry and structural biology of transcription factor IID (TFIID). *Annu Rev Biochem* **65:** 769–799.

Busch H, Reddy R, Rothblum L, Choi YC. 1982. SnRNAs, Sn RNPs, and RNA processing. *Annu Rev Biochem* **51**: 617–654.

Cai G, Imasaki T, Takagi Y, Asturias FJ. 2009. Mediator structural conservation and implications for the regulation mechanism. *Structure* **17**: 559–567.

Cairns BR. 2009. The logic of chromatin architecture and remodeling at promoters. *Nature* **461**: 193–198.

Candido EP, Dixon GH. 1972. Amino-terminal sequences and sites of in vivo acetylation of trout-testis histones 3 and lib 2. *Proc Natl Acad Sci* **69**: 2015–2019.

Carman GM, Henry SA. 2007. Phosphatidic acid plays a central role in the transcriptional regulation of glycerophospholipid synthesis in *Saccharomyces cerevisiae*. *J Biol Chem* **282**: 37293–37297.

Carninci P, Kasukawa T, Katayama S, Gough J, Frith MC, Maeda N, Oyama R, Ravasi T, Lenhard B, Wells C, et al. 2005. The transcriptional landscape of the mammalian genome. *Science* **309**: 1559–1563.

Carninci P, Sandelin A, Lenhard B, Katayama S, Shimokawa K, Ponjavic J, Semple CAM, Taylor MS, Engström PG, Frith MC, et al. 2006. Genome-wide analysis of mammalian promoter architecture and evolution. *Nature Genetics* **38**: 626–635.

Carthew RW, Sontheimer EJ. 2009. Origins and mechanisms of miRNAs and siRNAs. *Cell* **136**: 642–655.

Chang HY, Chi J-T, Dudoit S, Bondre C, van de Rijn M, Botstein D, Brown PO. 2002. Diversity, topographic differentiation, and positional memory in human fibroblasts. *Proc Natl Acad Sci* **99**: 12877–12882.

Chi SW, Zang JB, Mele A, Darnell RB. 2009. Argonaut HITs-CLIP decodes microRNA–mRNA interaction maps. *Nature* **460**: 479–486.

Chow J, Heard E. 2009. X inactivation and the complexities of silencing a sex chromosome. *Curr Opin Cell Biol* **21**: 359–366.

Chow JC, Ciaudo C, Fazzari MJ, Mise N, Servant N, Glass JL, Attreed M, Avner P, Wutz A, Barillot E, et al. 2010. LINE-1 activity in facultative heterochromatin formation during X chromosome inactivation. *Cell* **141**: 956–969.

Chukov S, Kurash JK, Wilson JR, Xiao B, Justin N, Ivanov GS, McKinney K, Tempst P, Prives C, Gamblin SJ, et al. 2004. Regulation of p53 activity through lysine methylation. *Nature* **432**: 353–360.

Churchman LS, Weissman JS. 2011. Nascent transcript sequencing visualizes transcription at nucleotide resolution. *Nature* **469**: 368–373.

Cinnamon E, Paroush Z. 2008. Context-dependent regulation of Groucho/TLE-mediated repression. *Curr Opin Genet Dev* **18**: 435–440.

Cline TW, Meyer BJ. 1996. Vive la difference: Males vs females in flies vs worms. *Annu Rev Genet* **30**: 637–702.

Cloos PAC, Christensen J, Agger K, Helin K. 2008. Erasing the methyl mark: Histone demethylases at the center of cellular differentiation and disease. *Genes Dev* **22**: 1115–1140.

Corden J, Wasylyk B, Buchwalder A, Sassoni-Corsi P, Kedinger C, Chambon P. 1980. Promoter sequences of eukaryotic protein-coding genes. *Science* **209**: 1406–1414.

Core LJ, Lis JT. 2008. Transcription regulation through promoter-proximal pausing of RNA polymerase II. *Science* **319**: 1791–1792.

Core LJ, Waterfall JJ, Lis JT. 2008. Nascent RNA sequencing reveals widespread pausing and divergent initiation at human promoters. *Science* **322**: 1845–1848.

Cramer P. 2010. Towards molecular systems biology of gene transcription and regulation. *Biol Chem* **391**: 731–735.

Cramer P, Bushnell DA, Kornberg RD. 2001. Structural basis of transcription: RNA polymerase II at 2.8 Ångstrom resolution. *Science* 292: 1863–1876.

Creyghton MP, Cheng AW, Welstead GG, Kooistra T, Carey BW, Steine EJ, Hanna J, Lodato MA, Frampton GM, Sharp PA, et al. 2010. Histone H3K27ac separates active from poised enhancers and predicts developmental state. *Proc Natl Acad Sci* 107: 21931–21936.

Cruz JA, Westhof E. 2009. The dynamic landscapes of RNA architecture. *Cell* 136: 604–609.

Darnell JE Jr. 1968. Ribonucleic acids from animal cells. *Bacteriol Rev* 32: 262–290.

Darnell JE Jr. 1975. The origin of mRNA and the structure of the mammalian chromosome. *Harvey Lect* 69: 1–47.

Darnell JE Jr. 1997. STATs and gene regulation. *Science* 277: 1630–1635.

Darnell J, Lodish H, Baltimore D. 1990. *Molecular cell biology*, 2nd ed. Scientific American Books/Freeman, New York.

Dean A. 2005. On a chromosome far, far away: LCRs and gene expression. *Trends Genet* 22: 38–45.

Deutscher MP. 2006. Degradation of RNA in bacteria: Comparison of mRNA and stable RNA. *Nucleic Acids Res* 34: 659–666.

Diep CQ, Tao X, Pilauri V, Losiewicz M, Blank TE, Hopper JE. 2008. Genetic evidence for sites of interaction between the Gal3 and Gal80 proteins of the *Saccharomyces cerevisiae* GAL gene switch. *Genetics* 78: 725–736.

D'Onofrio C, Colantuoni V, Cortese R. 1985. Structure and cell-specific expression of a cloned human retinol binding protein gene: The 5'-flanking region contains hepatoma specific transcriptional signals. *EMBO J* 4: 1981–1989.

Durrin LK, Mann RK, Kayne PS, Grunstein M. 1991. Yeast histone H4 N-terminal sequence is required for promoter activation in vivo. *Cell* 65: 1023–1031.

Dyson N. 1998. The regulation of E2F by pRB-family proteins. *Genes Dev* 12: 2245–2262.

Ebmeier CC, Taatjes DJ. 2010. Activator-Mediator binding regulates Mediator-cofactor interactions. *Proc Natl Acad Sci* 107: 11283–11288.

Egyhazi E. 1974. A tentative initiation inhibitor of chromosomal heterogeneous RNA synthesis. *J Mol Biol* 84: 173–183.

Evans RM. 1988. The steroid and thyroid hormone receptor superfamily. *Science* 240: 889–895.

Fire AZ. 2007. Gene silencing by double-stranded RNA. *Cell Death Differ* 14: 1998–2012. http://nobelprize.org/nobel_prizes/medicine/laureates/2006/fire_lecture.pdf.

Fire A, Albertson D, Harrison SW, Moerman DG. 1991. Production of antisense RNA leads to effective and specific innibition of gene expression in *C. elegans* muscle. *Development* 113: 503–514.

Fire A, Xu S, Montgomery M, Kostas S, Driver S, Mello C. 1998. Potent and specific genetic interferences by double-stranded RNA in *C. elegans*. *Nature* 391: 806–811.

Flanagan PM, Kelleher RJ III, Sayre MH, Tschochner H, Kornberg RD. 1991. A mediator required for activation of RNA polymerase II transcription in vitro. *Nature* 350: 436–438.

Fondell JD, Ge H, Roeder RG. 1996. Ligand induction of a transcriptionally active thyroid hormone receptor coactivator complex. *Proc Natl Acad Sci* 93: 8329–8333.

Fondell JD, Guermah M, Malik S, Roeder RG. 1999. Thyroid hormone receptor-associated proteins and general positive cofactors mediate thyroid hormone receptor function in the absence of the TATA box-binding protein-associated factors of TFIID. *Proc Natl Acad Sci* 96: 1959–1964.

Fuda NJ, Ardehali MB, Lis JT. 2009. Defining mechanisms that regulate RNA polymerase II transcription in vivo. *Nature* 461: 186–192.

Galindo MI, Pueyo JI, Fouix S, Bishop SA, Couso JP. 2007. Peptides encoded by short ORFs control development and define a new eukaryotic gene family. *PLoS Biol* **5:** 1052–1062.

Galli G, Guise J, Tucker PW, Nevins JR. 1988. Poly(A) site choice rather than splice site choice governs the regulated production of IgM heavy-chain RNA. *Proc Natl Acad Sci* **85:** 2439–2443.

Garneau NL, Wilusz J, Wilusz CJ. 2007. The highways and byways of mRNA decay. *Nat Rev Mol Cell Biol* **8:** 113–126.

Giguere V, Hollenberg SM, Rosenfeld MG, Evans RM. 1986. Functional domains of the human glucocorticoid receptor. *Cell* **46:** 645–652.

Gill G, Tjian R. 1992. Eukaryotic coactivators associated with the TATA box binding protein. *Curr Opin Genet Dev* **2:** 236–242.

Gilmour DS, Lis JT. 1986. RNA polymerase II interacts with the promoter region of the non-induced hsp70 gene in *Drosophila melanogaster* cells. *Mol Cell Biol* **6:** 3984–3989.

Gluzman Y, Shenk T. 1983. *Enhancers and eukaryotic gene expression.* Cold Spring Harbor Laboratory, Cold Spring Harbor, NY.

Gnatt AL, Cramer P, Fu J, Bushnell DA, Kornberg RD. 2001. Structural basis of transcription: An RNA polymerase II elongation complex at 3.3 Å resolution. *Science* **292:** 1876–1882.

Goñi JR, de la Cruz X, Orozco M. 2004. Triplex-forming oligonucletide sequences in the human genome. *Nucleic Acids Res* **32:** 354–360.

Gottschling D. 2006. Epigenetics: From phenomenon to field (a history of epigenetics at Cold Spring Harbor Symposia). In *Epigenetics* (ed. CD Allis et al.), pp. 2–7. Cold Spring Harbor Laboratory Press, Cold Spring Harbor, NY.

Graham FL, van der Eb AJ. 1973. A new technique for the assay of infectivity of human adenovirus 5 DNA. *Virology* **52:** 456–467.

Grayson DR, Costa RH, Xanthopoulos KG, Darnell JE Jr. 1988. A cell-specific enhancer of the mouse alpha 1-antitrypsin gene has multiple functional regions and corresponding protein binding sites. *Mol Cell Biol* **8:** 1055–1066.

Grossniklaus U, Paro R. 2006. Transcriptional silencing by Polycomb group proteins. In *Epigenetics* (ed. CD Allis et al.), pp. 211–230. Cold Spring Harbor Laboratory Press, Cold Spring Harbor, NY.

Guenther MG, Young RA. 2010. Repressive transcription. *Science* **329:** 150–151.

Guenther MG, Levine SS, Boyer LA, Jaenisch R, Young RA. 2007. A chromatin landmark and transcription initiation at most promoters in human cells. *Cell* **130:** 77–88.

Guidiyal M, Zamore PD. 2009. Small silencing RNAs: An expanding universe. *Nat Rev Genet* **10:** 94–107.

Guo S, Kemphues K. 1995. *par-1*, a gene for establishing polarity in *C. elegans* embryos, encodes a putative Ser/Thr kinase that is asymmetrically distributed. *Cell* **81:** 611–620.

Gurdon JB, Melton DA. 2008. Nuclear reprogramming in cells. *Science* **322:** 1811–1815.

Guttman M, Amit I, Garber M, French C, Lin MF, Feldser D, Huarte M, Zuk O, Carey BW, Cassady JP, et al. 2009. Chromatin signature reveals over a thousand highly conserved large non-coding RNAs in mammals. *Nature* **458:** 223–227.

Hamilton AJ, Baulcombe DC. 1999. A species of small antisense RNA in post-transcriptional gene silencing in plants. *Science* **286:** 950–952.

Han M, Grunstein M. 1988. Nucleosome loss activates yeast downstream promoters in vivo. *Cell* **55:** 1137–1145.

Hannon GJ. 2008. Small RNA silencing in germ and stem cells. *Cold Spring Harbor Symp Quant Biol* **73:** 283–290.

Harpold M, Wilson M, Darnell JE Jr. 1981. Chinese hamster poly(A)⁺ mRNA: Relationship to poly(A)⁻ sequences and relative conservation during mRNA processing. *Mol Cell Biol* 1: 188–198.

Hattori D, Chen Y, Matthews BJ, Salwinski L, Sabatti C, Grueber WB, Zipursky SL. 2009. Robust discrimination between self and non-self neurites requires thousands of Dscam1 isoforms. *Nature* **461:** 644–648.

Hebbes TR, Thorne AW, Crane-Robinson C. 1988. A direct link between core histone acetylation and transcriptionally active chromatin. *EMBO J* 7: 1395–1402.

Hebbes TR, Turner CH, Thorne AW, Crane-Robinson C. 1989. A "minimal epitope" anti-protein antibody that recognises a single modified amino acid. *Mol Immunol* **26:** 865–873.

Hengartner CJ, Thompson CM, Zhang J, Chao DM, Liao SM, Koleske AJ, Okamura S, Young RA. 1995. Association of an activator with an RNA polymerase II holoenzyme. *Genes Dev* **9:** 897–910.

Hermoso A, Aguilar D, Aviles FS, Querol E. 2004. TrSDB: A proteome data base of transcription factors. *Nucleic Acids Res* **32:** D171–D173.

Herr AJ, Baulcombe DC. 2004. RNA silencing in plants. *Cold Spring Harbor Symp Quant Biol* **69:** 363–370.

Hinnen A, Hicks JB, Fink GR. 1978. Transformation of yeast. *Proc Natl Acad Sci* **75:** 1929–1933.

Hofer E, Darnell JE Jr. 1981. The primary transcription unit of the mouse β-major globin gene. *Cell* **23:** 585–593.

Hollenberg SM, Weinberger C, Ong ES, Cerelli G, Oro A, Lebo R, Thompson EB, Rosenfeld MG, Evans RM. 1985. Primary structure and expression of a functional human glucocorticoid receptor cDNA. *Nature* **318:** 635–641.

Holt CE, Bullock SL. 2009. Subcellular mRNA localization in animal cells and why it matters. *Science* **326:** 1212–1216.

Horvitz HR. 2003. Worms, life and death (Nobel lecture). *Chembiochem* **4:** 697–711. http://nobelprize.org/nobel_prizes/medicine/laureates/2002/horvitz-lecture.html.

Huarte M, Guttman M, Feldser D, Garber M, Koziol MJ, Kenzelmann-Broz D, Khalil AM, Zuk O, Amit I, Rabani M, et al. 2010. A large intergenic noncoding RNA induced by p53 mediates global gene repression in the p53 response. *Cell* **142:** 409–419.

Jensen EV, DeSombre ER. 1972. Mechanism of action of the female sex hormones. *Annu Rev Biochem* **41:** 203–230.

Jiang F, Frey BR, Evans ML, Fried JC, Hopper JE. 2009. Gene activation by dissociation of an inhibitor from a transcriptional activation domain. *Mol Cell Biol* **29:** 5604–5610.

Juven-Gershon T, Hsu J-Y, Theisen JWM, Kadonaga JT. 2008. The RNA polymerase II core promoter—The gateway to transcription. *Curr Opin Cell Biol* **20:** 253–259.

Kagey MH, Newman JJ, Bilodeau S, Zhan Y, Orlando DA, van Berkum NL, Ebmeier CC, Goossens J, Rahl PB, Levine SS. 2010. Mediator and cohesin connect gene expression and chromatin architecture. **464:** 1082–1086.

Kanhere A, Viiri K, Araujo CC, Rasaiyaah J, Bouwman RD, Whyte WA, Pereira CF, Brookes E, Walter K, Bell GW, et al. 2010. Short RNAs are transcribed from repressed polycomb target genes and interact with Polycomb repressive complex-2. *Mol Cell* **38:** 675–688.

Kapranov P. 2009. From transcription start site to cell biology. *Genome Biol* **10:** 217. doi: 10.1186/gb-2009-10-4-217.

Khabar KS. 2010. Post-transcriptional control during chronic inflammation and cancer: A focus on AU-rich elements. *Cell Mol Life Sci* **67:** 2937–2955.

Khalil AM, Guttman M, Huarte M, Garber M, Raj A, Morales DR, Thomas K, Presser A, Berntein BE, van Oudenaarden A, et al. 2009. Many human large intergenic noncoding

RNAs associate with chromatin-modifying complexes and affect gene expression. *Proc Natl Acad Sci* **106:** 11675–11680.

Kim U-J, Han M, Kayne P, Grunstein M. 1988. Effects of histone H4 depletion on the cell cycle and transcription of *Saccharomyces cerevisiae. EMBO J* **7:** 2211–2219.

Kim JL, Nikolov DB, Burley SK. 1993. Co-crystal structure of TBP recognizing the minor groove of a TATA element. *Nature* **365:** 520–527.

Kim YJ, Bjorklund S, Li Y, Sayre MH, Kornberg RD. 1994. A multiprotein mediator of transcriptional activation and its interaction with the C-terminal repeat domain of RNA polymerase II. *Cell* **77:** 599–608.

Kim DH, Saetrom P, Snove O Jr, Rossi JJ. 2008. MicroRNA-directed transcriptional gene silencing in mammalian cells. *Proc Natl Acad Sci* **105:** 16230–16235.

King TJ, Briggs R. 1956. Serial transplantation of embryonic nuclei. *Cold Spring Harbor Symp Quant Biol* **21:** 271–290.

Kohoutek J. 2009. P-TEBb—The final frontier. *Cell Div* **4:** 19–33.

Kondo T, Hashimoto Y, Kato K, Inagaki S, Hayashi S, Kageyama Y. 2007. Small peptide regulators of actin-based cell morphogenesis encoded by a polycistronic mRNA. *Nat Cell Biol* **9:** 660–665.

Kondo T, Plaza S, Zanet J, Benrabah E, Valenti P, Hashimoto Y, Kobayashi S, Payre F, Kageyama Y. 2010. Small peptides switch the transcriptional activity of Shavenbaby during *Drosophila* embryogenesis. *Science* **329:** 336–339.

Kornberg RD. 1974. Chromatin structure: A repeating unit of histones and DNA. *Science* **184:** 868–871.

Kornberg RD. 2007. The molecular basis of eukaryotic transcription. *Proc Natl Acad Sci* **104:** 12955–12961.

Kornberg RD, Thomas JO. 1974. Chromatin structure: Oligomers of the histones. *Science* **184:** 865–868.

Kouzarides T, Berger S. 2006. Chromatin modifications and their mechanism of action. In *Epigenetics* (ed. CD Allis et al.), pp. 191–209. Cold Spring Harbor Laboratory Press, Cold Spring Harbor, NY.

Kruithof M, Chien FT, Routh A, Logie C, Rhodes D, van Noort J. 2009. Single-molecule force spectroscopy reveals a highly compliant helical folding for the 30-nm chromatin fiber. *Nat Struct Mol Biol* **16:** 534–540.

Kuechenmeister LJ, Anderson KL, Morrison JM, Dunman PM. 2009. The use of molecular beacons to directly measure bacterial mRNA abundances and transcript degradation. *J Microbiol Methods* **76:** 146–151.

Kuo MH, Brownell JE, Sobel RE, Ranalli TA, Cook RB, Edmondson DG, Roth SY, Allis CD. 1996. Transcription-linked acetylation by Gcn5p of histones H3 and H4 at specific lysines. *Nature* **383:** 269–272.

Lai E, Darnell JE Jr. 1991. Transcriptional control in hepatocytes: A window on development. *Trends Biochem Sci* **16:** 427–430.

Lamb P, McKnight SL. 1991. Diversity and specificity in transcriptional regulation: The benefits of heterotypic dimerization. *Trends Biochem Sci* **16:** 417–722.

Lee RC, Feinbaum RL, Ambros V. 1993. The *C. elegans* heterochromic gene *lin4* encodes small RNAs with antisense complementarity to *lin14. Cell* **75:** 843–854.

Lee Y, Ahn C, Han J, Choi H, Kim J, Yim J, Lee J, Provost P, Radmark O, Kim S, Kim VN. 2003. The nuclear RNase III Drosha initiates microRNA processing. *Nature* **425:** 415–419.

Lee W, Tillo D, Bray N, Morse RH, Davis RW, Hughes TR, Nislow C. 2007. A high-resolution atlas of nuleosome occupancy in yeast. *Nat Genet* **39:** 1235–1244.

Lee C, Li X, Hechmer A, Eisen M, Biggin MD, Venters BJ, Jiang C, Li J, Pugh BF, Gilmour DS. 2008. NELF and GAGA factor are linked to promoter-proximal pausing at many genes in *Drosophila*. *Mol Cell Biol* **28:** 3290–3300.

Lemaire MF, Thummel CS. 1990. Splicing precedes polyadenylation during *Drosophila* E74A transcription. *Mol Cell Biol* **10:** 6059–6063.

Levy DE, Darnell JE Jr. 2002. STATs: Transcriptional control and biologic impact. *Nat Rev Mol Cell Biol* **2:** 740–749.

Li Y, Flanagan PM, Tschochner H, Kornberg RD. 1994. RNA polymerase II initiation factor interactions and transcription start site selection. *Science* **163:** 805–807.

Li B, Carey M, Workman JL. 2007. The role of chromatin during transcription. *Cell* **128:** 707–719.

Licatalosi DD, Darnell RB. 2010. RNA processing and its regulation: Global insights into biological networks. *Nature* **11:** 75–87.

Licatalosi DD, Mele A, Fak JJ, Kayikci M, Chi SW, Clark TA, Schweitzer AC, Blume JE, Wang X, Darnell JC, Darnell RB. 2008. HITS-CLIP yields genome-wide insights into brain alternative RNA processing. *Nature* **456:** 464–469.

Lim LP, Lau NC, Weinstein EG, Abdelhakin A, Yetka S, Rhoades MW, Burge CB, Bartel DP. 2003. The microRNAs of *C. elegans*. *Genes Dev* **17:** 991–1008.

Lin R, Maeda S, Liu C, Karin M, Edgington TS. 2007. A large noncoding RNA is a marker for murine hepatocellular carcinomas and a spectrum of human carcinomas. *Oncogene* **26:** 851–858.

Lindbo JA, Silva-Rosales L, Proebsting WM, Doughery WG. 1993. Induction of a highly specific antiviral state in transgenic plants: Implications for regulation of gene expression and virus resistance. *Plant Cell* **5:** 1749–1759.

Lister R, Pelizzola M, Dowen RH, Hawkins RD, Hon G, Tonti-Filippini J, Nery JR, Lee L, Ye Z, Ngo QM, et al. 2009. Human DNA methylomes at base resolution show widespread epigenomic differences. *Nature* **462:** 315–322.

Liu J, Carmell MA, Rivas FV, Marsden CG, Thomson JM, Song J-J, Hammond SM, Joshua-Tor L, Hannon GJ. 2004. Argonaute 2 is the catalytic engine of mammalian RNAi. *Science* **305:** 1437–1441.

Liu WL, Coleman RA, Ma E, Grob P, Yang JL, Zhang Y, Dailey G, Nogales E, Tjian R. 2009. Structures of three distinct activator-TFIID complexes. *Genes Dev* **23:** 1510–1521.

Lodish H, Berk A, Matsudaira P, Kaiser CA, Krieger M, Scott MP, Zipursky L, Darnell J. 2003. *Molecular cell biology*, 5th ed. Freeman, New York.

Lodish H, Berk A, Kaiser CA, Krieger M, Scott MP, Bretscher A, Ploegh H, Matsudaira P. 2008. *Molecular cell biology*, 6th ed. Freeman, New York.

Longstaff M, Brigneti G, Boccard F, Chapman SN, Baulcombe DC. 1993. Extreme resistance to potato virus X infection in plants expressing a modified component of the putative viral replicase. *EMBO J* **12:** 379–386.

Lu LF, Thai TH, Calado DP, Chaudhry A, Kubo M, Tanaka K, Loeb GB, Lee H, Yoshimura A, Rajewsky K, Rudensky AY. 2009. Foxp3-dependent microRNA 155 confers competitive fitness to regulatory T cells by targeting SOCS1 protein. *Immunity* **30:** 80–91.

Lue NF, Kornberg RD. 1987. Accurate initiation at RNA polymerase II promoters in extracts from *Saccharomyces cerevisiae*. *Proc Natl Acad Sci* **84:** 8839–8843.

Luger K, Richmond TJ. 1998. DNA binding within the nucleosome core. *Curr Opin Struct Biol* **8:** 33–40.

Luger K, Mader AW, Richmond RK, Sargent DF, Richmond TJ. 1997. Crystal structure of the nucleosome core particle at 2.8 Å resolution. *Nature* **389:** 251–260.

Lyon MF. 1961. Gene action in the X chromosome of the mouse (*Mus musculus*). *Nature* **190:** 372–373.

Lyon MF. 1998. X-chromosome inactivation: A repeat hypothesis. *Cytogenet Cell Genet* **80:** 133–137.

Ma J, Ptashne M. 1987. Deletion analysis of GAL4 defines two transcriptional activating segments. *Cell* **48:** 847–853.

Maison C, Almouzni G. 2004. HP1 and the dynamics of heterochromatin maintenance. *Nat Rev Mol Cell Biol* **5:** 296–305.

Malik S, Roeder RG. 2000. Transcriptional regulation through Mediator-like coactivators in yeast and metazoan cells. *Trends Biochem Sci* **25:** 277–283.

Malone CD, Hannon GJ. 2009. Small RNAs as guardians of the genome. *Cell* **136:** 656–608.

Maniatis T, Tasic B. 2002. Alternative pre-mRNA splicing and proteosome expansion in metazoans. *Nature* **418:** 236–243.

Maquat LE, Kiledjian M. 2008a. RNA turnover in eukaryotes: Analysis of specialized and quality control RNA decay pathways. Preface. *Methods Enzymol* **449:** xvii–xviii.

Maquat LE, Kiledjian M. 2008b. RNA turnover in eukaryotes: Nucleases, pathways and analysis of mRNA decay. Preface. *Methods Enzymol* **448:** xxi–xxii.

Marcu MG, Doyle M, Bertolotti A, Ron D, Hendershot L, Neckers L. 2002. Heat shock protein 90 modulates the unfolded protein response by stabilizing IRE1a. *Mol Cell Biol* **22:** 8506–8513.

Margueron R, Reinberg D. 2011. The Polycomb complex PRC2 and its mark in life. *Nature* **469:** 343–349.

Margueron R, Justin N, Ohno K, Sharpe ML, Son J, Drury WJ III, Voigt P, Martin SR, Taylor WR, Marco VD, et al. 2009. Role of the Polycomb protein EEK in the propagation of repressive histone marks. *Nature* **461:** 762–767.

Matsui T, Segall J, Weil PA, Roeder RG. 1980. Multiple factors required for accurate initiation of transcription by purified RNA polymerase II. *J Biol Chem* **255:** 11992–11996.

Mavrich TN, Jiang C, Ioshikhes IP, Li X, Venters BJ, Zanton SJ, Tomsho LP, Qi J, Glaser RL, Schuster SC, et al. 2008. Nucleosome organization in the *Drosophila* genome. *Nature* **453:** 358–362.

McKenna NJ, Cooney AJ, DeMayo FJ, Downes M, Glass CK, Lanz RB, Lazar MA, Mangelsdorf DJ, Moore DD, Olin J, et al. 2009. Minireview: Evolution of NURSA, the nuclear receptor signaling atlas. *Mol Endocrinol* **23:** 740–746.

Mercer TR, Dinger ME, Mattick JS. 2009. Long non-coding RNAs: Insights into functions. *Nat Rev Genet* **10:** 155–159.

Merton RK. 1965. *On the shoulders of giants: A Shandean postscript.* The Free Press, New York [republished 1993, University of Chicago Press, Chicago].

Merton RK. 1973. *The sociology of science: Theoretical and empirical investigations.* University of Chicago Press, Chicago.

Méthot N, Basler K. 1999. Hedgehog controls limb development by regulating the activities of distinct transcriptional activator and repressor forms of *Cubitus interruptus*. *Cell* **96:** 819–831.

Méthot N, Basler K. 2001. An absolute requirement for *Cubitus interruptus* in Hedgehog signaling. *Development* **128:** 733–742.

Miele A, Dekker J. 2008. Long-range chromosomal interactions and gene regulation. *Mol Biol Syst* **4:** 1046–1057.

Mikkelsen TS, Ku M, Jaffe DB, Issac B, Lieberman E, Giannoukos G, Alvarez P, Brockman W, Kim T-K, Koche RP, et al. 2007. Genome-wide maps of chromatin state in pluripotent and lineage-committed cells. *Nature* **448:** 553–560.

Milani P, Chevereau G, Vaillant C, Audit B, Haftek-Terreau Z, Marilley M, Bouvet P, Argoul F, Arneodo A. 2009. Genome-scale identification of nucleosome positions in *S. cerevisiae*. *Proc Natl Acad Sci* **106:** 22257–22262.

Miska EA, Alvarez-Saavedra E, Abbot AL, Lau NC, Hellman AB, McGonagle SM, Bartel DP, Ambros V, Horvitz HR. 2007. Most *Caenorhabditis elegans* microRNAs are individually not essential for development or viability. *PLoS Genet* **3:** e215. doi: 10.1371/journal.pgen.0030215.

Montzka KA, Steitz JA. 1988. Additional low-abundance human small nuclear ribonucleoproteins: U11, U12, etc. *Proc Natl Acad Sci* **85:** 885–889.

Moore MJ, Proudfoot NJ. 2009. Pre-mRNA processing reaches back to transcription and ahead to translation. *Cell* **136:** 688–700.

Moore MJ, Query CC, Sharp PA. 1993. Splicing of precursors to mRNAs by the spliceosome. In *The RNA world* (ed. R Gesteland, JF Atkins), pp. 303–357. Cold Spring Harbor Laboratory Press, Cold Spring Harbor, NY.

Morimoto RI. 1993. Cells in stress: Transcriptional activation of heat shock genes. *Science* **259:** 1409–1410.

Morrison AJ, Shen X. 2005. DNA repair in the context of chromatin. *Cell Cycle* **4:** 568–571.

Naar AM, Boutin J-M, Lipkin SM, Yu VC, Holloway JM, Glass CK, Rosenfeld MG. 1991. The orientation and spacing of core DNA-binding motifs dictate selective transcriptional responses to three nuclear receptors. *Cell* **65:** 1267–1279.

Nakajima N, Horikoshi M, Roeder RG. 1988. Factors involved in specific transcription by mammalian RNA polymerase II: Purification, genetic specificity and TATA box-promoter interactions of TFIID. *Mol Cell Biol* **8:** 4028–4040.

Nevins JR, Chen-Kiang S. 1981. Processing of adenovirus nuclear RNA to mRNA. *Adv Virus Res* **26:** 1–35.

Nikolov DB, Hu S-H, Lin J, Gasch A, Hoffmann A, Horikoshi M, Chua N-H, Roeder RG, Burley SK. 1992. Crystal structure of TFIID TATA-box binding protein. *Nature* **360:** 40–46.

Nikolov DB, Chen H, Halay ED, Lusheva AA, Hisatake K, Lee DK, Roeder RG, Burley SK. 1995. Crystal structure of a TFIIB-TBP-TATA-element ternary complex. *Nature* **377:** 119–128.

Nilsen TW, Graveley BR. 2010. Expansion of the eukaryotic proteome by alternative splicing. *Nature* **463:** 457–463.

Nonet ML, Young RA. 1989. Intragenic and extragenic suppressors of mutations in the heptapeptide repeat domain of *Saccharomyces cerevisiae* RNA polymerase II. *Genetics* **123:** 715–724.

Nott A, Hir HL, Moore MJ. 2004. Splicing enhances translation in mammalian cells: An additional function of the exon junction complex. *Genes Dev* **18:** 210–222.

Nüsslein-Volhard C, Wieschaus E. 1980. Mutations affecting segment number and polarity in *Drosophila*. *Nature* **287:** 795–801.

Okada Y, Yamagata K, Hong K, Wakayama T, Zhang Y. 2010. A role for the elongator complex in zygotic paternal genome demethylation. *Nature* **463:** 554–558.

Olins AL, Olins DE. 1974. Spherical chromatin units (v bodies). *Science* **183:** 330–332.

Ørom UA, Derrien T, Beringer M, Gumireddy K, Gardini A, Bussotti G, Lai F, Zytnicki M, Notredame C, Huang Q, et al. 2010. Long noncoding RNAs with enhancer-like function in human cells. *Cell* **143:** 46–58.

Orlando V, Paro R. 1993. Mapping Polycomb-repressed domains in the bithorax complex using in vivo formaldehyde cross-linked chromatin. *Cell* **75:** 1187–1198.

Padgett RA, Konarska MM, Grabowski PJ, Hardy SF, Sharp PA. 1984. Lariat RNAs as intermediates and products in the splicing of messenger RNA precursors. *Science* **225:** 898–903.

Palstra R-J, Tolhuis B, Splinter E, Nijmeijer R, Grosveld F, de Laat W. 2003. *Nature Genetics* **35:** 190–194.

Panne D, Maniatis T, Harrison SC. 2007. An atomic model of the interferon-β enhanceosome. *Cell* **129:** 1111–1123.

Papin JA, Hunter T, Palsson BO, Subramanian S. 2005. Reconstruction of cellular signalling networks and analysis of their properties. *Nat Rev Mol Cell Biol* **6:** 99–111.

Park JM, Werner J, Kim JM, Lis JT, Kim YJ. 2001. Mediator, not holoenzyme, is directly recruited to the heat shock promoter by HSF upon heat shock. *Mol Cell* **8:** 9–19.

Parker R, Song H. 2004. The enzymes and control of eukaryotic mRNA turnover. *Nat Struct Mol Biol* **11:** 121–127.

Pasquinelli AE, Reinhart BJ, Slack F, Martindate MQ, Kuroda MI, Maller B, Hayward DC, Ball EE, Degnan B, Muller P, et al. 2000. Conservation of the sequence and temporal expression of *let7* heterochronic regulatory RNA. *Nature* **408:** 86–89.

Payvar F, DeFranco D, Firestone GL, Edgar B, Wrange O, Okret S, Gustafsson JA, Yamamoto KR. 1983. Sequence-specific binding of glucocorticoid receptor to MTV DNA at sites within and upstream of the transcribed region. *Cell* **35:** 381–392.

Penman S, Scherrer K, Becker Y, Darnell JE. 1963. Polyribosomes in normal and poliovirus infected HeLa cells and their relationship to messenger RNA. *Proc Natl Acad Sci* **49:** 654–662.

Peterlin BM, Price DH. 2006. Controlling the elongation phase of transcription with P-TEFb. *Mol Cell* **23:** 297–305.

Phatnani HP, Greenleaf AL. 2006. Phosphorylation and functions of the RNA polymerase II CTD. *Genes Dev* **20:** 2922–2936.

Ponting CP, Oliver PL, Reik W. 2009. Evolution and functions of long noncoding RNAs. *Cell* **136:** 629–641.

Probst AV, Dunleavy E, Almouzni G. 2009. Epigenetic inheritance during the cell cycle. *Nat Rev Mol Cell Biol* **10:** 192–206.

Puckett L, Darnell JE. l977. Essential factors in the kinetic analysis of RNA synthesis in HeLa cells. *J Cell Phys* **90:** 521–534.

Puckett L, Chambers S, Darnell JE. l975. Short-lived messenger RNA in HeLa cells and its impact on the kinetics of accumulation of cytoplasmic polyadenylate. *Proc Natl Acad Sci* **72:** 389–393.

Queen C, Baltimore D. 1983. Immunoglobulin gene transcription is activated by downstream sequence elements. *Cell* **33:** 741–748.

Rachez C, Lemon BD, Suldan Z, Bromleigh V, Gamble M, Naar AM, Erdjument-Bromage H, Tempst P, Freedman LP. 1999. Ligand-dependent transcription activation by nuclear receptors requires the DRIP complex. *Nature* **398:** 824–828.

Rasmussen EB, Lis JT. 1995. Short transcripts of the ternary complex provide insight into RNA polymerase II elongational pausing. *J Mol Biol* **252:** 522–535.

Rea S, Eisenhaber F, O'Carroll D, Strahl BD, Sun ZW, Sahmid M, Opravil S, Mechtler K, Ponting CP, Allis CD, Jenuwein T. 2000. Regulation of chromatin structure by site-specific histone H3 methyltransferases. *Nature* **406:** 593–596.

Reina JH, Hernandez N. 2007. On a roll for new TRF targets. *Genes Dev* **21:** 2855–2860.

Reinhart BJ, Slack FJ, Basson M, Pasquinelli AE, Bettinger JC, Rougvie AE, Horvitz HR, Ruvkun G. 2000. The 21-nucleotide let-7 RNA regulates deveopmental timing in *Caenorhabditis elegans*. *Nature* **403:** 901–906.

Richard P, Manley JL. 2009. Transcription termination by nuclear RNA polymerases. *Genes Dev* **23:** 1247–1269.

Riddle DL, Bluenthal T, Meyer BJ, eds. 1997. *C. elegans II*. Cold Spring Harbor Laboratory Press, Cold Spring Harbor, NY.

Ringrose L, Paro R. 2007. Polycomb/Trithorax response elements and epigenetic memory of cell identity. *Development* **134:** 223–232.

Rinn JL, Kertesz M, Wang JK, Squazzo SL, Xu X, Brugmann SA, Goodnough LH, Helms JA, Farnham PJ, Segal E, Chang HY. 2007. Functional demarcation of active and silent chromatin domains in human HOX loci by noncoding RNAs. *Cell* **129:** 1311–1323.

Rinn JL, Wang JK, Allen N, Brugmann SA, Mikels AJ, Liu H, Ridky TW, Stadler HS, Nusse R, Helms JA, Chang HY. 2008. A dermal HOX transcriptional program regulates site-specific epidermal fate. *Genes Dev* **22:** 303–307.

Robinson PJ, An W, Routh A, Martino F, Chapman L, Roeder RG, Rhodes D. 2008. 30 nm chromatin fibre decompaction requires both H4-K16 acetylation and linker histone eviction. *J Mol Biol* **381:** 816–825.

Rocheleau CE, Downs WD, Lin R, Wittmann C, Bei Y, Cha YH, Ali M, Priess JR, Mellow CC. 1997. Wnt signaling and an APC-related gene specify endoderm in early *C. elegans* embryos. *Cell* **90:** 707–716.

Roeder RG. 1996. The role of general initiation factors in transcription by RNA polymerase II. *Trends Biochem Sci* **21:** 327–335.

Roeder RG. 2005. Transcriptional regulation and the role of diverse coactivators in animal cells. *FEBS Lett* **579:** 909–915.

Rosenfeld MG, Mermod J-J, Amara SG, Swanson LW, Sawchenko PE, Rivier J, Vale WW, Evans RM. 1983. Production of a novel neuropeptide encoded by the calcitonin gene via tissue-specific RNA processing. *Nature* **304:** 129–135.

Rosenfeld MG, Lunyak VV, Glass CK. 2006. Sensors and signals: A coactivator/corepressor/epigenetic code for integrating signal-dependent programs of transcriptional response. *Genes Dev* **20:** 1405–1428.

Rougvie AE, Lis JT. 1988. The RNA polymerase II molecule at the 5′ end of the uninduced hsp70 gene of *D. melanogaster* is transcriptionally engaged. *Cell* **54:** 795–804.

Ruskin B, Krainer AR, Maniatis T, Green MR. 1984. Excision of an intact intron as a novel lariat structure during pre-mRNA splicing in vitro. *Cell* **38:** 317–331.

Ruthenburg AJ, Li H, Patel DJ, Allis CD. 2007. Multivalent engagement of chromatin modifications by linked binding modules. *Nat Rev Mol Cell Biol* **8:** 983–994.

Ruvkun G, Giusto J. 1989. The *C. elegans* heterochonic gene *lin 14* encodes a nuclear protein that forms a temporal developmental switch. *Nature* **338:** 313–319.

Saha A, Wittmeyer J, Cairns BR. 2006. Chromatin remodeling: The industrial revolution of DNA around histones. *Nat Rev Mol Cell Biol* **7:** 437–447.

Salditt-Georgieff M, Darnell JE Jr. 1982. Further evidence that the majority of primary nuclear RNA transcripts in mammalian cells do not contribute to mRNA. *Mol Cell Biol* **2:** 701–707.

Salditt-Georgieff M, Jelinek W, Darnell JE, Furuichi Y, Morgan M, Shatkin A. 1976. Methyl labeling of HeLa cell HnRNA: A comparison with mRNA. *Cell* **7:** 227–237.

Salditt-Georgieff M, Harpold M, Chen-Kiang S, Darnell JE Jr. 1980. The addition of 5′ cap structures occurs early in hnRNA synthesis and prematurely terminated molecules are capped. *Cell* **19:** 69–78.

Salditt-Georgieff M, Harpold M, Wilson M, Darnell JE Jr. 1981. Large heterogeneous nuclear ribonucleic acid has three times as many 5′ caps as polyadenylic acid segments, and most caps do not enter polyribosomes. *Mol Cell Biol* **1:** 179–187.

Schermelleh L, Spada F, Easwaran HP, Zolghadr K, Margot JB, Cardoso MC, Leonhardt H. 2005. Trapped in action: Direct visualization of DNA methyltransferase activity in living cells. *Nat Methods* **2:** 751–756.

Schlessinger J. 2004. Common and distinct elements in cellular signaling via EGF and FGF receptors. *Science* **306:** 1506–1507.

Schmeing TM, Ramakrishnan V. 2009. What recent ribosome structures have revealed about the mechanism of translation. *Nature* **461:** 1234–1242.

Schroder M, Kauman RJ. 2005. The mammalian unfolded protein response. *Annu Rev Biochem* **74:** 739–789.

Schutt C, Nothiger R. 2000. Structure, function and evolution of sex-determining systems in Dipteran insects. *Development* **127:** 667–677.

Sehgal PG, Derman E, Molloy GR, Tamm I, Darnell JE. 1976. 5,6-Dichloro-1-Beta-D-ribofur-anosylbenzimidazole inhibits initiation of nuclear heterogeneous RNA chains in HeLa cells. *Science* **194:** 431–433.

Seila AC, Calabrese JM, Levine SS, Yeo GW, Rahl PB, Flynn RA, Young RA, Sharp PA. 2008. Divergent transcription from active promoters. *Science* **322:** 1849–1851.

Seila AC, Core LJ, Lis JT, Sharp PA. 2009. Divergent transcription: A new feature of active promoters. *Cell Cycle* **8:** 2557–2564.

Sheng W, Yan H, Rausa FM III, Costa RH, Liao X. 2004. Structure of the hepatocyte nuclear factor 6α and its interaction with DNA. *J Biol Chem* **279:** 33928–33936.

Shi Y. 2007. Histone lysine demethylases: Emerging roles in development, physiology and disease. *Nat Rev Genet* **8:** 829–833.

Shi Y, Matson C, Mulligan P, Whetstine JR, Cole PA, Casero RA, Shi Y. 2004. Histone demethylation mediated by the nuclear amine oxidase homolog LSD1. *Cell* **119:** 941–953.

Shi Y, Giammartino DC, Taylor D, Sarkeshik A, Rice WJ, Yates JR III, Frank J, Manley JL. 2009. Molecular architecture of the human pre-mRNA 3′ processing complex. *Mol Cell* **33:** 365–376.

Sikder HA, Devlin MK, Dunlap S, Ryu B, Alani RM. 2003. Id proteins in cell growth and tumorigenesis. *Cancer Cell* **3:** 525–530.

Singh J, Padgett RA. 2009. Rates of in situ transcriptional splicing in large human genes. *Nat Struct Mol Biol* **16:** 1128–1133.

Small EM, Olson EN. 2011. Pervasive roles of microRNAs in cardiovascular biology. *Nature* **469:** 336–342.

Solomon MJ, Larsen PL, Varshavsky A. 1988. Mapping protein–DNA interactions in vivo with formaldehyde: Evidence that histone H4 is retained on a highly transcribed gene. *Cell* **53:** 937–947.

Soshnikova N, Duboule D. 2009. Epigenetic temporal control of mouse Hox genes in vivo. *Science* **324:** 1320–1323.

Spies N, Nielsen CB, Padgett RA, Burge CB. 2009. Biased chromatin signatures around poly-adenylation sites and exons. *Mol Cell* **36:** 245–254.

St Johnston D, Nüsslein-Volhard C. 1992. The origin of pattern and polarity in the *Drosophila* embryo. *Cell* **68:** 201–219.

Sugiyama T, Cam H, Verdel A, Moazed D, Grewal SI. 2005. RNA-dependent RNA polymerase is an essential component of a self-enforcing loop coupling heterochromatin assembly to siRNA production. *Proc Natl Acad Sci* **102:** 152–157.

Sulston JE. 2003. *Caenorhabditis elegans*: The cell lineage and beyond (Nobel lecture). *Chembiochem* **4:** 688–696. http://nobelprize.org/nobel_prizes/medicine/laureates/2002/sulston-lecture.html.

Taatjes DF, Schneider-Poetsch T, Tjian R. 2004. Distinct conformational states of nuclear receptor-bound CRSP-Med complexes. *Nat Struct Mol Biol* **11:** 664–671.

Taft RJ, Glazov EA, Cloonan N, Simons C, Stephen S, Faulkner GJ, Lassmann T, Forrest AR, Grimmond SM, Schroder K, et al. 2009. Tiny RNAs associated with transcription start sites in animals. *Nat Genet* **41:** 572–578.

Takada R, Nakatani Y, Hoffmann A, Kokubo T, Hasegawa S, Roeder RG, Horikoshi M. 1992. Identification of human TFIID components and direct interaction between a 250-kDa polypeptide and the TATA box-binding protein (TFIID tau). *Proc Natl Acad Sci* **89**: 11809–11813.

Takagi Y, Kornberg RD. 2006. Mediator as a general transcription factor. *J Biol Chem* **281**: 80–89.

Takahashi K, Yamanaka S. 2006. Induction of pluripotent stem cells from mouse embryonic and adult fibroblast cultures by defined factors. *Cell* **126**: 663–676.

Tamm I, Kiku T, Darnell JE Jr, Salditt-Georgieff M. 1980. Short capped hnRNA precursor chains in HeLa cells: Continued synthesis in presence of 5,6-dichloro-l-β-D-ribofuranosylenzimidazole. *Biochemistry* **19**: 2743–2748.

Tarn WY, Steitz JA. 1996. A novel spliceosome containing U11, U12, and U5 snRNPs excises a minor class (AT-AC) intron in vitro. *Cell* **84**: 801–811.

Tarn WY, Steitz JA. 1997. Pre-mRNA splicing: The discovery of a new splicesome doubles the challenge. *Trends Biochem Sci* **22**: 132–137.

Taverna SD, Li H, Rutheburg AJ, Allis CD, Patel DJ. 2007. How chromatin-binding modules interpret histone modifications: Lessons from professional pocket pickers. *Nat Struct Mol Biol* **14**: 1025–1040.

Thanos D, Maniatis T. 1995. Virus induction of human IFNβ gene expression requires the assembly of an enhanceosome. *Cell* **83**: 1091–1100.

Tian D, Sun S, Lee JT. 2010. The long noncoding RNA, Jpx, is a molecular switch for X chromosome inactivation. *Cell* **143**: 390–403.

Tsai MC, Manor O, Wan Y, Mosammaparast N, Wang JK, Lan F, Shi Y, Segal E, Chang HY. 2010. Long noncoding RNA as modular scaffold of histone modification complexes. *Science* **329**: 689–693.

Ule J, Jensen KB, Ruggiu M, Mele A, Ule A, Darnell RB. 2003. CLIP identifies nova-regulated RNA networks in the brain. *Science* **302**: 1212–1215.

Ule J, Ule A, Spencer J, Williams A, Hu J-S, Cline M, Wang H, Clark T, Fraser C, Ruggiu M, et al. 2005. Nova regulates brain-specific splicing to shape the synapse. *Nat Genet* **37**: 844–852.

Umesono K, Murakami KK, Thompson CC, Evans RM. 1991. Direct repeats as selective response elements for the thyroid hormone, retinoic acid, and vitamin D$_3$ receptors. *Cell* **65**: 1255–1266.

Vagin VV, Sigova A, Li C, Seitz H, Gvozdev V, Zamore PD. 2006. A distinct small RNA pathway silences selfish genetic elements in the germline. *Science* **313**: 320–314.

Venkataraman K, Brown KM, Gilmartin GM. 2005. Analysis of a noncanonical polyA site reveals a tripartite mechanism for vertebrate polyA site recognition. *Genes Dev* **19**: 1315–1327.

Vermeulen M, Mulder KW, Denissov S, Pijnappel WW, van Schaik FM, Varier RA, Baltissen MP, Stunnenberg HG, Mann M, Timmers HT. 2007. Selective anchoring of TFIID to nucleosomes by trimethylation of histone H3 lysine 4. *Cell* **131**: 58–69.

Vousden KH, Prives C. 2009. Blinded by the light: The growing complexity of p53. *Cell* **137**: 413–431.

Wada T, Takagi T, Yamaguchi Y, Ferdous A, Imai T, Hirose S, Sugimoto S, Yano K, Hartzog GA, Winston F, et al. 1998. DSIF, a novel transcription elongation factor that regulates RNA polymease II processivity, is composed of human Spt4 and Spt5 homologs. *Genes Dev* **12**: 343–356.

Wada Y, Ohta Y, Xu M, Tsutsumi S, Minami T, Inoue K, Komura D, Kitakami J, Oshida N, Papantonis A, et al. 2009. A wave of nascent transcription on activated human genes. *Proc Natl Acad Sci* **106**: 18357–18361.

Wahl MC, Will CL, Luhrmann R. 2009. The spliceosome: Design principles of a dynamic RNP machine. *Cell* **136:** 701–718.

Walker MD, Adlund T, Boulet AM, Rutter WJ. 1983. Cell-specific expression controlled by the 5′-flanking region of insulin and chymotrypsin. *Nature* **306:** 557–561.

Wang GI, Semenza GL. 1995. Purification and characterization of hypoxia-inducible factor 1. *J Biol Chem* **270:** 1230–1237.

Wang Y, Wysocka J, Savegh J, Lee YH, Perlin JR, Leonelli L, Sonbuchner LS, McDonald CH, Cook RG, Dou Y, et al. 2004. Human PAD4 regulates histone arginine methylation levels via demethylimination. *Science* **306:** 279–283.

Wang D, Bushnell DA, Westover KD, Kaplan CD, Kornberg RD. 2006. Structural basis of transcription: Role of the trigger loop in substrate specificity and catalysis. *Cell* **127:** 941–954.

Wang ET, Sandberg R, Luo S, Khrebtukova I, Zhang L, Mayr C, Kingsmore SF, Schroth GP, Burge CG. 2008. Alternative isoform regulation in human tissue transcriptomes. *Nature* **456:** 470–476.

Weber J, Jelinek W, Darnell JE. 1977. The definition of a large viral transcription unit late in Ad2 infection of HeLa cells: Mapping of nascent RNA molecules labeled in isolated nuclei. *Cell* **10:** 811–816.

Weber J, Blanchard J-M, Ginsberg HM, Darnell JE Jr. 1980. Order of polyadenylic acid addition and splicing events in early adenovirus mRNA formation. *J Virol* **33:** 286–291.

Weil PW, Luse DS, Segall J, Roeder RG. 1979. Selective and accurate initiation of transcription at the Ad2 major late promoter in a soluble system dependent on purified RNA polymerase II and DNA. *Cell* **18:** 469–484.

Weinmaster G. 2000. Notch signal transduction: A real rip and more. *Curr Opin Genet Dev* **10:** 363–369.

Westaway SK, Abelson J. 1995. Splicing of tRNAs. In *tRNA: Structure, biosynthesis and function* (ed. D Soll, UL RajBhandary), pp. 79–92. American Society for Microbiology, Washington, DC.

Wightman B, Burglin TR, Gatto J, Arasu P, Revkun G. 1991. Negative regulatory sequences in the *lin14* 3′-untranslated region are necessary to generate a temporal switch during *C. elegans* development. *Genes Dev* **5:** 1813–1824.

Wightman B, Ha I, Ruvkun G. 1993. Posttranscriptional regulation of the heterochronic gene *lin-14* by *lin-4* mediates temporal pattern formation in *C. elegans. Cell* **75:** 855–862.

Wightman R, Bell R, Reece RJ. 2008. Localization and interaction of the proteins constituting the *GAL* genetic switch in *Saccharomyces cerevisiae. Eukaryot Cell* **7:** 2061–2068.

Wigler M, Silverstein S, Lee LS, Pellicer A, Cheng Y, Axel R. 1977. Transfer of purified herpes virus thymidine kinase gene to cultured mouse cells. *Cell* **11:** 223–232.

Wigler M, Sweet R, Sim GK, Wold B, Pellicer A, Lacy E, Maniatis T, Silverstein S, Axel R. 1979. Transformation of mammalian cells with genes from prokaryotes and eukaryotes. *Cell* **16:** 444–452.

Williams AH, Liu N, van Rooij E, Olson EN. 2009. MicroRNA control of muscle development and disease. *Curr Opin Cell Biol* **21:** 461–469.

Wilson D, Charoensawan V, Kummerfeld SK, Teichmann SA. 2008. DBD—Taxonomically broad transcription factor predictions: New content and functionality. *Nucleic Acids Res* **36:** D88–D92.

Wilusz CJ, Wormington M, Peltz SW. 2001. The cap-to-tail guide to mRNA turnover. *Nat Rev Mol Cell Biol* **2:** 237–246.

Wingender E, Chen X, Hehl R, Karas H, Liebich I, Matys V, Meinhardt T, Pruss M, Reuter I, Schacherer F. 2000. TRANSFAC: An integrated system for gene expression regulation. *Nucleic Acids Res* **28:** 316–319.

Workman JL. 2006. Nucleosome displacement in transcription. *Genes Dev* **20:** 2009–1017.

Workman JL, Roeder RG. 1987. Binding of transcription factor TFIID to the major late promoter during in vitro nucleosome assembly potentiates subsequent initiation by RNA polymerase II. *Cell* **51:** 613–622.

Wutz A. 2007. Xist function: Bridging chromatin and stem cells. *Trend Genet* **23:** 457–464.

Xanthopoulos KG, Prezioso VR, Chen WS, Sladek FM, Cortese R, Darnell JE Jr. 1991. The different tissue transcription patterns of genes for HNF-1, C/EBP HNF-3, and HNF-4, protein factors that govern liver-specific transcription. *Proc Natl Acad Sci* **88:** 3807–3811.

Xiao C, Calado DP, Galler G, Thai TH, Patterson HC, Wang J, Rajewsky N, Bender TP, Rajewsky K. 2007. MiR-150 controls B cell differentiation by targeting the transcription factor c-Myb. *Cell* **131:** 146–159.

Yamanaka S. 2009. A fresh look at IPS cells. *Cell* **137:** 13–17.

Young RA. 1991. RNA polymerase II. *Annu Rev Biochem* **60:** 689–715.

Zhan X-L, Clemens JC, Neves G, Hattori D, Flanagan JJ, Hummel T, Vasconcelos ML, Chess A, Zipursky SL. 2004. Analysis of Dscam diversity in regulating axon guidance in *Drosophila* mushroom bodies. *Neuron* **43:** 673–686.

Zhang D, Paley AJ, Childs G. 1998. The transcriptional repressor ZFM1 interacts with and modulates the ability of EWS to activate transcription. *J Biol Chem* **273:** 18086–18091.

Zhang L, Xu T, Maeder C, Bud LO, Shanks J, Nix J, Guthrie C, Pleiss JA, Zhao R. 2009. Structural evidence for consecutive Hel308-like modules in the spliceosomal ATPase Brr2. *Nat Struct Mol Biol* **16:** 731–739.

Zinzen RP, Girardot C, Gagneur J, Braun M, Furlong EEM. 2009. Combinatorial binding predicts spatio-temporal *cis*-regulatory activity. *Nature* **462:** 65–70.

Zwaka TP. 2010. Unraveling the score of the enhancer symphony. *Proc Natl Acad Sci* **107:** 21240–21241.

6

RNA and the Beginning of Life

THERE IS MUCH LEFT TO ANSWER regarding the role of RNA in present-day biology. But with current momentum and ever-improving technology, there remains little doubt that the many roles of RNA in gene function and gene regulation will be revealed in increasing detail.

Certainly, the greatest remaining challenge that all of the recent progress in RNA studies has left unsolved is the role of RNA in the origin of life and the evolution of cells. And despite encouraging progress, many would be surprised to see these questions resolved soon. These are hardly new topics (Haldane 1929; Oparin 1976). Speculation regarding Archaean geology, atmospheric conditions, and physical/chemical events that could have led to producing the necessary biological building blocks for life is more than 60 years old, predating most modern molecular biology achievements (for a recent review, see Leach et al. 2006). Even Charles Darwin in a letter to his eminent friend and supporter, the botanist Joseph Hooker, speculated that the first spark of life may have taken place in a "warm little pond, with all sorts of ammonia and phosphoric salts, lights, heat, electricity, etc. present, so that a *proteine* [Darwin's spelling] compound was chemically formed ready to undergo still more complex changes" (Browne 2002).

By 1960, the centrality to life of macromolecules—not only protein and DNA but finally RNA—became incontrovertible. Ideas regarding the chemical evolution of life took on a new flavor. Clearly, most of the chemistry in cells depended on proteins, but ordered amino acids capable of catalysis could hardly have arisen de novo. Such order required an informational molecule.

If the contest as to the first informational molecule were between DNA and RNA, RNA would win hands down. Even today, beginning DNA replication requires a primer, almost always furnished by RNA. Moreover, the building blocks for DNA—deoxyribonucleotides—derive through the action of ribonucleotide reductase on ribonucleotide diphosphates: rCDP, rADP, rGDP, and rUDP yield dCDP, dADP, dGDP, and dUDP, respectively. dUDP (as dUMP) is then methylated by thymidine synthase to form dTMP. All of the deoxynucleotides are then converted to triphosphates for incorporation into DNA. All present-day organisms use these enzymes, speaking strongly for the presence during evolution of RNA before DNA.

By 1956, tobacco mosaic virus (TMV) RNA was proven to be an informational molecule that could, at least in the infected cell, direct its own replication. When the messenger RNA (mRNA) idea was "invented" by François Jacob and Jacques Monod and then mRNA was actually discovered in 1961, the case was settled—RNA is not only in wide use in today's cells as an informational molecule, but it is also the responsible agent for information transfer to protein. DNA is "only" the stable storehouse of genetic information that must be converted into RNA to perform operational genetics.

In the late 1960s, Francis Crick, Leslie Orgel, and Carl Woese (Woese 1967; Crick 1968; Orgel 1968) all wrote persuasively that RNA was the best candidate for the "first informational molecule." Their conclusions drew mainly on the then recently recognized roles of mRNA, transfer RNA (tRNA), and ribosomal RNA (rRNA) in translation of DNA information into proteins.

Beyond the Watson-Crick base-pairing properties of DNA and RNA, little was appreciated at that time regarding RNA as a chemically active entity. But during the late 20th and early 21st centuries, explosions revealing the central roles of RNA in present-day macromolecular synthesis and advances in understanding the catalytic capacity of RNA have led to a much sharper focus on RNA in prebiotic and precellular chemistry (Joyce 2002; de Pouplana 2004; Gesteland et al. 2006; Atkins et al. 2011; Deamer and Szostak 2011). Tom Cech and Sidney Altman began this revolution in the early and mid 1980s, showing the first catalytic properties of RNA (for review, see Altman 1990; Cech 1990) (see Chapter 4). These groundbreaking observations included not only RNA chain scission and transesterification, but also, under proper conditions, chain elongation by the *Tetrahymena* rRNA intron. And in the last decade, all of the recently discovered RNAs discussed in Chapter 5 have come into view.

All of these advances stimulate optimism regarding several current lines of prebiotic research: (1) the chemical origin of biomolecules under plausible prebiotic conditions and the evolution of RNA as the first informational macromolecule, (2) geological studies on ancient but accurately aged rocks to recover chemical "fossils" and physically observable microfossils; and (3) the definition, by sequence analysis, of the three cellular kingdoms on the planet and the use of this information to illuminate the origin of cells.

PREBIOTIC CHEMISTRY: BUILDING BLOCKS

The oldest type of research dealing with prebiotic questions naturally concerns the possible sources of basic macromolecular building blocks. Several lines of imaginative synthetic organic chemistry are moving in the direction of describing how amino acids and nucleic acid constituents could have arisen on the Archaean Earth.

Amino Acids

The famous experiments of Stanley Miller in 1953 (Miller 1953) showed that, given an atmosphere containing H_2, methane, and ammonia, electrical sparks simulating lightning could generate amino acids. These early experiments have been questioned as relevant to early life because the Archaean atmosphere probably did not contain the concentration of methane used in the experiments, although atmospheric methane almost certainly did exist during part of the Hadean–Archaean epochs (Zahnle et al. 2011).

Although there is no proof of how amino acids might have become available, there are several major possibilities. Formation of amino acids at hydrothermal and volcanic vents is generally agreed to be likely (for review, see Huber and Wachtershauser 2006). Possibly the most abundant early source for amino acids indigenous to Earth was volcanic hydrothermal vents. In addition, stellar and interstellar synthesis of organic molecules seems well established, and the presence of amino acids and hundreds of other organic molecules in meteorites is proven (Bernstein 2006; Schmitt-Kopplin et al. 2010). Thus for the purposes of biologists/biochemists trying to deal with prebiotic problems, the presence of at least some amino acids on the Archaean Earth is generally taken for granted. Moreover, plausible explanations based on differential solubility have recently been advanced for the choice of L-amino acids versus D-amino acids as well as for the choice of D-ribose over L-ribose (Breslow and Cheng 2009; Breslow et al. 2010).

Nucleotides

Evidence for the presence of nucleic acid precursors on the Archaean Earth is much different from that for amino acids. Miller's original experiments produced adenine, and continued efforts with collaborators eventually produced all of the nucleobases under plausible prebiotic conditions (Robertson and Miller 1995; Levy et al. 1999; Nelson et al. 2001). For some time, however, attempts to show pentose (especially ribose) formation were unrewarding.

Many organic chemists have considered whether evolutionary precursors with some structural similarities to RNA constituents or polymeric RNA precursors arose before RNA (for review, see Benner 2006, 2009; Englehardt and Hud 2011; Robertson and Joyce 2011). These considerations included earlier, easier-to-make sugars and nucleobases. One appealing aspect of the logic in these experiments is that because evolution of life follows a path from the simple to the more complex but still workable solutions, why not expect the same of organic molecules? Reading these discussions leads the reader to believe that although the field is a difficult one, progress is very definite.

For example, in 2006, Steve Benner (2006) emphasized the importance of stabilization by borate salts of ribose that can be formed by glyceraldehyde reactions (the so-called formose reactions). Glycoaldehyde and glyceraldehyde spontaneously form in space and most likely did so in the probable Archaean atmosphere (Zahnle et al. 2011). Together with the ready availability of borates on the Earth's surface, ribose formation therefore seemed plausible. After discussing these considerations on possible ribose formation, Benner concludes his 2006 review by saying that "the RNA first hypothesis appears to be more defensible today than it was a decade ago" (Benner 2006). But how ribose and nucleobases might have been put together had not been solved.

A recent report by John Sutherland and colleagues at the University of Manchester in England (Powner et al. 2009) has created quite some optimism in prebiotic circles (Szostak 2009). The Sutherland group reported the test tube synthesis of pyrimidine nucleotides under conditions that might have existed on the Archaean Earth. Rather than going step by step to make a nucleic acid base first, then a sugar, and then attempting to unite them by a phosphodiester bond, the Sutherland group assembled carbon- and nitrogen-containing potential nucleoside precursors—cyanamide, cyanoacetylene, glycoaldehyde, and glyceraldehyde—along with inorganic phosphate. After several steps of incubation under differing temperatures and salt conditions, all plausible on Archaean Earth, they succeeded in producing cytidine monophosphate. They proposed probable chemical steps to this success and isolated many of the proposed intermediates to confirm their ideas (Fig. 6-1A,B) (Powner et al. 2009). The presence of phosphate was crucial to several steps during the formation of the nucleotide.

In addition, the Sutherland laboratory found that UV irradiation of CMP (cytidine 5′ monophosphate) produced UMP (uridine 5′ monophosphate) (Fig. 6-1C). For at least two of the four recognized ribonucleotides, there now seemed to be a plausible (if difficult) route of prebiotic synthesis of nucleotides. Because purine and purine-like compounds were likely available (Oro 1961; Levy et al. 1999), it now seemed more reasonable that all four nucleotides could have been available in the Archaean.

The point of this brief discussion is not to consider prebiotic synthesis as in any way settled, but to point out that investigators in the field place the precursors of RNA high on the list of the necessary "first" ingredients of life.

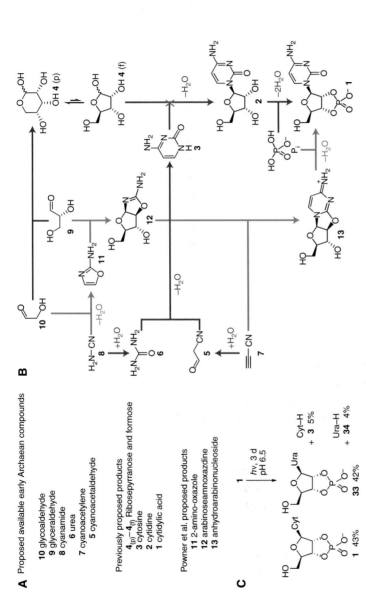

A Proposed available early Archaean compounds

10 glycoaldehyde
9 glyceraldehyde
8 cyanamide
6 urea
7 cyanoacetylene
5 cyanoacetaldehyde

Previously proposed products

4(p)–4(f) Ribosepyrranose and formose
3 cytosine
2 cytidine
1 cytidylic acid

Powner et al. proposed products

11 2-amino-oxazole
12 arabinoseamnoxazdine
13 anhydroarabinonucleoside

Figure 6-1. Possible routes to pyrimidine nucleotides on the Archaean Earth as summarized by Powner et al. (2009). Compounds widely believed to have been available on the Archaean Earth are listed in *A*, and the numbers are then used in *B* and *C* to follow the reactions. Cyanamide (8) and urea (6) are taken to be interconvertible in H_2O, as are cyanoacetylene (7) and cyanoacetaldehyde (5). (*B*, blue lines) Interactions known to occur. According to Powner et al. (2009), the problem in the blue pathway is that condensation of ribose (4) with cytosine (3) to produce cytidine (2) does not occur at an appreciable rate (red X). Powner et al. (2009) then obtained cytidylic acid (1) through reactions diagrammed by the green pathway. (*C*) Cytidylic acid (1) is converted to uridylic acid (33) by limited ultraviolet (*hv*) light exposure for 3 d. Percentages are yields after irradiation. Some phosphates are lost, yielding cytosine (3) and uridine (34). (Adapted, with permission, from Powner et al. 2009 [©Macmillan].)

Furthermore, it is clear that many talented chemists have turned to, and will probably increasingly turn to, the task of searching for plausible steps to "life in a test tube" that could have occurred here on Earth, as well as collaborations involving explorations of Mars for hints regarding how and if such chemistry occurred elsewhere (McLoughlin et al. 2007; for up-to-date details and a discussion of important problems not yet solved by organic chemistry, see Atkins et al. 2011; Deamer and Szostak 2011).

SELF-REPLICATION OF RNA

Years before any knowledge of how ribonucleotides might be formed on ancient Earth, other scientists began to investigate polymerization of RNA-like molecules and, later, self-replicating polynucleotide systems.

Despairing that although nucleic acid bases might be formed, Leslie Orgel concluded that no clear path to ribonucleotides could be envisioned. He and colleagues in the late 1970s began using nucleoside phosphorimidazolides as activated nucleotide analogs. Under the correct Mg^{++} concentrations, these compounds polymerized in a template-dependent fashion, granting some success to years of arduous efforts (for review, see van Roode and Orgel 1980; Robertson and Joyce 2011). Therefore, at least if plausible schemes for making ribose and ribonucleotides could ever be discovered, dimerization and ultimately "spontaneous" polymerization might be possible.

Reactions That Can Occur with Already Formed Polynucleotides

The first success was when Tom Cech and colleagues showed that the self-excising *Tetrahymena* intron (or IVS, for intervening sequence) normally undergoes further reactions, it can cyclize, clearing off 15 nucleotides (Fig. 4-35) or 19 nucleotides to form a linear L-19 molecule. In an appropriate salt concentration and temperature, the L-19 molecule can act as either a nuclease or a polymerase that can add a number of nucleotides (Fig. 6-2A,B) (Zaug and Cech 1986). The full-length IVS, in addition to cyclizing with a loss of 15 nucleotides, can oligomerize, showing a ligase ability (Fig. 6-2C,D) (Zaug and Cech 1985; Doudna and Cech 2002). These experiments, aside from their monumental importance in simply showing the enzymatic capacity of RNA, focused attention on the likely fact that, as is the case for protein enzymes, a precise folded physical structure of a ribozyme underlies its catalytic potential. This has been amply proven by later structural studies (for review, see Burke 2004; Hougland et al. 2006; Pyle 2010).

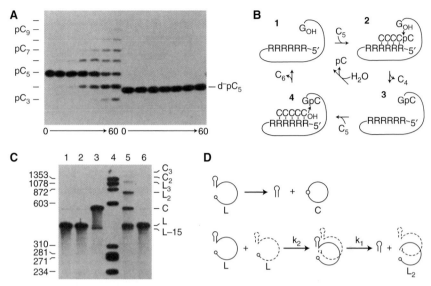

Figure 6-2. *Tetrahymena* rRNA intron is an enzyme. After release of the *Tetrahymena* pre-rRNA intron, the IVS (intervening sequence) or intron undergoes secondary reactions, finally cleaving off the 19 nucleotides at the 3′ end of the IVS to produce the linear L minus 19, or L−19. (*A*) Using the 3′-G_{OH}, the L−19 RNA is still an active enzyme, either as a ribonuclease or as a polymerase. When incubated (0 to 60 min) with L−19 RNA, ^{32}P-labeled C_5 (pC_5) reacts to yield both shorter and longer chain lengths (*left* half of autoradiogram). If the reaction is run to exhaustion of pC_5 substrate, chains as long as 30 Cs can be observed. The ^{32}P-labeled deoxycytidine chain (d-pC_5) is not a substrate (*right* half of autoradiogram). (*B*) Model of enzymatic reactions in *A*. Steps 1, 2, and 3 result in cleavage to yield C_4 and regenerate the enzyme (L−19). Step 4 results in adding one C residue to the 3′ end of the ^{32}P-labeled 5′-pC_5. Repetition can either shorten or lengthen the original ^{32}P-labeled pC_5 chains. (*C*) Cyclization and/or oligomerization (ligation) of the *Tetrahymena* IVS. Autoradiogram of electrophoretic separation of ^{32}P-labeled full-length IVS (L) (~400+ nucleotides) after various treatments, showing only a trace of L−15 linear molecule. (Lanes *1,6*) Starting material, (lane *2*) no cyclization conditions, (lane *3*) cyclization conditions allow circularization with loss of 15 nucleotides (circles, marked as C on right margin, move more slowly during electrophoresis), (lane *4*) size markers, (lane *5*) cyclization and oligomerization are essentially the same reaction (conditions that allow cyclization also show ligation to form circles [C] as well as oligomers, the dimer [L_2], and a small amount of a trimer [L_3]. (*D*) Interpretation of oligomerization (formation of dimers and trimers) of IVS. In addition to a single circle, two full-length IVSs (*bottom*) can interact so that the 3′ OH of one attacks the cyclization site of its partner, forming a longer linear molecule (marked on gel as L2). (*A,B,* Reprinted, with permission, from Zaug and Cech 1986 [©AAAS]; *C,D,* reprinted, with permission, from Zaug and Cech 1985 [©AAAS].)

Taking off from Cech's results, Jack Szostak and David Bartel studied the evolution of ribozymes by starting with a large series of machine-made random polyribonucleotides to mimic what might evolve in undirected synthesis. They isolated molecules (*ribozymes*) that were capable of "ligase" action (i.e., joining template-aligned oligonucleotides in 3′–5′ phosphodiester linkages) (Bartel and Szostak 1993; Ekland et al. 1995; Ekland and Bartel 1996). Additional work on RNA molecules with this ligase activity led to true polymerase capacity, wherein individual nucleoside triphosphates could be added to preexisting oligonucleotides in a template-dependent fashion (for review, see Burke 2004). Gerald Joyce and colleagues have made the latest advance in this field of RNA-directed RNA synthesis by using a designed set of polynucleotides to form a truly self-replicating system. The system originally requires six designed polyribonucleotides (two double-stranded target "genomes" [four molecules] and two "polymerases"). Given this set of molecules, cross-replication by the two polymerases of each of the two chains in the target "genomes" can be performed. With these cross-replicating molecules, serial dilutions (to simulate evolutionary time) were carried out, and the products were further incubated. The final products after many dozens of dilutions were extremely efficient, with the greatest rate of self-replication of ~one molecule per hour (Lincoln and Joyce 2009).

What do all of the presently discussed prebiotic "evolutionary" experiments prove? Progress in formation of necessary small-molecule precursors for both amino acids and nucleotides under prebiotic conditions is certainly encouraging. How the first oligonucleotides could have arisen, however, is still a major question (Ferris 2006; Robertson and Joyce 2011). But given a pool of oligonucleotides, the recent experiments certainly show that oligoribonucleotides could quite reasonably develop by selection into self-replicating molecules. In 2006, Joyce and Orgel countered the pessimists who view these experiments as parlor tricks: "Nature did not have the opportunity to conduct carefully arranged evolution experiments using highly purified reagents but did have the luxury of *much greater reaction volumes* and *much more time*" (Joyce and Orgel 2006) (my emphasis).

In addition to test tube evolution of self-replicating RNAs, other small RNAs (ribozymes) have been selected in in vitro evolution systems that perform reactions that resemble the tasks now carried out by tRNAs (Lohse and Szostak 1996; Lee et al. 2000; for review, see Breaker 2006; Breaker 2011; Yarus 2011). These molecules were selected from large libraries of RNAs ~100 nucleotides long, so-called *aptamers*. For example, a ribozyme of 110 nucleotides will catalyze the addition of either leucine or phenylalanine to its own 5′ terminus, forming an anhydride linkage. Still another ribozyme catalyzes peptide-bond formation between methionine and a phenylalanine that is tethered to the 5′ end of the ribozyme, resulting in a dipeptide. Another ribozyme will add aminoacyl adenylates to its 2′- or 3′-hydroxyl terminus, more like a primitive tRNA.

Although none of the ribozyme reactions mimic exactly the actual steps that occur during polypeptide synthesis on today's ribosomes, they again illustrate that given "endless" time ($\sim$1 billion years), RNA polymers showing the abilities required for replication and peptide synthesis could have evolved (Breaker 2006; Noller 2011; Yarus 2011). These chemical evolution experiments will certainly be continued and promise eventual success in more closely mimicking present-day directed peptide synthesis.

PRESENT-DAY RNA MACHINES SUGGEST AN RNA WORLD

The pivotal role of RNA in protein synthesis discovered between 1955 and 1966 first stimulated the hypothesis that RNA may have evolved to make polypeptides before the evolution of DNA (Woese 1967; Crick 1968; Orgel 1968). The recently uncovered structural and functional details of peptide synthesis by tRNA and ribosomes strongly support these early predictions, as do several other ribonucleoprotein (RNP) complexes discussed in this section.

rRNA Performs Peptidyl Synthesis

Harry Noller (2011) reviewed early evidence in the 1970s and 1980s that small chemical changes in bacterial rRNA greatly decreased protein synthesis, suggesting a direct participation of rRNA in protein synthesis. In addition, by the early 1990s, Noller et al. (1992) showed that most of the protein of the *Escherichia coli* ribosome could be removed and the remaining rRNA would perform the peptidyl synthetase reaction, extending the idea that peptidyl synthesis was performed at or close to defined sites in 23S rRNA (Noller 1991).

The recent striking advances in ribosome structure and function based on crystallography have consolidated the idea that RNA alone acted at least for peptide synthesis in an RNA world (Nissen et al. 2000; Schmeing and Ramakrishnan 2009; Voorhees et al. 2009). Granted that the present-day ribosome has dozens of proteins that help to maintain the present-day rRNAs in proper configuration, the catalysis of peptide linkage is nevertheless carried out in the interior of the large ribosomal subunit in an essentially protein-free environment (Fig. 6-3). Moreover, the weak interaction of only a three-base complementarity between tRNA and mRNA functions because of stabilizing tRNA:rRNA interactions while the codon:anticodon pairing between mRNA and tRNA is checked for a correct match, allowing peptidyl synthesis to proceed with fidelity. Thus, all of the specificity during protein synthesis resides

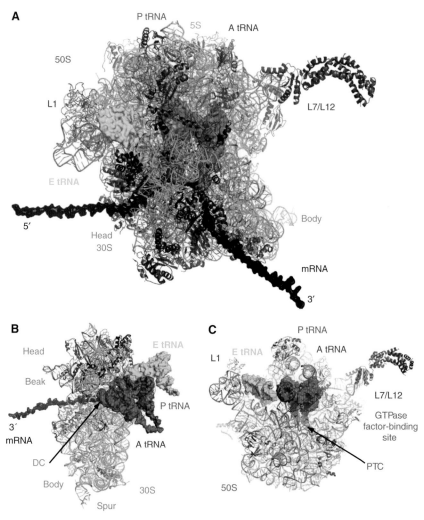

Figure 6-3. Peptidyl transferase reaction is peformed in an RNA environment. (*A*) RNA portion of the 70S ribosome showing that the interface between the 50S and 30S subunits consists mainly of rRNA ([orange] 50S rRNA, [aqua] 30S rRNA). Structural landmarks are designated, and the three tRNAs in positions A ([purple] entry), P ([green] peptidyl), and E ([yellow] exit) reveal that the 3′ ends of the A and P tRNAs lie very close together, whereas the E tRNA, now without its amino acid but as a peptidyl tRNA, is relatively far away. The L1 and L7/L12 ribosomal proteins are modeled in, but most protein structures are absent. The mRNA is black. (*B,C*) Separate views of the 30S and 50S subunits. (DC [blue]) Site of decoding, (PTC [brown]) peptidyl transfer center. (Reprinted, with permission, from Schmeing and Ramakrishnan 2009 [©Macmillan].)

in actions of RNA molecules. Although proteins are crucial to stabilize the RNA in the modern ribosome, the 1968 intuition of RNA as the primordial informational macromolecule in cells seems increasingly reasonable.

Pre-rRNA Processing Requires Small Nucleolar RNAs

Before leaving the ribosome, consider how the rRNA itself is formed: This is perhaps the prime extant example of an essential early life function that presumably was and still is performed by a specific set of RNAs. In the modern world, all of the rRNA (eukaryotic, archaeal, and bacterial) is transcribed as a precursor molecule (Chapter 4, pp. 154–155) that must be processed to be functional. This process includes methylation of sugars and pseudouridylation at specific sites as well as cleavage to release the final pieces of RNA found in the ribosome.

In eukaryotes, there are several hundred small RNAs named snoRNAs (small *nucle*olar RNAs) because many were first isolated from nucleoli, where eukaryotic ribosome manufacture occurs. The snoRNAs are complementary to pre-rRNAs at specific sites and have been proven to be necessary for specific steps in the maturation of pre-rRNAs (Fig. 6-4). Like most (if not all) noncoding RNAs that have known functions in modern cells, the snoRNAs operate as RNP complexes (for review, see Filipowicz and Pogacic 2002; Tycowski et al. 2006). Some archaea have small RNAs like snoRNAs, but bacteria do not. The argument has been made that the oldest form of processing pre-rRNA used short RNAs, and that these were lost in rapidly growing single cells such as bacteria and some archaea (Jeffares et al. 1995). This conjecture is one of several (spliceosomal splicing being another) that raise the possibility that some RNA biochemistry in eukaryotes reflects earlier evolution than does RNA biochemistry in modern bacteria and archaea.

Other Ribonucleoprotein Machines

In addition to the ribosome and the events of pre-rRNA processing, that by themselves essentially make the case for an RNA world, several other cellular machines obligatorily contain RNA to function. Several examples illustrate that, although RNA retains a critical role in performing modern-day cell functions, it is always now in conjunction with proteins that, as is the case with ribosomes, have a structural role.

RNase P1, A Ribozyme That Is a Nuclease

Recall that the first-discovered present-day RNA enzyme, RNase P1, can function as a pure RNA in the right ionic environment (high Mg^{++})

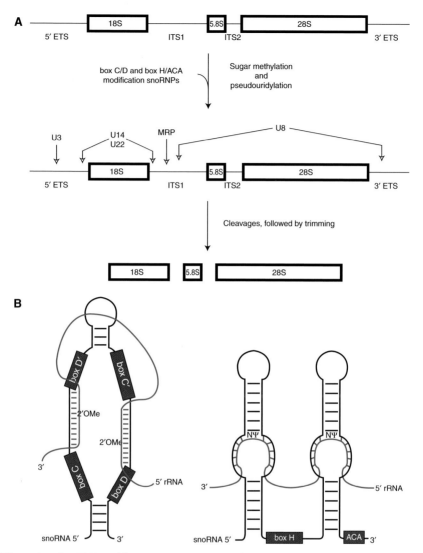

Figure 6-4. SnoRNAs guide pre-rRNA processing. (*A*) Steps during typical eukaryotic pre-rRNA processing. (Boxes) The three finished rRNAs: 18S, 5.8S, and 28S separated by internal transcribed spacers (ITSs) in the pre-rRNA (external transcribed spacer, ETS). Modifications of sugars and uridines are assisted by particular snoRNPs. U3, U14, U22, and U8 yeast particles that have mammalian counterparts, whereas MRP is mammalian. All of these act in cleaving the pre-rRNA but can act at a distance (arrows point to regions of association); exonucleolytic "trimming" to borders makes precise sites at ends of finished products. The 5.8S is hydrogen-bonded to the 28S in the ribosome and derives from processing a larger precursor (32S in mammals). (*B*) Examples of base-pair interactions in snoRNAs. Some (box C/D snoRNAs; *left*) have long (~20-bp) regions of perfect complementarity to a region in pre-rRNA. SnoRNAs that direct pseudouridylation (*right*) have short self-complementary regions that are separated. (Modified from Tycowski et al. 2006 [©CSHLP].)

(Guerrier-Takada et al. 1983) (see Chapter 4, pp. 202–204). However, inside of cells of all types—bacteria, archaea, and eukaryotes—this pre-tRNA processing enzyme is held in proper conformation by at least one protein.

The Signal Recognition Particle

In archaea, bacteria, and eukaryotes, a protein/RNA complex called SRP (signal recognition protein) binds proteins destined for secretion as they are being formed. Secretion occurs through the *translocon*, a multiprotein complex embedded in the plasma membrane or the endoplasmic reticulum (ER). One subunit in the SRP of both the bacterial and eukaryotic structure is a short RNA. The bacterial molecule (4.5S RNA) has a segment with sequence similarity to the larger 7S RNA of the eukaryotic SRP. In the SRP of all cells, there is also a GTPase that, by cleaving GTP, has a role in the exit of the protein chain through to the outside of the cell in prokaryotes or in eukaryotes into the lumen of the ER as it is being synthesized. The bacterial 4.5S RNA by itself can bind to the surface of a ribosome. This RNA-binding function presumably operates as the SRP picks up the amino-terminal amino acid segment—the signal sequence—just as it emerges at the ribosome surface (Poritz et al. 1990; Keenan et al. 2001).

Telomerase

Replication of the ends of linear eukaryotic chromosomes requires a special complex. During replication, the 5′-to-3′ growth of one chain of the DNA duplex (the *leading strand*) can proceed to the chromosome end, but during synthesis of the *lagging strand*, the terminal gap cannot be filled. This job is completed by a protein/RNA enzyme complex. The protein enzyme, telomerase, is structurally related to reverse transcriptases (RTs) of tumor viruses. The complex also contains a short RNA molecule. Telomerase can copy the RNA molecule into DNA, and the DNA is then ligated to the preexisting gap on the unfinished chromosome end (Blackburn 2006; Qiao and Cech 2008). The use of an RT to accomplish this task is interesting because, as we note below, group II introns that are self-splicing and can enter the DNA genome encode an RT. In the hypothetical transition from an RNA world to a DNA world, an RT would have been greatly beneficial, if not obligatory (Darnell and Doolittle 1986).

This discussion of present-day RNA/protein complexes also suggests another point regarding early chemical evolution. Ribozymes may have originally been the only enzymes. Proteins by themselves replaced most of these ribozymes, *but not all.* Some of the functions originally performed by ribozymes have, during the course of evolution, been too central or too complicated to have been subsumed by proteins (Cech 2009). These include protein synthesis itself and protein secretion, RNA processing, the initial priming step in DNA synthesis, and the concluding step in the synthesis of linear chromosomes by telomerase.

Small RNAs: Carryovers from an RNA World?

In Chapter 5, we discussed some of the many varieties of small RNAs with negative regulatory functions in eukaryotic cells.

A number of small RNAs were discovered in bacteria before the intense interest in such eukaryotic RNAs (for review, see Storz and Gottesman 2006; Breaker 2011). Many of the small bacterial RNAs are antisense to mRNA and function to inhibit translation or to speed mRNA decay. Many of these *riboswitches* are imperfectly paired with the mRNAs that they affect, reminiscent of microRNAs (miRNAs) in mammals. Some of the bacterial small RNAs (e.g., 6S RNA) (Wassarman and Storz 2006) and others (DsrA, RprA, and OxyS RNAs) affect transcription by interfering with the major σ factor (σ^{70}) that functions in transcription during logarithmic growth. In some cases, the σ factor that functions in the stationary phase, σ^{S}, is up-regulated by these small RNAs. Mammalian transcription has also been recently reported to be affected by small RNAs copied from repetitive sequences in mammalian cells (Mariner et al. 2008).

Many bacterial short RNAs are bound by a protein called Hfq (for "a *h*ost *f*actor that assists in the replication of the RNA bacteriophage Qβ"). The crystal structure of Hfq shows an RNA-binding domain that closely resembles that of the RNA-binding domain of the so-called Sm proteins that function in eukaryotic spliceosomes (Sauter et al. 2003). Archaea also have such proteins (Mura et al. 2003). The Sm proteins, by binding to the U1, U2, U4, U5, and U6 small nuclear RNAs, undoubtedly have an important role in determining spliceosomal function in all eukaryotes.

The bacterial Hfq is thought to assist small RNAs in pairing to an mRNA site with which the small RNA has imperfect homology. Bacterial small RNAs can also be encoded by viruses and plasmids but function by binding bacterial mRNAs. Although most of the bacterial small RNAs are not directly comparable to the small RNAs derived in eukaryotic cells by dedicated RNA-processing pathways, it seems that all cell types may, in fact, use RNA:RNA binding and small RNA:protein binding for information regulation. All of this new evidence provokes the speculation that such RNP functions are a holdover from the earliest time of cellular life or even of precellular existence.

Conclusion

There is a virtually universal belief among scientists that RNA chemistry early in the Earth's history was involved in starting life. There is no evidence and no convincing ideas as to exactly what this means. A functioning cell with only an RNA genome and functioning three-dimensional (3D) ribozymes certainly seems impossible, if by *cell* we mean a growing, dividing, genetically stable structure. But the invention of a functional protein synthesis system doubtless depended on RNA, as it still does. When the origin of cells is discussed below, we return to the question of, and refer to speculation regarding, properties of the *precellular* state.

THE GEOLOGICAL RECORD AND EARLY CELLULAR LIFE

Nineteenth-century and early 20th-century geology and paleontology, of course, provided ample evidence for proof of the evolution of plants and animals. Striking recent discoveries of "intermediate" forms continue to fill in an ever more complete fossil record of organismal evolution. For example, the fossil *Tiktaalik* was discovered by Neil Shubin and colleagues in 2006. This ~360-million-year-old fish-like animal had advanced fins containing bones resembling wrists that could likely propel it onto land and a head that could rotate (unlike fish). Clearly, paleontology remains a lively and fertile field (Daeschler et al. 2006; Shubin et al. 2006) in charting the evolution of multi-cellular organisms.

Geology and, in particular, Charles Lyell's 1830 volume *Principles of Geology*, was an important early influence on Darwin's thinking. Janet Browne, the Darwin biographer, quotes from Darwin's notebooks written after his ship the Beagle's visit to the Galápagos: "Animals differ in different countries in exact proportion to the time they have been separated. Countries longest separated, greatest differences" (Browne 1955).

Modern geology has also had a decisive role in explaining more precisely the present-day geographic distribution of many plants and animals. The wide acceptance (only in the 1960s) of plate tectonics solidified modern evidence for continent formation after the beginning breakup more than 200 million years ago of the single giant landmass, Pangaea. Geologic isolation explains a great deal regarding development of different species—Old World and New World monkeys, for example.

Thus, geology and historical biology have been entwined for more than 150 years and continue to be closely related. In fact, geological research offers the *only* source of hard evidence regarding life's origins and, in particular, the origins of cellular life. As discussed below, sequence comparisons using nucleic acids or proteins stimulate much speculation regarding the earliest evolution but yield no definitive answers. In contrast, two kinds of geological findings have great relevance for questions regarding early life on Earth: chemical and microscopic examination of accurately dated rock samples.

Dating of Geological Samples

By now, geophysicists have devised an ever-increasing armamentarium of physical isotopic techniques for accurately dating rock samples. A very readable summary of these techniques is available in G. Brent Dalrymple's book, *Ancient Earth, Ancient Skies: The Age of Earth and its Cosmic Surroundings* (2004). When care is taken in sample collecting, modern techniques of rock dating are accurate within tens of millions of years. Anything *within* rock samples up to ~4 billion years old can be accurately dated.

Microfossils Interpreted as Bacteria or Archaea

By the mid 1970s, thin sections of rocks were examined—by either transmission electron microscopy (TEM) or surface replicate techniques—for "microfossils" (Knoll and Barghoorn 1977). Structures the size and shape of microorganisms were observed and widely reported as bacteria (summarized in Schopf 2006).

Present-day collections of mats of microbial colonies (especially cyanobacteria) consisting of bacteria, secreted extracellular material, and fine sand form layered shoreline structures called stromatolites (Westall et al. 2006). Ancient layered stromatolite-like structures that became exposed in geologic outcrops were logically also thought to be formed by microorganisms. Extensive microscopic studies led to the proposal that microbes of several different physical types could be seen in stromatolite-like rock formations as old as 3.5 billion years. These spherical structures were originally proposed to be cyanobacteria (for review, see Schopf 2006). This is now discredited because cyanobacteria are aerobic oxygen producers credited with producing atmospheric oxygen and geologists agree that high levels of oxygen only developed ~2.4 billion years ago (Holland 2006). It remains possible (as Schopf emphasizes) that the spheroidal structures observed might be some other fossilized bacteria. However, it has also been pointed out that metamorphic changes in rocks can produce nonbiogenic defects that are indistinguishable from what appear to be biogenic structural residues (Brasier et al. 2006; McLoughlin et al. 2008).

Apprised of the care that must be exercised in judging microfossils to be biogenic, geologists have continued to use microscopic data from ancient rocks. Comparisons to modern stromatolites that are undoubtedly caused by layers of bacterial extracellular and cellular deposits have led a number of investigators examining similar ancient formations to conclude that anaerobic organisms *did* exist on Earth 3–3.5 billion years ago (Allwood et al. 2009). Such anoxygenic photosynthetic organisms *do* exist even today. Among the microscopic characteristics thought most positively to identify presumed microfossils are long, fiber-like structures scattered among which are unit-shaped (round or tubular) structures that are interpreted to have once been embedded among the fibrous material. It is certain that modern microbes and their extracellular products form such structures (for review, see Westall 2009). It has also been established by Raman spectroscopy that there is car-bonaceous material in the layers of the ancient purported stromatolites. Moreover, the $^{12}C/^{13}C$ ratio is high, suggesting that it was derived from living organisms (see below) (Tice and Lowe 2004; Westall 2009). Because of the different environments from which these samples most likely originally came, some are thought to be hydrothermal in origin, and all are thought to be anaerobic single-cell organisms similar to organisms that exist today (Rasmussen 2000; Westall et al. 2006; Allwood et al. 2009; Westall 2009).

Acritarchs: "Large" Microfossils

Although most of the bacteria-like microfossils discussed above are ~10 μm or smaller in diameter (as are the vast majority of extant bacteria and ar-chaea), older, well-preserved structures that are much larger (~50 μm in diameter) have been found with TEM in rocks 1.8 billion years old (summar-ized in Knoll et al. 2006). In their review, Knoll et al. (2006) suggest that these large *acritarchs* could be some kind of protist or protist precursor (i.e., a fore-runner to eukaryotes). They propose several features that such fossils might possess to merit this claim: layered and decorated ("spiny") surface structure, ruptured resting "cells" releasing vegetative forms, or specific geochemistry suggesting chemical synthesis known to be performed only by eukaryotes. This has become an area of active interest recently because similar well-preserved fossils were reported in sedimentary rock formations that are 3.2 billion years old (Javaux et al. 2010; for review, see Buick 2010).

All microfossil evidence is stated with circumspection, but the consensus among geobiologists seems to be that microbes—probably mostly if not only bacteria—existed 3–3.5 billion years ago. Furthermore, oxygenation of the atmosphere occurred by the action of cyanobacteria 2.4 billion years ago. As we see in the next section, geochemistry on the new large microfossils could possibly push the appearance of eukaryotes back to Archaean time (possibly as early as ~3 billion years ago) (Buick 2010).

Geochemistry and Early Cellular Life

Perhaps less ambiguous conclusions regarding the earliest cellular life have come from examination of "chemical fossils." To simplify, there are two main tests applied to geological samples to determine their biogenicity: (1) First is the analysis of carbon isotopes in geologic samples. ^{12}C and ^{13}C isotope ratios in nonbiological samples have changed only slightly from ~ 3.5 billion years ago to the present (Smith 2007). The enzymes in all present-day and presumably ancient aerobic and anaerobic organisms fractionate the two isotopes, preferring the lighter ^{12}C to ^{13}C. Thus, any putative ancient biogenic sample should have a higher $^{12}C/^{13}C$ isotope ratio than a nonbiogenic sample. (2) Compounds that, as far as anyone can tell, had to be synthesized by living organisms can now be analyzed in femtomole (and less) amounts. Large polycyclic lipids are the most useful biogenic marker molecules for stating with quite reasonable certainty that an ancient sample contains evidence of former living organisms (Dutkiewicz et al. 2006; Summons et al. 2006).

Possibly the most informative samples for using this chemical approach have come from deep drilling ("coring") and recovery of rock samples 2–3 billion years old. Roger Summons and colleagues (Waldbauer et al. 2009) have gone to great pains to be assured that samples uncontaminated with younger material have been obtained. Recently described samples from South Africa that span from 2.7 to 2.4 billion years ago contain lipids that can be synthesized by bacteria and archaea. But, strikingly, in samples that are 2.6 billion years old, there are lipids that today are synthesized only by eukaryotes. Others have contested this possibility (Rasmussen et al. 2008). But the Summons group and a group headed by Roger Buick have produced strong chemical evidence for early eukaryotic biosynthesis. These samples date considerably earlier than any microfossil evidence for eukaryotes (Dutkiewicz et al. 2006; George et al. 2008), except possibly for the recently described 3-billion-year-old acritarchs discussed in the previous section (Javaux et al. 2010).

To biologists interested in "final" answers to questions such as the age of the oldest cells and when eukaryotes arrived on the scene—the appropriate attitude would seem to be to wait and see. Life is very probably at least 3 billion years old, perhaps older, and the exact time of arrival of eukaryotes has to be judged as uncertain at present but could have been well over 2.4 billion years ago.

MEMBRANES ARE NECESSARY FOR CELL FORMATION

As we have seen, prebiotic chemistry is a fascinating but difficult subject if inarguable conclusions are the only satisfaction. Aside from the yet-unsolved problems of organic chemistry and the extent of functions in an RNA world, there are other problems. All present-day life operates in spaces constrained by lipid-containing membranes. To corral any successful early chemical operations on the way to forming a cell, cell membranes would appear to be a necessity. Only a few intrepid researchers have attempted to avoid guesswork regarding this ancient era and waded in to consider how cellular membranes might have formed in order to allow "internal" primitive nucleic acid chemistry to occur. Jack Szostak and colleagues have embarked on such an exploration to describe a *protocell.*

The spontaneous formation of vesicles by amphiphilic lipids has been known for some time (Hargreaves and Deamer 1978; for review, see Chen and Walde 2011). Szostak and colleagues started with lipid mixtures that would form vesicles and found that a clay-like mineral would greatly accelerate vesicle formation (Hanczyc et al. 2003). It had earlier been established that such a clay, montmorillonite, would adsorb oligoribonucleotides. The lipid vesicles can grow by further lipid incorporation and with slight agitation can "divide" (be fractured) without losing their contents.

The Szostak group then found that charged molecules such as nucleotides could enter these vesicles at elevated temperatures. Moreover, if the vesicles contained a partially double-stranded nucleic acid template, a nonenzymatic primer extension reaction occurred within the vesicle (Mansy et al. 2008). Vesicles can be stable to temperatures of up to 100°C so that strand separation and nonenzymatic multiplication of templates in such vesicles can be envisioned (Mansy and Szostak 2008).

Whether these experiments accurately model early prebiotic events seems irrelevant at this point. The experiments do show that imaginative research is being directed toward very daunting challenges in grappling with the unknown, but just possibly not unknowable, prebiotic and precellular world.

INITIAL SEQUENCE COMPARISONS AND EVOLUTIONARY HISTORY: CARL WOESE AND THREE KINGDOMS OF LIFE

Although establishing cellular identity and species relatedness by sequence comparisons seems child's play today, this situation could not be imagined in the early 1970s. One impatient visionary, Carl Woese, did not wait for extensive DNA sequence data. Woese had been attracted in the late 1960s (Woese 1967) to the origin of cells and the relationship among different cells based on the universal genetic code. He decided in the early 1970s that even primitive sequence comparisons of some universal cellular entity should provide deep insight into the origin of cells. He chose bacterial 16S rRNA because all cells possess two-subunit ribosomes and purification was simple for the 16S rRNA from many different bacteria (and later the 18S small-subunit rRNA from eukaryotes).

Following Fred Sanger's lead of making tryptic peptide maps to sequence proteins, production of overlapping oligonucleotide maps of short RNAs had allowed sequencing of a few RNAs. Sequencing of the first tRNA was by Robert Holley and colleagues (Holley et al. 1965). The 120-nucleotide sequence of a human 5S rRNA had been accomplished by Sherman Weissman and colleagues (Forget and Weissman 1967), and a 119-nucleotide sequence of *E. coli* 5S rRNA was generated by Bart Barrell and George Brownlee in the Sanger laboratory (Brownlee et al. 1967).

But the sequencing of an RNA molecule ∼1600 nucleotides long by fragmentation and oligonucleotide sequencing was not feasible (and, of course, DNA sequencing was not yet available). Woese and colleagues did, however, achieve reproducible oligonucleotide patterns—*fingerprints*—from the total small rRNA after complete or partial ribonuclease T1 digestion. This enzyme cleaves after the GMP in RNA leaving a set of oligonucleotides terminated by a single G, some of which could be cleanly separated and individually sequenced. Woese, first with his graduate student George Fox and then other colleagues, made T1-produced oligonucleotide maps with 16S rRNA from 27 bacterial species (Woese et al. 1975) and found a striking conservation of many oligonucleotides, several of which were widely shared. (More than 50% of the bacterial species examined contained a collection of the same oligonucleotides.) There seemed no doubt that this group of commonly studied bacteria that included a broad spectrum of organisms, some disease causing and some not, had maintained a close molecular relationship throughout their evolution.

When Woese and colleagues branched out to some unusual bacteria, they made a great discovery. When the 16S rRNAs of 10 methanogenic (methane-producing) organisms were analyzed, there was conservation of oligonucleotide pattern among the 10. However, there were striking differences between the previous group of 27 "ordinary" bacteria and the methanogens (Fox et al. 1977).

Furthermore, the 18S small-subunit rRNAs of three eukaryotes—a yeast (*Saccharomyces cerevisiae*), a plant (*Lemna minor*, the duckweed), and cultured mouse L cells—were all similar and clearly distinguishable from those of the bacteria and the methanogens. Heretofore, the methanogens had been considered bacteria, but Woese and Fox stated that "These 'bacteria' appear to be no more related to typical bacteria than they are to eukaryotic organisms" (Woese and Fox 1977).

By the following year, extreme halophiles (salt loving) and various thermoacidophiles (capable of growth up to 100°C) had been isolated and their 16S rRNA fingerprinted. These unusual organisms obviously belonged with the original 10 methanogens (although they did not produce methane). Woese and colleagues proposed a separate kingdom of life, the Archaebacteria (Woese et al. 1978). This was originally thought to be an unwarranted and unnecessary claim by classical evolutionists. Ernst Mayr, for example, was such a holdout (Mayr 1998). But when DNA sequencing was first developed in the late 1970s and early 1980s, many more rRNAs were sequenced. Woese's position that there are three kingdoms—Bacteria, Archaea, and Eukarya (Fig. 6-5)—became widely accepted. From rRNA sequencing alone, at least 52 groups (possibly phyla) of bacteria and at least two archaeal groups, Crenarchaeota and Euryarchaeota, are recognized (Walsh and Doolittle 2005).

It had originally been hoped that rRNA alone would serve to root a "universal tree of life," and the idea of a "last universal common ancestor" (LUCA) to all cells became the Holy Grail. But from rRNA genes alone, this did not seem to be either reasonable or possible, for among other reasons there was no way to determine whether the archaea or the bacteria were more ancient.

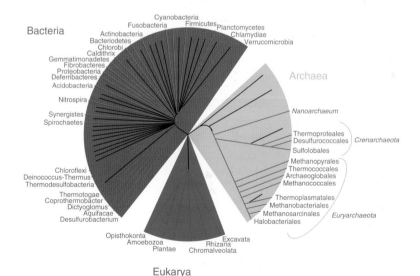

Figure 6-5. Diagrammatic representation of the organization of life into three domains based on SSU (16S small-subunit) rRNA gene sequence similarity. (Blue branches) Groups of organisms that have been cultured, (black branches) groups only known from culture-independent environmental studies. (Reprinted, with permission, from Walsh and Doolittle 2005 [©Elsevier].)

WHOLE-GENOME SEQUENCING AND ORIGIN
OF CELLULAR KINGDOMS

From the 1990s until the present, as whole-genome sequences of hundreds of bacteria and dozens of archaea, plants, and animals were assembled, it seemed reasonable that the pathway of cellular evolution would become transparent. Many convictions regarding protein families—both orthologs and paralogs—were confirmed. But, alas, the hopes of discerning the earliest events in cellular history have foundered.

Lateral Gene Transfer

In recent years, the most intense discussions in the "tree of life" community have been concerned with *lateral gene transfer* among and between bacteria and archaea. Recent whole-genome sequencing results ($\sim$200 bacteria and 20 archaea) have made it clear that many protein sequences are shared between the bacteria and archaea (Table 6-1) (Doolittle 2005; Walsh and Doolittle 2005). For example, of orthologous proteins, 100% of which were in all archaea, 28% were also found in bacteria. Thus, it is not possible to rigidly tell *species*, strictly defined by separate genomes, apart from one another. Over time, genes appear to have moved freely between all bacteria and archaea. Ford Doolittle describes a "web of life" among single cells and not a "tree of life" (Fig. 6-6) (Doolittle 2009; Doolittle and Zhaxybayeva 2009). It is now

Table 6-1. Comparison of proteins in bacteria and archaea suggests lateral gene transfer

No. of orthologs present in	Archaeal genomic signature			Bacterial genomic signature		
	0% bacteria	≤10% bacteria	≤20% bacteria	0% archaea	≤10% archaea	≤20% archaea
100% archaea	28.1, sd = 4.1	43.2, sd = 4.2	48.3, sd = 4.8	–	–	–
≥90% archaea	48.0, sd = 4.6	77.4, sd = 7.6	89.2, sd = 10	–	–	–
≥80% archaea	64.5, sd = 6.1	105.7, sd = 11.13	123, sd = 14.6	–	–	–
100% bacteria	–	–	–	13.6, sd = 1.7	15.7, sd = 1.8	17.1, sd = 1.8
≥90% bacteria	–	–	–	32.3, sd = 2.4	40.7, sd = 3.6	43.2, sd = 3.5
≥80% bacteria	–	–	–	47.8, sd = 5.3	64.9, sd = 7.4	70.5, sd = 7.9

Many proteins found in all archaea are also found in many bacteria. The reverse is also true. The complete sequences of 21 archaeal and 196 bacterial genomes were available for comparison. (Reprinted, with permission, from Walsh and Doolittle 2005 [©Elsevier].)

Phylogenetic relationships   Taxonomic rank arbitrary   Chimeric history
unknown

Figure 6-6. The problem with "species designations" (taxonomics) in view of extensive lateral gene transfer within and between bacteria and archaea. Traditional evolutionary thinking (applied to bacteria) supposed that given sufficient details (including sequences), individual species (*left*) could be identified or that if sufficient ancestral differences and similarities could be discerned by arbitrary rules, individual species could be arranged in groups in a tree-like fashion (*center*). However, if genes are swapped promiscuously (*right*), no such species designations or tree-like arrangements are possible because of a "chimeric" history of the prokaryotic organisms. (Reprinted from Doolittle and Zhaxy-bayeva 2009 [©CSHLP].)

widely accepted that sequencing of bacterial and archaeal genomes will not be able—now or ever—to lead to an incontrovertible conclusion regarding the earliest forms of life on the planet.

In addition to the sequencing of individual recognized and cultured organisms, so-called environmental sequencing—taking DNA for sequencing from soil, sea, or ocean water—shows that microbial life is both unbelievably prolific and varied. Presumably, if this is true for today's single cells, it might well have been true billions of years ago. Perhaps we can never know, barring discovery of some revelatory geochemical biomarker, which if any present-day microbes resemble the most ancient single cells.

Protein Sequence Comparisons and Evolutionary History

What, therefore, has genomic sequencing told us regarding the origin of the eukaryotic cell? First of all, as eukaryotic organisms increased in complexity, gene duplication (probably even whole-genome duplication) led to larger genomes. But the startling fact was that much larger genomes did *not* mean vast differences in protein-coding genes. Many protein families did duplicate and account for some increase in genes that encode proteins. Both orthologs (similarly functioning genes with recognizably similar sequences) and paralogs (recognizable as originating from the same sequence family but drifted in both sequence and function) are widely recognized. But most multicellular plants, animals, and fungi have 15,000–25,000 protein-coding loci. (These loci are frequently simply called *genes* [see Chapter 5] but because of alternate splicing and alternate transcription start sites (TSSs) they give rise to many more mutable mRNA products than 15,000–25,000. However, there is an economy of space to use *genes* to mean "protein-coding loci.")

An equally surprising more recent result has come from sequencing free-living single-celled eukaryotes. Many of these organisms also have ~15,000 genes. For example, the protist *Naegleria gruberi* (Fritz-Laylin et al. 2010) and the choanoflagellate *Monosiga brevicollis* (King et al. 2008) each has ~15,000 genes. By comparing representative organisms across all of the clades of eukaryotes, Fritz-Laylin et al. (2010) and Sebé-Pedrós et al. (2010, 2011) offer evidence that the great majority of eukaryotes have descended from organisms(s) that possessed virtually the entire eukaryotic array of genes (Fig. 6-7). Possibly of great importance in considering the origin of eukaryotes is that between 25% and 40% of the eukaryotic gene repertoire is *not* related to that of bacteria or archaea by any sophisticated data-searching method, but it includes proteins that are present throughout all of the eukaryotes. The origin of these eukaryotic-specific proteins remains unknown (Fritz-Laylin et al. 2010; also see Hartman and Fedorov 2002; Fedorov and Hartman 2004; Aravind et al. 2006; Doolittle 2009). For the remaining ~60%–75% of eukaryotic proteins encoded in eukaryotic *nuclear* genes, there is at least some relationship to proteins encoded by either bacteria or archaea (Aravind et al. 2006). Both the chloroplast and (especially) mitochondria have reduced genomes compared to bacteria. Many of the nuclear eukaryotic genes resembling those found in bacteria and archaea could logically have derived from endosymbionts or from lateral gene transfer at an early time.

Archaeal–Eukaryotic Similarity in Macromolecular Synthetic Machinery

Perhaps the single most striking observations to come from the protein sequence comparisons relevant to cellular evolution of the three kingdoms is the great similarity of many proteins concerned with macromolecular synthesis in archaea and eukaryotes (for review, see Walsh and Doolittle 2005; Cann 2008). Some recent analyses place the eukaryotic proteins that engage in macromolecular synthesis closer to those of Crenarchaeota than to those of Euryarchaeota (Cox et al. 2008; Foster et al. 2009).

All of the archaeal and eukaryotic RNA polymerases are multi-subunit enzymes—10 or so subunits of similar proteins—in contrast to the four subunits ($\alpha_2\beta\beta'$) in all bacteria. In addition, several members of the general transcription factors identified in eukaryotes (TBP, TFIIB, TFIIE) are also present and function similarly in archaea but are not present in bacteria. Both archaea

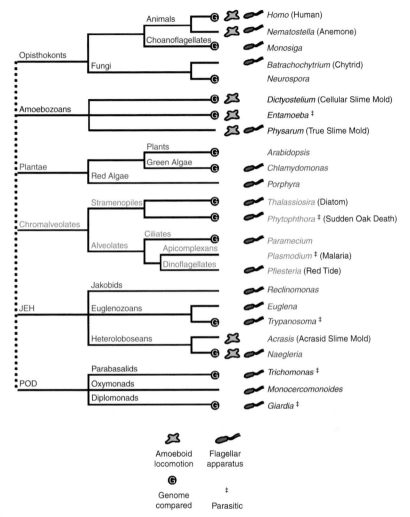

Figure 6-7. Consensus cladogram of selected eukaryotes including animals, plants, and single-celled organisms. Comparative analyses of shared proteins in six major groups with diverse molecular phylogenies were carried out. The dotted polytomy indicates uncertainty regarding the order of early branching events. The circled "G" indicates groups in which at least one complete genomic sequence was available. Extensive protein-coding sequence was available for all entries. Representative taxa are shown on the *right*, with glyphs indicating flagellar- and/or actin-based amoeboid movement. (Although commonly referred to as *amoeboid*, *Trichomonas* does not undergo amoeboid locomotion.) Sequence comparisons among organisms listed here show virtually all eukaryotic-specific proteins to be present in some member of each clade indicated at the *left* in different colors. (Reprinted, with permission, from Fritz-Laylin et al. 2010 [ⓒElsevier].)

and eukaryotes use TA-rich transcription start sites. Furthermore, histone-like proteins in some archaea form dimers spontaneously and then bind DNA as tetramers, resembling the two tetramers present in octomeric eukaryotic nucleosomes (Fig. 6-8) (for review, see Reeve et al. 2004). In addition, although archaean rRNA sequences set archaea apart from both bacteria and eukaryotes, the ribosomal proteins of eukaryotes and prokaryotes are clearly quite similar and easily distinguishable from those of bacteria.

Finally, although all DNA polymerases have some similarities in crystal structure, there are clear sequence distinctions; thus DNA polymerases are divisible into subfamilies. Both eukaryotic and archaeal enzymes fall into the same DNA polymerase subfamily (Delaye et al. 2001; Walsh and Doolittle 2005). Not only are the DNA polymerases similar in archaea and eukaryotes, but other proteins concerned with DNA replication are also similar. For example, in eukaryotes, six different MCM (*minichromosome maintenance*) proteins form a ring that surrounds DNA and serves a helicase function. Archaea have a single protein with helicase activity that forms a six-membered ring (Pape et al. 2003). Bacteria do not encode such a protein. Furthermore, the clamp that holds DNA polymerase on DNA is composed of six structural units that are similar in all organisms. However, the eukaryotic and archaeal clamps have three similar dimers, whereas bacteria have two similar trimers to comprise the six similar units (Fig. 6-9) (Matsumiya et al. 2001; Cann 2008).

Archaea have a primitive form of a divided cell cycle with a limited period of DNA replication before division, as is the case with eukaryotes. In contrast, bacteria simply complete chromosome replication, divide, and immediately resume DNA replication. Moreover, the archaean *Sulfolobus acidocaldarius*

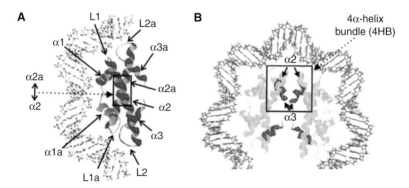

Figure 6-8. Crystal structures of methanococcal histone-like proteins. These proteins form dimers and, in the presence of DNA, nucleosome-like tetramers. (*A*) Dimeric structure of the histone fold domains (the helical regions are noted). (*B*) Arrows point to regions of interaction that lead to tetramer formation. (Reprinted, with permission, from Reeve et al. 2004 [©Biochemical Society].)

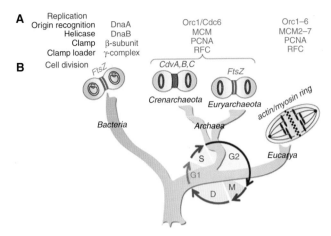

Figure 6-9. Three cellular domains emphasizing the similarity of eukaryotes and archaea: Distribution of similar proteins concerned with DNA replication and cell division. (*A*) Proteins concerned with DNA replication are very similar in archaea and eukaryotes (red). (Orc) Protein complex that locates DNA replication origins, (MCM) proteins that form a six-membered circular helicase that surrounds DNA and participates in replication fork movement, (PCNA) clamp that binds the replicative polymerase to the DNA to ensure processive copying, (RFC [replication factor complex]), a DNA-binding protein complex required for replication (see text for details). (*B*) Growing bacteria have a virtually continuous cell cycle. The archaean *Sulfolobus acidocaldarius* was recently shown to have a cell cycle separated into phases similar to those of eukaryotes. All archaea and bacteria have proteins that form a ring that participates in cell division (FtsZ). Eukaryotes use a more elaborate structure consisting of tubulins and associated proteins to divide. (Reprinted, with permission, from Cann 2008 [©National Academy of Sciences]; see also Duggin et al. 2008.)

expresses genes in a tightly controlled fashion around the cell cycle. Like eukaryotes, genes for proteins required in archaeal DNA replication are transcribed just before the "S phase" (Lundgren and Bernander 2007).

All of these remarkable similarities between archaea and eukaryotes led some years ago to the conclusion that there was no single root to a tree of life but, rather, that bacteria and archaea should be set apart, with eukaryotes being placed on a branch along with archaea on an *unrooted* tree (Fig. 6-10). This seems to be a reasonable conclusion. No matter how closely the proteins and the processes of eukaryotic macromolecular synthesis are related to archaeal proteins, it remains unclear how or whether archaea evolved directly into eukaryotes.

Carl Woese has written persuasively (at least to some scientists) that the concept of pervasive lateral gene transfer, detected by sequence analysis of extant bacteria and archaea, should be extended back to precellular time.

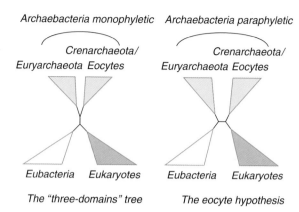

Figure 6-10. Possible origins of eukaryotes according to a "three-domains" tree (*left*) or a tree recognizing a special relationship between Crenarchaeota (eocytes) and eukaryotes (*right*). (Reprinted, with permission, from Foster et al. 2009 [©The Royal Society]; see also Cox et al. 2008.)

The precellular gene pool in this hypothesis contained perhaps everything that was required for the epoch in which actual cells capable of real Darwinian evolution took over (Woese 2004).

THE ORIGIN OF EUKARYOTIC CELLS

Despite the puzzle provoked by lateral gene transfer, the overwhelmingly interesting problem of the origin of eukaryotic cells remains. DNA sequencing (whole-genome sequencing included) has clearly left the problem unsolved.

Endosymbiosis

Modern ideas regarding eukaryotic cell origins originated with the proposal by Lynn Margulis more than three decades ago (Margulis 1975; for review, see Margulis and Schwartz 1998) that mitochondria and chloroplasts are *endosymbionts*. An ancestral eukaryotic precursor was envisioned as forming a commensal dependence with a bacterium, eventually ending with engulfment of the bacterium by the eukaryotic precursor and the persistence of at least some portion of the bacterial genome. And this proposal has correctly become embedded in any story of eukaryotic cell origin.

All known eukaryotic cells, including single cells (e.g., single-cell fungi) and all cells of multicellular organisms, have mitochondria. Recently some anaerobic eukaryotes (typified by *Giardia*) were thought to lack mitochondria and were referred to as Archezoa (Martin 1999). All such organisms have now been shown to harbor nuclear genes, the sequences of which suggest that they came, at some time in the past, from mitochondria. In addition, these organisms have membrane-bound organelles such as hydrogenosomes that make ATP or mitosomes that import proteins as mitochondria do. The idea has now been abandoned that eukaryotes thought to lack mitochondria exemplified primitive amitochondrial eukaryotes of the sort that could have engulfed a bacterium (for review, see Embley and Martin 2006).

Sequencing of hundreds of mitochondrial DNAs and comparing the sequences to those of bacterial genomes strongly suggests that an α-proteobacterium, an organism capable of aerobic respiration, was the progenitor of all mitochondria (Esser et al. 2004). Similar results show that chloroplasts in plants and single-cell light-responsive eukaryotes such as *Chlamydomonas* were likely derived from cyanobacteria. But, as noted above, there are many more "bacteria-like" encoding regions in the nuclear DNA of eukaryotes than remain in the relatively compact mitochondrial DNA. Likewise, chloroplast DNA is not by any means a full-length cyanobacterial genome (for review, see Martin 1999). Thus, all of the extant eukaryotes have one or more organelles that are of endosymbiotic origin—but the origin of the nucleus remains unsolved.

Two Possible Evolutionary Pathways to a Eukaryote

Perhaps the majority of scientists wish to deal with problems of eukaryotic origins using verifiable facts obtained from studying modern cells, and they do not concentrate on how any type of cell arose from a hypothetical RNA world or, indeed, a precellular non-Darwinian world with free sampling of a developing gene pool. To this group, such speculation is, at the moment (and may be forever), a dead end. This understandable philosophy continues the tradition of "prokaryote" (bacteria and archaea) to eukaryote—from "simple" to complex—that has reigned for 100 yr or more. And, of course, this may be correct. The fundamental tenet of this approach is that crossing the precellular/cellular divide occurred *once*—bacteria emerged and then gave rise to archaea—or at most twice—bacteria and archaea each managed to emerge and one or both then gave rise to eukaryotes (Martin 1999; Cavalier-Smith 2006; Koonin 2006; Martin and Koonin 2006). Because extensive gene sequence comparisons suggest that all eukaryotic organisms have many nuclear genes resembling each lineage—bacterial and archaeal—the task, according to this line of thought, is to get an example of each type of these two genomes together as the logical precursor to the eukaryotic cell. All eukaryotes have mitochondria, courtesy of an α-proteobacterium, and the macromolecular synthetic capacity of eukaryotes is closely related to that of archaea. So why not hypothesize a fusion/union of a cell from each of these two kingdoms to comprise the eukaryotic ancestor and assume that any eukaryotic-specific proteins have simply evolved independently so as not to be recognizable to be of bacterial/archaeal origin?

There are two major stumbling blocks to simply giving in to this seemingly reasonable logic. One is that all eukaryotes have an internal structure unlike that of either archaea or bacteria. This is not an insuperable problem because there are examples of genes that encode similar proteins (e.g., tubulin) that are found in bacteria. In addition, there are bacteria (Planctomycetes and Verrucomicrobia) that have internal compartmentalization resembling that of eukaryotes. *Gemmatata obscuriglobus*, a Planctomycete, has a nucleoid containing its chromosome separate from the cytoplasm (Fuerst and Webb 1991). Both Planctomycetes and Verrucomicrobia have structural proteins that shape internal membranes (for review, see Fuerst 2005; Santarella-Mellwig et al. 2010). *G. obscuriglobus* even appears to be capable of endocytosis (Lonhienne et al. 2010). Therefore, without a doubt, some present-day bacteria show that the bacterial kingdom has candidates to contribute protein and even internal structures to the possible origins of eukaryotes. However, these bacteria are not proteobacteria, and they are not archaea.

To some scientists, the bigger problem is that of information storage in the DNA of eukaryotes and the retrieval of information in a useful form (as mRNA). And in this age of molecular genetics, these are the chief defining characteristics that most clearly separate eukaryotes and single-celled nonnucleated bacteria and archaea.

Eukaryotes universally have nuclei in which their chromosomal DNA is housed and in which mRNA is formed, whereas their protein synthesis is cytoplasmic. All eukaryotes that have been examined have some primary RNA transcripts that must be deprocessed into mRNA using similar final biochemistry. There is, in every eukaryotic mRNA, a $5'$ m^7G cap and a $3'$-terminal poly(A), and *one protein is translated from each*. There are internal ribosome entry sites (IRESs) but independent ribosome entry is required to translate each coding region. There are no operons like those that exist in bacteria and archaea, where one primary transcript is translated into two or more different proteins by ribosome entry and polar ($5'$ to $3'$) transcription (for review, see Hellen and Sarnow 2001).

Most important for the question of evolutionary origin is that at least some transcripts in all eukaryotes have introns that must be removed. And all of the eukaryotes examined (whether single celled or multicellular), use very large *spliceosomal* machinery ($\sim$200 proteins in humans and five small nuclear RNAs, universally) to make an mRNA. Thus, uncovering the origin of spliceosomal splicing is critical for understanding the origin of the eukaryotic cell.

Indeed, the present-day existence of spliceosomal RNA splicing raises the possibility, not immediately refutable, that the late stage of precellular evolution may have included such splicing. After all, splicing is a property of RNA chemistry and might have evolved in an RNA world.

Some scientists attuned to the problems in crossing the precellular/cellular barrier in the first place are not willing to be forced into the formerly uncontested "prokaryotic"-to-eukaryotic mold (Fedorov et al. 2003; Fedorov and Hartman 2004; Woese 2004; Collins and Penny 2005; Jeffares et al. 2006; Kurland et al. 2006; Poole and Penny 2006). Rather, they hold out for an early, perhaps quite primitive and long-extinct "protoeukaryote" (Fritz-Laylin et al. 2010), certainly sharing a close kinship with archaea but perhaps already having a nucleus and possibly performing at least protein-assisted RNA-mediated RNA splicing, as the cell that *engulfed* (or, at any rate, joined with) the α-proteobacterium to the ultimate great benefit of the protoeukaryote. The creation of an energy factory—a mitochondrion—was the union that made eukaryotes successful. This clan of investigators rests its case on a hypothetical cell—a premitochondrial protoeukaryote, no example of which has ever been detected.

The foregoing paragraphs are a personal redaction of a most scholarly summary of the field by Francisco Rodríguez-Trelles, Rosa Tarríoa, the eminent evolutionist Francisco Ayala (Rodríguez-Trelles et al. 2006), and Carl Woese (2004). With this discussion as a backdrop, some of the current published thinking regarding cellular origins in general and eukaryotic origins in particular is outlined below.

ARCHAEAN–α-PROTEOBACTERIAL UNION: THE MARTIN-KOONIN MODEL

Perhaps the most straightforward and parsimonious proposal regarding the origin of a eukaryotic cell from the union of an archaean and a bacterium has been described by William Martin and Eugene Koonin (2006). Earlier, less specific but similar ideas were also put forward (e.g., Doolittle 1991; Palmer and Logsdon 1991; Stoltzfus et al. 1994). Starting with the knowledge that so many genes directing macromolecular synthesis have sequence and functional similarity in archaea and eukaryotes, they propose that an archaeon somehow fused with/incorporated the α-proteobacterium that was the precursor to a mitochondrion. As explained later, they speculate that this fusion offers a possible explanation for the frequent necessity of splicing in the ultimately derived eukaryote.

But the assumption of an archaean built like today's archaeal cells capable of ingesting another single-celled organism such as an α-proteobacterium has no obvious basis (Poole and Penny 2006; also see discussion of Koonin 2006). Furthermore, the only known present-day multigenomic noneukaryotic commensals are two related proteobacteria that both live independently within a eukaryotic cell (a "bacteriome") in the gut of the mealy bug (von Dohlen et al. 2001).

Mobile Bacterial Elements as Likely Precursors to Many (Perhaps All) Introns in Eukaryotic Cells

A second major inventive element of the Koonin-Martin proposal is directed at answering how a nucleus that performs pre-mRNA splicing arose. This part of their proposal derives from the mode of action of group II introns that exist as mobile genetic elements in some α-proteobacteria and cyanobacteria as well as in organelles thought to be derived from such bacteria—the mitochondria of fungi and the chloroplast of plants. Martin and Koonin propose, as have others, that these mobile elements are the sole source of introns in eukaryotes (Keating et al. 2010).

After the discovery of nuclear splicing by spliceosomes in the late 1970s and early 1980s, reports of introns in mitochondrial and chloroplast mRNAs soon appeared (Michel and Dujon 1983). Using sequence comparisons and predicted secondary structures of primary RNA transcripts, François Michel and Bernard Dujon recognized that two classes of introns were found in mitochondria and chloroplast mRNAs. One of these had similarities to the *Tetrahymena* intron studied by Cech and colleagues and was designated as *group I*. A larger number of introns, however, had very different sequences with easily recognizable, highly conserved predicted secondary structural domains, and these were designated as *group II introns* (Fig. 6-11).

The original discovery by Cech and coworkers of self-splicing of the ribosomal intron and of transesterification without intervention of protein stimulated attempts to determine the removal mechanisms of the group II mitochondrial and chloroplast introns. In a few cases, self-splicing of the type-II intron did occur. It was clear, however, that the two types of self-splicing introns differed fundamentally. In all of the type-II introns, there were five conserved domains predicted to form very similar secondary structures. In at least 25% of group II introns, there were coding sequences (ORFs) between domains III and V of the conserved structural domains (Fig. 6-11B). One of the encoded proteins strongly resembled the sequence of known reverse transcriptases (RTs). The other protein, when produced, was found to have a structural role in intron folding that greatly facilitated splicing of the mitochondrial chloroplast introns in vitro and was called *maturase*. Mutational analysis of the conserved elements in these introns showed that the structural elements as well as the two encoded proteins were necessary for splicing inside of the cell (for review, see Lambowitz and Zimmerly 2004, 2011; Pyle and Lambowitz 2006; Pyle 2010).

Quite early in the study of group II introns, it was recognized that the excision reaction that removed group II introns resembled the spliceosomal reaction that removes eukaryotic nuclear introns. The removed intron in both cases had a lariat structure—a branched circle with a dangling $3'$ tail (Fig. 6-11). (Of course, the group I intron [IVS in Cech's original nomenclature] has a short nucleotide stretch that cleaves itself from the main portion of the IVS as a lariat.) Second, a very compelling point of similarity was the presence in group II introns of five presumably folded double-stranded structural domains, and in the eukaryotic nucleus there was the participation of five spliceosomal RNAs—U1, U2, U4, U5, and U6—that also were discovered to have double-stranded structures (Keating et al. 2010). Advanced crystallographic studies have shown that the active center for splicing by group II introns is in domain five, and proper folding is assisted by other domains plus the encoded proteins. The ultimate details of spliceosomal splicing have yet to be solved, but interaction between base-paired RNA structures during spliceosomal action is established (e.g., U4 and U6 interaction is disrupted and replaced by U2 and U6 interaction to create the "active" spliceosome) (see Fig. 5-28).

In 1993, using nucleic acid probes taken from the mitochondrial and/or chloroplast group II introns, Jean-Luc Ferat and François Michel (1993) uncovered group II introns in α-proteobacteria and cyanobacteria. Martin and Koonin (2006) and others (Pyle 2010) take the close resemblance between group II introns and spliceosomal introns in the mechanism of intron removal, plus the finding of group II introns in bacteria, as conclusive evidence that all of the present-day eukaryotic introns are derived from endosymbiotic bacteria.

In addition to the apparently similar mechanisms of intron removal between group II introns and spliceosomal introns, group II introns exist in the RNA transcripts of mobile DNA elements in bacteria (and in plastids). It is hypothesized that a possible mechanism for mobility of these DNA elements operates through the RNA copy of the element. The protruding $3'$ tail of the lariat created by group II intron removal is thought to be capable of attacking DNA through transesterification, after which the encoded RT of the group II intron can insert a DNA copy of the RNA intron into the genomic DNA of the host. In their hypothetical α-proteobacterial–archaeal hybrid, Koonin and Martin hypothesize, as do others (Lambowitz and Zimmerly 2004; Pyle and Lambowitz 2006), that widespread insertion of these mobile DNA elements would have occurred during evolution after mitochondrial establishment in the early eukaryotic precursor.

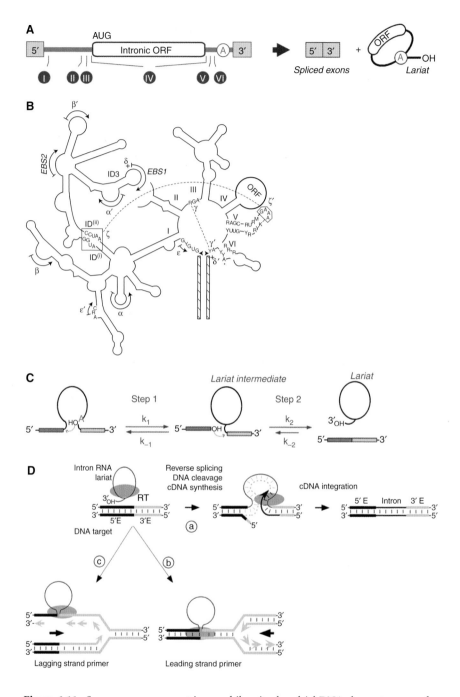

Figure 6-11. Sequence arrangement in a mobile mitochondrial DNA element proposed as the source of eukaryotic introns. (*A*) Site into which the intron RNA last caused entry is marked by 5′ and 3′ boxes (gray). (I, II, III, V, and VI) Conserved RNA structural domains found in all of the group II introns. Note that domain IV is an open reading frame (ORF) that often encodes an RT and also a *maturase*, i.e., a protein that acts to hold an RNA copy in position for "self-splicing." (*Legend continued on facing page.*)

Because these elements produce a means of being removed from any RNA transcript across their integration sites—they are self-splicing—they would initially cause no harm. However, it is conjectured that because the elements do not encode necessary (or even useful) functions, they would eventually disintegrate and no longer be capable of self-excision from RNA. This would kill any gene into which such an element had integrated. Koonin and Martin offer a plausible if difficult series of steps by which eventually a separate compartment for translation—a cytoplasm—would be walled off from the transcribed genome, to protect the precursors to mRNAs in what became the nucleus. Evolution of spliceosomes would have occurred by capturing the small RNAs of spliceosomes from the disintegrating and fragmented group II introns, and genes would have had to arise to produce the many proteins of the spliceosome.

In a limited support of this latter possibility, archaea and bacteria both have genes that encode a possible forerunner of a spliceosomal protein domain, the Sm domain (see Chapter 5). In eukaryotic spliceosomes, all of the individual small RNAs (U1, U2, U4, U5, and U6) are bound by the multiple Sm proteins (Khusial et al. 2005). Although both archaea and bacteria have Sm-like proteins, these proteins do not have a role in group II intron splicing, but they do bind RNA (Toro et al. 2001). The Martin-Koonin proposal shows a great awareness of all of the problems, both molecular genetic and in structural cell biology. The proposal lacks (as the authors realize) any immediate way to acquire confirmatory evidence, but it does provide a logical solution as to how eukaryotes arose from presently understood extant organisms without recourse to precellular times.

Anthony Poole and David Penny together and separately (among many others) argue against direct α-proteobacterial–archaeal fusion, holding out for a nucleated eukaryotic precursor that had phagocytic ability and contained macromolecular synthesis capabilities such as those of archaea as the entity that engulfed the α-proteobacterium (Poole and Penny 2006, 2007).

So where does this leave us at present? In the absence of any decisive answer to the question of a protoeukaryotic or archaeal precursor that joined with an α-proteobacterium, a wide gap remains in the implications of these

---

Figure 6-11. (*Continued*) (*B*) Secondary structure of conserved group II intron sequences, showing domains I, II, III, IV, V, and VI as stem loops. (*C*) Entire intron is excised by the transesterification reactions of the 2′ OH of the adenosine (red A) and the 3′ OH of the upstream host sequence. These reactions leave a free 3′ OH at the downstream end of the host intron. The intron, containing the coding sequence, forms a lariat. (*D*) Excised intron lariat can reenter the host genome (either the mitochondrial or bacterial genome) by the 3′ OH of the lariat attacking *DNA*. With the aid of the RT, the lariat can be copied back into DNA, with DNA repair completing the integration of the intron at a new site. (*A,C,D*, Reprinted from Pyle and Lambowitz 2006 [©CSHLP]; *B*, modified, with permission, from Michel and Ferat 1995 [©Annual Reviews Inc.].)

two classes of ideas for what actually happened during the early evolution of eukaryotes. Traditionalist thinking is that prokaryotes—now, one must logically say bacteria and archaea—came before eukaryotes by at least tens to hundreds of millions of years. This seems certainly to be the case, but that does not mean that prokaryotes designed like the *extant* bacteria and archaea *must* have given rise to eukaryotes.

SPLICING AND TRANSITION FROM THE RNA TO THE RNP TO THE CELLULAR WORLD

For many scientists (see Rodríguez-Trelles et al. 2006), spliceosomal splicing and RNA chemistry have refocused attention to a different stage of evolution, namely, the development of a precellular state dominated by RNA or RNPs. In this precellular state, piecing together information was likely originally necessary. This opens the possibility that some form of splicing may have been a part of the RNA/RNP world and that a protoeukaryote had such capability.

No conclusion can be made nor need be made on the timescale of the presumed precellular-to-cellular events (i.e., the exit from the RNA/RNP world). If the protoeukaryote emerged 1 billion years later than bacteria and archaea, is that impossible? Certainly, the acquisition of a mitochondrion was necessary for robust eukaryotic success. But nothing known at present bars a protoeukaryote, before the mitochondrial acquisition, from sampling the same precellular RNA/RNP gene pool as did archaea and bacteria before taking off successfully after the engulfment of an α-proteobacterium. In this formulation, the precellular gene pool may have included some type of splicing ability that would have been available to become amplified after α-proteobacterial engulfment.

However sophisticated genomic studies of present-day organisms may become, it is by no means clear that we will learn how a cell or many different primordial cells jumped the RNA/RNP gap to become functioning growing entities capable of Darwinian selection. Carl Woese, a major moving force in all of modern molecular evolutionary studies (Woese 2004), seems to be on safe ground in breaking with the notion of a single primordial DNA-containing *cellular* ancestor, a LUCA that was the Darwinian precursor to all present-day cells. It remains a possibility that all three present successful cellular domains each succeeded in crossing the line from a precursor state to success as contenders in a Darwinian world.

Furthermore, who is to say how many *potential* "kingdom-forming" cell types did originally emerge, only to become extinct? After all, extinction has been the fate of most organisms ever to evolve on the planet. The only widely accepted conclusion is that today there are three functioning cellular kingdoms: Bacteria, Archaea, and Eukarya.

A major point of the present discussion is that the discovery of discontinuous protein coding and splicing in 1977 and all of the studies on RNA chemistry that followed have had a huge impact on the thinking of researchers struggling to understand not only the evolution of the earliest cell but also of the earliest eukaryotic cell.

SOME FINAL THOUGHTS: THE PRIMACY OF RNA

What do we make of RNA at the conclusion of all of our foregoing discussions? To begin, the origin of life seems routed through RNA. The overwhelming consensus by knowledgeable chemists is that studying how RNA arrived on the scene is an increasingly tractable subject—nucleic acid bases and ribose are "easy" to imagine in the prebiotic world. And if Sutherland and colleagues (Powner et al. 2009) and efforts stimulated by their recent work succeed, a realistic path to nucleotides will soon be described.

Oligonucleotide assembly is more problematic, but in combination with mineral catalysts, not completely hopeless. Given mixtures of long-enough oligoribonucleotides, it is established that ribozymes will form. And, at least under laboratory conditions, ribozymes with oligonucleotide ligase abilities will "evolve." Furthermore, with designed ribozymes, it is a proven fact that RNA self-replication will ensue. Ribozymes with many different catalytic properties including some related to peptide synthesis have been produced in the laboratory.

The next steps require an ambitious and impatient collection of RNA polymers to "recognize" the value of even short peptides, "spontaneously" composed from available amino acids. This recognition of the value of peptides would encourage the enormously critical step of directed catalysis of useful peptides—the forerunner of protein synthesis.

Advance to this stage is what many investigators consider the *RNA world*. How far such chemical evolution could go in forming even more RNA catalysts and, indeed, short peptide catalysts is obviously only a guess. Acknowledging a stage at which this level of sophistication could have been reached, the necessity arises of adopting chemical compartmentation in lipid vesicles or some other sort of appropriately "sticky" mineral substrate to lead in the direction of integration into "protocell(s)."

Discussion to this point is thought in many quarters, perhaps correctly, to be so much idle speculation. Speculation beyond this point invites, possibly requires, some whimsy.

Having gotten this far, RNA would have "recognized" that the "heavy lifting" of inventing coded information to make useful peptides, however short, was too tiresome to continue to invent. How complicated this protocell had to become before this recognition occurred is, of course, an even deeper mystery.

But the imperative arrived that information embodied in RNA sequences was too precious to lose, and a more stable information storehouse was demanded—DNA had to be invented. Affirmation or rejection of this story represents potential experimental tasks that 21st-century prebiotic chemistry/biochemistry must attempt.

Given a seething, writhing molecular mass, how an actual cell was formed seems buried, perhaps forever, in a past that has left, at best, only traces of chemical clues that geochemists may be best at analyzing.

However, making a cell today is at least a tractable laboratory exercise (Gibson et al. 2010). How many proteins it takes to make a functioning cell is a project already on the drafting boards in several laboratories. Will this inform us regarding early evolution? Quite possibly not. It is still a long way from making such a jerry-rigged laboratory organism to understanding the decisive, true, Darwinian steps to three, and only three, cellular kingdoms that without argument have been around for at least $\sim 1.8-2.7$ billion years. But as difficult as the task of evolving a eukaryotic cell may seem, it is certainly no more difficult than the solution to the earlier chemical events leading up to a cell.

Although there is an inevitable hallucinatory atmosphere to thoughts regarding early evolution, what seems universally accepted is that RNA is at the center of all of the problems and possibilities for further enlightenment.

Moving to modern times, it took the epoch-making discoveries of Watson and Crick and Sanger to finally break the back of opposition to the central importance of macromolecules. The idea of a linear relationship between genes and proteins demanded emphasis on how linear genetic information could be translated into linear polypeptides. Only the successful prosecution of that problem finally brought RNA to a central role in contemporary biology. The experimental success that followed in the pursuit of how RNA is synthesized and of the chemical abilities of RNA constitutes many of the most fundamental advances of the last five decades. Technical and mechanical success using DNA, that durable, pliable storehouse of RNA blueprints, has figured very prominently in the success of experiments designed to learn about RNA. But very often in the era of recombinant DNA and rapid DNA sequencing, the underlying motive for experiments is to learn about RNA. And in this new century, it seems as if the cornucopia of new RNAs has certainly not been exhausted.

From the narrow perspective of the RNA biochemist, it is possible to ask, "Why all the furor in the biochemical world of the past regarding DNA and protein?" They are both simply the handmaidens that have allowed RNA, the first informational molecule, to have its way with this planet.

REFERENCES

Allwood AC, Grotzinger JP, Knoll AH, Burch IW, Anderson MS, Coleman ML, Kanik I. 2009. Controls on development and diversity of Early Archean stromatolites. *Proc Natl Acad Sci* **106:** 9548–9555.

Altman S. 1990. Nobel lecture. Enzymatic cleavage of RNA by RNA. *Biosci Rep* **10:** 317–337.

Aravind L, Iyer LM, Koonin EV. 2006. Comparative genomics and structural biology of the molecular innovations of eukaryotes. *Curr Opin Struct Biol* **16:** 409–419.

Atkins JF, Gesteland RE, Cech TR, eds. 2011. *RNA worlds: From life's origins to diversity in gene regulation.* Cold Spring Harbor Laboratory Press, Cold Spring Harbor, NY.

Bartel DP, Szostak JW. 1993. Isolation of new ribozymes from a large pool of random sequences. *Science* **261:** 1411–1418.

Benner SH. 2006. Setting the stage: The history, chemistry, and geobiology behind RNA. In *The RNA world*, 3rd ed. (ed. RF Gesteland et al.), pp. 1–21. Cold Spring Harbor Laboratory Press, Cold Spring Harbor, NY.

Benner SH. 2009. Understanding nucleic acids using synthetic chemistry. *Acc Chem Res* **37:** 784–797.

Bernstein M. 2006. Prebiotic materials from on and off the early Earth. *Philos Trans R Soc Lond B Biol Sci* **361:** 1689–1704.

Blackburn E. 2006. Telomerase RNA. In *The RNA world*, 3rd ed. (ed. RF Gesteland, et al.), pp. 419–436. Cold Spring Harbor Laboratory Press, Cold Spring Harbor, NY.

Brasier M, McLoughlin N, Green O, Wacey D. 2006. A fresh look at the fossil evidence for early Archaean cellular life. *Philos Trans R Soc Lond B Biol Sci* **361:** 887–902.

Breaker RR. 2006. Riboswitches and the RNA world. In *The RNA world*, 3rd ed. (ed. RF Gesteland et al.), pp. 89–103. Cold Spring Harbor Laboratory Press, Cold Spring Harbor, NY.

Breaker RR. 2011. Riboswitches and the RNA world. In *RNA worlds: From life's origins to diversity in gene regulation* (ed. JF Atkins et al.), pp. 63–79. Cold Spring Harbor Laboratory Press, Cold Spring Harbor, NY.

Breslow R, Cheng ZL. 2009. On the origin of terrestrial homochirality for nucleosides and amino acids. *Proc Natl Acad Sci* **106:** 9144–9146.

Breslow R, Levine M, Cheng ZL. 2010. Imitating prebiotic homochirality on Earth. *Orig Life Evol Biosph* **40:** 11–26.

Browne J. 1996. *Charles Darwin: Voyaging* [quote from Darwin's Notebooks B15, D23], p. 363. Princeton University Press, Princeton, NJ.

Browne J. 2002. *Charles Darwin, the power of place*, p. 392. Princeton University Press, Princeton, NJ.

Brownlee GG, Sanger F, Barrell BG. 1967. Nucleotide sequence of a 5S ribosomal RNA from *E. coli. Nature* **215:** 735–736.

Buick R. 2010. Ancient acritarchs. *Nature* **463:** 885–886.

Burke DH. 2004. Ribozyme-catalyzed genetics. In *The genetic code and the origin of life* (ed. L Ribas de Pouplana), pp. 48–64. Kluwer Academic/Plenum Publishers, New York.

Cann IKO. 2008. Cell sorting protein homologs reveal an unusual diversity in archaeal cell division. *Proc Natl Acad Sci* **105:** 18653–18654.

Cavalier-Smith T. 2006. Cell evolution and Earth history: Stasis and revolution. *Philos Trans R Soc Lond B Biol Sci* **361:** 969–1006.

Cech TR. 1990. Nobel lecture. Self-splicing and enzymatic activity of an intervening sequence RNA from *Tetrahymena. Biosci Rep* **10:** 239–261.

Cech TR. 2009. Crawling out of the RNA world. *Cell* **136:** 599–603.

Chen IA, Walde P. 2011. From self assembled vesicles to protocells. In *The origin of life* (ed. D Deamer, JW Szostak), pp. 179–193. Cold Spring Harbor Laboratory Press, Cold Spring Harbor, NY.

Collins L, Penny D. 2005. Complex spliceosomal organization ancestral to extant eukaryotes. *Mol Biol Evol* **22:** 1053–1066.

Cox CJ, Foster PG, Hirt RP, Harris SR, Embley TM. 2008. The archaebacterial origin of eukaryotes. *Proc Natl Acad Sci* **105:** 20356–20361.

Crick FHC. 1968. The origin of the genetic code. *J Mol Biol* **38:** 367–379.

Daeschler EB, Shubin NH, Jenkins FA Jr. 2006. A Devonian tetrapod-like fish and the evolution of the tetrapod body plan. *Nature* **44:** 757–763.

Darnell JE, Doolittle WF. 1986. Speculations on the early course of evolution. *Proc Natl Acad Sci* **83:** 1271–1275.

Deamer D, Szostak JW, eds. 2011. *The origins of life.* Cold Spring Harbor Laboratory Press, Cold Spring Harbor, NY.

Delaye L, Vazquez H, Lazcano A. 2001. The cenancestor and its contemporary biological relics: The case of nucleic acid polymerases. In *First steps in the origin of life in the universe* (ed. J Chela-Floes et al.), pp. 223–230. Kluwer Academic Publisher, Dordrecht, The Netherlands.

de Pouplana LR, ed. 2004. *The genetic code and the origin of life.* Kluwer Academic/Plenum Publishers, New York.

Doolittle WF. 1991. The origin of introns. *Curr Biol* **1:** 145–146.

Doolittle RF. 2005. Evolutionary aspects of whole genome biology. *Curr Opin Struct Biol* **15:** 248–253.

Doolittle WF. 2009. The practice of classification and the theory of evolution, and what the demise of Charles Darwin's tree of life hypothesis means for both of them. *Philos Trans R Soc Lond B Biol Sci* **364:** 2221–2228.

Doolittle WF, Zhaxybayeva O. 2009. On the origin of prokaryotic species. *Genome Res* **19:** 744–756.

Doudna JA, Cech TR. 2002. The chemical repertoire of natural ribozymes. *Nature* **418:** 222–228.

Duggin IG, McCallum SA, Bell SD. 2008. Chromosome replication dynamics in the archaeon *Sulfolobus acidocaldarius. Proc Natl Acad Sci* **105:** 16737–16742.

Dutkiewicz A, Volk H, George SC, Ridley J, Buick R. 2006. Biomarkers from Huronian oil-bearing fluid inclusions: An uncontaminated record of life before the Great Oxidation Event. *Geology* **34:** 437–440.

Ekland EH, Bartel DP. 1996. RNA-catalysed RNA polymerization using nucleoside triphosphates. *Nature* **382:** 373–376.

Ekland EH, Szostak JW, Bartel DP. 1995. Structurally complex and highly active RNA ligases derived from random sequences. *Science* **269:** 364–370.

Embley TM, Martin W. 2006. Eukaryotic evolution, changes and challenges. *Nature* **440:** 623–630.

Englehardt AE, Hud NV. 2011. Primitive genetic polymers. In *The origins of life* (ed. D Deamer, JW Szostak), pp. 207–227. Cold Spring Harbor Laboratory Press, Cold Spring Harbor, NY.

Esser C, Ahmadinejad N, Wiegand C, Rotte C, Sebastiani F, Gelius-Dietrich G, Henze K, Kretschmann E, Richly E, Leister D, et al. 2004. A genome phylogeny for mitochondria among α-proteobacteria and a predominantly eubacterial ancestry of yeast nuclear genes. *Mol Biol Evol* **21:** 1643–1660.

Fedorov A, Hartman H. 2004. What does the microsporidian *E. cuniculi* tell us about the origin of the eukaryotic cell? *J Mol Evol* **59:** 695–702.

Fedorov A, Roy S, Cao X, Gilbert W. 2003. Phylogenetically older introns strongly correlate with module boundaries in ancient proteins. *Genome Res* **13:** 1155–1157.

Ferat J-L, Michel F. 1993. Group II self-splicing introns in bacteria. *Nature* **364:** 358–361.

Ferris JP. 2006. Montmorillonite-catalysed formation of RNA oligomers: The possible role of catalysis in the origins of life. *Philos Trans R Soc Lond B Biol Sci* **361:** 1777–1786.

Filipowicz W, Pogacic V. 2002. Biogenesis of small nucleolar ribonucleoproteins. *Curr Opin Cell Biol* **14:** 319–327.

Forget BG, Weissman SM. 1967. Nucleotide sequence of KB cell 5S RNA. *Science* **158:** 1695–1699.

Foster PG, Cox CJ, Embley TM. 2009. The primary divisions of life: A phylogenomic approach employing composition-heterogeneous methods. *Philos Trans R Soc Lond B Biol Sci* **364:** 2197–2207.

Fox GE, Magrum LJ, Balch WE, Wolfe RS, Woese CR. 1977. Classification of methanogenic bacteria by 16S ribosomal rRNA characterization. *Proc Natl Acad Sci* **74:** 4537–4541.

Fritz-Laylin LK, Prochnik SE, Giner ML, Dacks JB, Carpenter ML, Field MC, Kuo A, Paredez A, Chapman J, Pharm J, et al. 2010. The genome of *Naegleria gruberi* illuminates early eukaryotic versatility. *Cell* **140:** 631–642.

Fuerst JA. 2005. Intracellular compartmentation in planctomycetes. *Annu Rev Microbiol* **59:** 299–328.

Fuerst JA, Webb RI. 1991. Membrane-bounded nucleoid in the eubacterium *Gemmatata obscuriglobus. Proc Natl Acad Sci* **88:** 8184–8188.

George SC, Volk H, Dutkiewicz A, Ridley J, Buick R. 2008. Preservation of hydrocarbons and biomarkers in oil trapped inside fluid inclusion for >2 billion years. *Geochim Cosmochim Acta* **72:** 844–870.

Gesteland RF, Cech TR, Atkins JF, eds. 2006. *The RNA world*, 3rd ed. Cold Spring Harbor Laboratory Press, Cold Spring Harbor, NY.

Gibson DG, Glass JL, Lartigue C, Noskov VN, Chuang RY, Algire MA, Benders GA, Montague MG, Ma L, Moodie MM, et al. 2010. Creation of a bacterial cell controlled by a chemically synthesized genome. *Science* **329:** 52–56.

Guerrier-Takada C, Gardiner K, Marsh T, Pace N, Altman S. 1983. The RNA moiety of ribonuclease P is the catalytic subunit of the enzyme. *Cell* **35:** 849–857.

Haldane JBS. 1929. The origin of life. Reprinted in *On being the right size* (ed. JM Smith). Oxford University Press, New York.

Hanczyc MM, Fujikawa SM, Szostak JW. 2003. Experimental models of primitive cellular compartments: Encapsulation, growth, and division. *Science* **302:** 618–622.

Hargreaves WR, Deamer DW. 1978. Liposomes from ionic single-chain amphiphiles. *Biochemistry* **17:** 3759–3768.

Hellen CUT, Sarnow P. 2001. Internal ribosome entry sites in eukaryotic mRNA molecules. *Genes Dev* **15:** 1593–1612.

Holland HD. 2006. The oxygenation of the atmosphere and oceans. *Philos Trans R Soc Lond B Biol Sci* **361:** 903–915.

Holley RW, Apgar J, Everett G, Madison JT, Marquisee M, Merrill S, Penswick JR, Zamir A. 1965. Structure of ribonucleic acid. *Science* **147:** 1462–1465.

Hougland JL, Picitille JR, Forconi M, Lee J, Herchlag D. 2006. How the group I intron works: A case study of RNA structure and function. In *The RNA world*, 3rd ed. (ed. RF Gesteland et al.), pp. 133–205. Cold Spring Harbor Laboratory Press, Cold Spring Harbor, NY.

Huber C, Wachtershauser G. 2006. α-Hydroxy and α-amino acids under possible Hadean, volcanic origin-of-life conditions. *Science* **324:** 630–632.

Javaux EJ, Marshall CP, Bekker A. 2010. Organic-walled microfossils in 3.2-billion-year-old shallow-marine silliciastic deposits. *Nature* **463:** 934–938.

Jeffares DG, Poole AM, Penny D. 1995. Pre-rRNA processing and the path from the RNA world. *Trends Biochem Sci* **20:** 298–299.

Jeffares DC, Mourier T, Penny D. 2006. The biology of intron gain and loss. *Trends Genet* **22:** 16–22.

Joyce GF. 2002. The antiquity of RNA-based evolution. *Nature* **418:** 214–221.

Joyce GF, Orgel LE. 2006. Progress toward understanding the origins of the RNA world. In *The RNA world*, 3rd ed. (ed. RF Gesteland et al.), pp. 23–56. Cold Spring Harbor Laboratory Press, Cold Spring Harbor, NY.

Keating KS, Toor N, Perlman PS, Pyle AM. 2010. A structural analysis of the group II intron active site and implications for the spliceosome. *RNA* **16:** 1–9.

Keenan RJ, Freymann DM, Stroud RM, Walter P. 2001. The signal recognition particle. *Annu Rev Biochem* **70:** 755–775.

Khusial P, Plaag R, Zieve GW. 2005. LSm proteins form heptameric rings that bind to RNA via repeating motifs. *Trends Biochem Sci* **30:** 522–528.

King N, Westbrook MJ, Young SL, Kuo A, Abedin M, Chapman J, Fairclough S, Hellsten U, Isogai Y, Letunic I, et al. 2008. The genome of the choanoflagellate *Monosiga brevicollis* and the origin of metazoans. *Nature* **451:** 783–788.

Knoll AH, Barghoorn ES. 1977. Archean microfossils showing cell division from the Swaziland System of South Africa. *Science* **198:** 396–398.

Knoll AH, Javaux EJ, Hewitt D, Cohen P. 2006. Eukaryotic organisms in Proterozoic oceans. *Philos Trans R Soc Lond B Biol Sci* **361:** 1023–1038.

Koonin EV. 2006. The origin of introns and their role in eukaryogenesis: A compromise solution to the introns-early versus introns-late debate? *Biol Direct* **1:** 22. doi: 10.1186/1745-6150-1-22.

Kurland CG, Collins LJ, Penny D. 2006. Genomics and the irreducible nature of eukaryote cells. *Science* **312:** 1011–1014.

Lambowitz AM, Zimmerly S. 2004. Mobile group II introns. *Annu Rev Genet* **38:** 1–35.

Lambowitz AM, Zimmerly S. 2011. Group II introns: Mobile ribozymes that invade DNA. In *RNA worlds: From life's origins to diversity in gene regulation* (ed. JF Atkins et al.), pp. 103–122. Cold Spring Harbor Laboratory Press, Cold Spring Harbor, NY.

Leach S, Smith IW, Cockell CS. 2006. Introduction: Conditions for the emergence of life on the early Earth. *Philos Trans R Soc Lond B Biol Sci* **361:** 1675–1679.

Lee N, Bessho Y, Wei K, Szostak JW, Suga H. 2000. Ribozyme-catalyzed tRNA aminoacylation. *Nat Struct Biol* **7:** 28–33.

Levy M, Miller SL, Oro J. 1999. Production of guanine from NH_4CN polymerizations. *J Mol Evol* **49:** 165–168.

Lincoln TA, Joyce GF. 2009. Self-sustained replication of an RNA enzyme. *Science* **323:** 1229–1232.

Lohse PA, Szostak JW. 1996. Ribozyme-catalyzed amino-acid transfer reactions. *Nature* **381:** 442–444.

Lonhienne TG, Sagulenko E, Webb RI, Lee KC, Franke J, Devos DP, Nouwens A, Carroll BJ, Fuerst JA. 2010. Endocytosis-like protein uptake in the bacterium *Gemmata obscuriglobus*. *Proc Natl Acad Sci* **107:** 12883–12888.

Lundgren M, Bernander R. 2007. Genome-wide transcription map of an archaeal cell cycle. *Proc Natl Acad Sci* **104:** 2939–2944.

Mansy SS, Szostak JW. 2008. Thermostability of model protocell membranes. *Proc Natl Acad Sci* **105:** 13351–13355.

Mansy SS, Schrum JP, Krishnamurthy M, Tobe S, Treco DA, Szostak JW. 2008. Template-directed synthesis of a genetic polymer in a model protocell. *Nature* **454:** 122–125.

Margulis L. 1975. Symbiotic theory of the origin of eukaryotic organelles; criteria for proof. *Symp Soc Exp Biol* **29:** 21–38.

Margulis L, Schwartz KV. 1998. *Five kingdoms: An illustrated guide to the phyla of life on Earth*, 3rd ed. Freeman, New York.

Mariner PD, Walters RD, Espinoza CA, Drullinger LF, Wagner SD, Kugel JF, Goodrich JA. 2008. Human Alu RNA is a modular transacting repressor of mRNA transcription during heat shock. *Mol Cell* **29:** 499–509.

Martin W. 1999. A briefly argued case that mitochondria and plastids are descendants of endo-symbionts, but that the nuclear compartment is not. *Philos Trans R Soc Lond B Biol Sci* **266:** 1387–1395.

Martin W, Koonin EV. 2006. Introns and the origin of nucleus–cytosol compartmentalization. *Nature* **440:** 41–45.

Matsumiya S, Ishino Y, Morikawa K. 2001. Crystal structure of an archaeal DNA sliding clamp: Proliferating cell nuclear antigen from *Pyrococcus furiosus*. *Protein Sci* **10:** 17–23.

Mayr E. 1998. Two empires or three? *Proc Natl Acad Sci* **95:** 9720–9723.

McLoughlin N, Brasier MD, Wacey D, Green OR, Perr RS. 2007. On biogenicity criteria for endolithic microborings on early Earth and beyond. *Astrobiology* **7:** 10–26.

McLoughlin N, Wilson LA, Brasier MD. 2008. Growth of synthetic stromatolites and wrinkle structures in the absence of microbes—Implications for the early fossil record. *Geobiology* **6:** 95–105.

Michel F, Dujon B. 1983. Conservation of RNA secondary structures in two intron families including mitochondrial-, chloroplast- and nuclear-encoded members. *EMBO J* **2:** 33–38.

Michel F, Ferat J-L. 1995. Structure and activities of group II introns. *Annu Rev Biochem* **64:** 435–461.

Miller SL. 1953. A production of amino acids under possible primitive Earth conditions. *Science* **117:** 528–529.

Mura C, Phillips M, Kozhukhovsky A, Eisenberg D. 2003. Structure and assembly of an augmented Sm-like archaeal protein 14-mer. *Proc Natl Acad Sci* **100:** 4539–4544.

Nelson KE, Robertson MP, Levy M, Miller SL. 2001. Concentration by evaporation and the prebiotic synthesis of cytosine. *Orig Live Evol Biosph* **31:** 221–229.

Nissen P, Hansen J, Ban N, Moore PB, Steitz TA. 2000. The structural basis of ribosome activity in peptide bond synthesis. *Science* **289:** 920–930.

Noller HF. 1991. Ribosomal RNA and translation. *Annu Rev Biochem* **60:** 191–227.

Noller HF. 2011. Evolution of protein synthesis from an RNA world. In *RNA worlds: From life's origins to diversity in gene regulation* (ed. JF Atkins et al.), pp. 141–144. Cold Spring Harbor Laboratory Press, Cold Spring Harbor, NY.

Noller HF, Hoffarth V, Zimniak L. 1992. Unusual resistance of peptidyl transferase to protein extraction procedures. *Science* **256:** 1416–1419.

Oparin AI. 1976. Evolution of the concepts of the origin of life, 1924–1974. (A lecture read at the International Seminar, Origin of Life, August 1974, Moscow.) *Orig Life* **7:** 3–8.

Orgel LE. 1968. Evolution of the genetic apparatus. *J Mol Biol* **38:** 381–393.

Oro J. 1961. Mechanism of synthesis of adenine from hydrogen cyanide under possible primitive earth conditions. *Nature* **191:** 1193–1194.

Palmer JD, Logsdon JM Jr. 1991. The recent origins of introns. *Curr Opin Genet Dev* **1:** 470–477.

Pape T, Meka H, Chen S, Vicentini G, van Heel M, Onesti S. 2003. Hexameric ring structure of the full-length archaeal MCM protein complex. *EMBO Rep* **4:** 1079–1083.

Poole AM, Penny D. 2006. Evaluating hypotheses for the origin of eukaryotes. *BioEssays* **29:** 74–84.

Poole A, Penny D. 2007. Engulfed by speculation. *Nature* **447:** 913.

Poritz MA, Bernstein HB, Strub K, Zopf D, Wilhelm H, Walter P. 1990. An *E. coli* ribonucleoprotein containing 4.5S RNA resembles mammalian signal recognition particle. *Science* **250:** 1111–1117.

Powner MW, Gerland B, Sutherland JD. 2009. Synthesis of activated pyrimidine ribonucleotides in prebiotically plausible conditions. *Nature* **459:** 239–242.

Pyle AM. 2010. The tertiary structure of group II introns: Implications for biological function and evolution. *Crit Rev Biochem Mol Biol* **45:** 215–232.

Pyle AM, Lambowitz AM. 2006. Group II introns: Ribozymes that splice RNA and invade DNA. In *The RNA world*, 3rd ed. (ed. RF Gesteland et al.), pp. 469–505. Cold Spring Harbor Laboratory Press, Cold Spring Harbor, NY.

Qiao F, Cech TR. 2008. Triple-helix structure in telomerase RNA contributes to catalysis. *Nat Struct Mol Biol* **15:** 634–640.

Rasmussen B. 2000. Filamentous microfossils in a 3,235-million-year-old volcanogenic massive sulphide deposit. *Nature* **405:** 676–679.

Rasmussen B, Fletcher IR, Brocks J, Kilburn MR. 2008. Reassessing the first appearance of eukaryotes and cyanobacteria. *Nature* **455:** 1101–1104.

Reeve JN, Bailey KA, Li W-T, Marc F, Sandman K, Soares DJ. 2004. Archaeal histones: Structures, stability and DNA binding. *Biochem Soc Trans* **32:** 227–230.

Robertson MP, Joyce GF. 2011. The origins of the RNA world. In *RNA worlds: From life's origins to diversity in gene regulation* (ed. JF Atkins et al.), pp. 21–42. Cold Spring Harbor Laboratory Press, Cold Spring Harbor, NY.

Robertson MP, Miller SL. 1995. An efficient prebiotic synthesis of cytosine and uracil. *Nature* **375:** 772–774.

Rodríguez-Trelles F, Tarrio R, Ayala FJ. 2006. Origins and evolution of spliceosomal introns. *Annu Rev Genet* **40:** 47–76.

Santarella-Mellwig R, Franke J, Jaedicke A, Gorjanacz M, Bauer U, Budd A, Mattaj IW, Devos DP. 2010. The compartmentalized bacteria of the Planctomycetes-Verrucomicrobia-Chlamydiae superphylum have membrane coat-like proteins. *PLoS Biol* **8:** 1–11.

Sauter C, Bazquin J, Suck D. 2003. Sm-like proteins in Eubacteria: the crystal structure of the Hfq protein from *Escherichia coli*. *Nucleic Acids Res* **31:** 4091–4098.

Schmeing TM, Ramakrishnan V. 2009. What recent ribosome structures have revealed about the mechanism of translation. *Nature* **461:** 1234–1242.

Schmitt-Kopplin P, Gabelica Z, Gougeon RD, Fekete A, Kanawati B, Harir M, Gebefuegi I, Eckel G, Hertkorn N. 2010. High molecular diversity of extraterrestrial organic matter in Murchison meteorite revealed 40 years after its fall. *Proc Natl Acad Sci* **107:** 2763–2768.

Schopf JW. 2006. Fossil evidence of Archaean life. *Philos Trans R Soc Lond B Biol Sci* **361:** 869–885.

Sebé-Pedrós A, Roger AJ, Lang FB, King N, Ruiz-Trillo I. 2010. Ancient origin of the integrin-mediated adhesion and signaling machinery. *Proc Natl Acad Sci* **107:** 10142–10147.

Sebé-Pedrós A, de Mendoza A, Lang BF, Degnan BM, Ruiz-Trillo I. 2011. Unexpected repertoire of metazoan transcription factors in the unicellular holozoan Capsaspora owczarzaki. *Mol Biol Evol* **28:** 1241–1254.

Shubin NH, Daeschler EB, Jenkins FA Jr. 2006. The pectoral fin of *Tiktaalik roseae* and the origin of the tetrapod limb. *Nature* **440:** 764–771.

Smith BN. 2007. Natural abundance of the stable isotopes of carbon in biological systems. *Bioscience* **22:** 226–231.

Stoltzfus A, Spencer PF, Zuker M, Logsdon JM, Doolittle WF. 1994. Testing the exon theory of genes. The evidence from protein structure. *Science* **265:** 202–207.

Storz G, Gottesman S. 2006. Versatile roles of small RNA regulators in bacteria. In *The RNA world*, 3rd ed. (ed. RF Gesteland et al.), pp. 567–594. Cold Spring Harbor Laboratory Press, Cold Spring Harbor, NY.

Summons RE, Bradley AS, Jahnke LL, Waldbauer JR. 2006. Steroids, triterpenoids and molecular oxyen. *Philos Trans R Soc Lond B Biol Sci* **361:** 951–968.

Szostak JW. 2009. Origins of life: Systems chemistry on early Earth. *Nature* **459:** 171–172.

Tice MM, Lowe DR. 2004. Photosynthetic microbial mats in the 3,416-Myr-old ocean. *Nature* **431:** 549–552.

Toro I, Thore S, Mayer C, Basquin J, Seraphin B, Suck D. 2001. RNA binding in an Sm core domain: X-ray structure and functional analysis of an archael Sm protein complex. *EMBO J* **20:** 2293–2303.

Tycowski K, Kolev NG, Conrad NK, Fok V, Steitz JA. 2006. The ever-growing world of small nuclear ribonucleoproteins. In *The RNA world*, 3rd ed. (ed. RF Gesteland et al.), pp. 327–368. Cold Spring Harbor Laboratory Press, Cold Spring Harbor, NY.

van Roode JH, Orgel LE. 1980. Template-directed synthesis of oligoguanylates in the presence of metal ions. *J Mol Biol* **144:** 579–585.

von Dohlen CD, Kohler S, Alsop ST, McManus WR. 2001. Mealybug β-proteobacterial symbionts contain γ-proteobacterial symbionts. *Nature* **412:** 433–436.

Voorhees RM, Weixlbaumer A, Loakes D, Kelley AC, Ramakrishnan V. 2009. Insights into substrate stabilization from snapshots of the peptidyl transferase center of the intact 70S ribosome. *Nat Struct Mol Biol* **16:** 528–533.

Waldbauer JR, Sherman LS, Sumner DY, Summons RE. 2009. Late Archean molecular fossils from the Transvaal Supergroup record the antiquity of microbial diversity and aerobiosis. *Precambrian Res* **169:** 28–47.

Walsh DA, Doolittle WF. 2005. The real 'domains' of life. *Curr Biol* **15:** R237–R240.

Wassarman KM, Storz G. 2006. 6S RNA regulates *E. coli* polymerase activity. *Cell* **101:** 613–623.

Westall F. 2009. Life on an anaerobic planet. *Science* **323:** 471–472.

Westall F, de Ronde CEJ, Southam G, Grassineau N, Colas M, Cockell C, Lammer H. 2006. Implications of a 3.472–3.333 Gyr-old subaerial microbial mat from the Barberton greenstone belt, South Africa for the UV environmental conditions on the early Earth. *Philos Trans R Soc Lond B Biol Sci* **361:** 1857–1876.

Woese C. 1967. *The origin of the genetic code*. Harper & Row, New York.

Woese CR. 2004. A new biology for a new century. *Microbiol Mol Biol Rev* **68:** 173–186.

Woese CR, Fox GE. 1977. Phylogenetic structure of the prokaryotic domain: The primary kingdoms. *Proc Natl Acad Sci* **74:** 5088–5090.

Woese CR, Fox GE, Zablen L, Uchida T, Bonen L, Pechman K, Lewis BJ, Stahl D. 1975. Conservation of primary structure in 16S ribosomal RNA. *Nature* **254:** 83–85.

Woese CR, Magrum LJ, Fox GE. 1978. Archaebacteria. *J Mol Evol* **11:** 245–251.

Yarus M. 2011. Getting past the RNA world: The initial Darwinian ancestor. In *RNA worlds: From life's origins to diversity in gene regulation* (ed. JF Atkins et al.), pp. 43–50. Cold Spring Harbor Laboratory Press, Cold Spring Harbor, NY.

Zahnle K, Schaefer L, Fegley B. 2011. Earth's earliest atmospheres. In *The origins of life* (ed. D Deamer, JW Szostak), pp. 49–65. Cold Spring Harbor Laboratory Press, Cold Spring Harbor, NY.

Zaug AJ, Cech TR. 1985. Oligomerization of intervening sequence RNA molecules in the absence of proteins. *Science* **229:** 1060–1064.

Zaug AJ, Cech TR. 1986. The intervening sequence RNA of *Tetrahymena* is an enzyme. *Science* **231:** 470–475.

Sources

Atkins JF, Gesteland RF, Cech TR, eds. 2011. *RNA worlds: From life's origins to diversity in gene regulation*. Cold Spring Harbor Laboratory Press, Cold Spring Harbor, NY.

Brenner S. 2001. *My life in science* (as told to Lewis Wolpert) (ed. EC Friedberg, E Lawrence). BioMed Central, London.

Cairns J, Stent GS, Watson JD, eds. 1966. *Phage and the origins of molecular biology*. Cold Spring Harbor Laboratory, Cold Spring Harbor, NY.

Carlson EA. 2004. *Mendel's legacy: The origin of classical genetics*. Cold Spring Harbor Laboratory Press, Cold Spring Harbor, NY.

Cavalier-Smith T, Brasier M, Embley TM, eds. 2006. Major steps in cell evolution: Palaeontological, molecular and cellular evidence of their timing and global effects. Papers from a discussion meeting at the Royal Society, September 26–27, 2005, London, United Kingdom. *Philos Trans R Soc Lond B Biol Sci* **361:** 843–1083.

Chargaff E. 1963. *Essays on nucleic acids*. Elsevier, New York.

Crick F. 1988. *What mad pursuit: A personal view of scientific discovery*. Basic Books, New York.

Dalrymple GB. 2004. *Ancient Earth, ancient skies: The age of Earth and its cosmic surroundings*. Stanford University Press, Stanford, CA.

De Pouplana LR, ed. 2004. *The genetic code and the origin of life*. Kluwer Academic/ Plenum Publishers, New York.

Deamer D, Szostak JW, eds. 2010. *The origins of life*. Cold Spring Harbor Laboratory Press, Cold Spring Harbor, NY.

Dickerson RE. 2005. *Present at the Flood: How structural molecular biology came about*. Sinauer, Sunderland, MA.

Fruton JS. 1999. *Proteins, enzymes, genes: The interplay of chemistry and biology*. Yale University Press, New Haven, CT.

Gesteland RF, Cech TR, Atkins JF, eds. 2006. *The RNA world*, 3rd ed. Cold Spring Harbor Laboratory Press, Cold Spring Harbor, NY.

Inglis J, Sambrook J, Witkowski J, eds. 2003. *Inspiring science: Jim Watson and the age of DNA*. Cold Spring Harbor Laboratory Press, Cold Spring Harbor, NY.

Jacob F. 1973. *The logic of life: A history of heredity*. Pantheon Books, New York.

Jacob F. 1988. *The statue within.* Basic Books, New York.

Judson HF. 1996. *The eighth day of creation: Makers of the revolution in biology* (expanded edition). Cold Spring Harbor Laboratory Press, Cold Spring Harbor, NY.

Knoll AH. 2003. *Life on a young planet: The first three billion years of evolution on Earth.* Princeton University Press, Princeton, NJ.

Kay LE. 2000. *Who wrote the book of life? A history of the genetic code.* Stanford University Press, Stanford, CA.

Landecker H. 2007. *Culturing life: How cells became technologies.* Harvard University Press, Cambridge, MA.

Levene PA, Bass LW. 1931. *Nucleic acids.* Chemical Catalog Company, New York.

McCarty M. 1985. *The transforming principle: Discovering that genes are made of DNA.* Norton, New York.

McElroy WD, Glass B, eds. 1957. *A symposium on the chemical basis of heredity.* Johns Hopkins Press, Baltimore, MD.

Miller JH, Reznikoff WS. 1978. *The operon.* Cold Spring Harbor Laboratory, Cold Spring Harbor, NY.

Olby R. 2009. *Francis Crick: Hunter of life's secrets.* Cold Spring Harbor Laboratory Press, Cold Spring Harbor, NY.

Rheinberger H-J. 1997. *Toward a history of epistemic things: Synthesizing proteins in the test tube.* Stanford University Press, Stanford, CA.

Ridley M. 2006. *Francis Crick: Discoverer of the genetic code.* HarperCollins, New York.

Singh S. 2004. *Big Bang: The origin of the universe.* HarperCollins, NY.

Stetten DeW, Carrigan WT, eds. 1984. *NIH: An account of research in its laboratories and clinics.* Academic Press, New York.

Tanford C, Reynolds J. 2003. *Nature's robots: A history of proteins.* Oxford University Press, New York.

Ullmann A, ed. 2003. *Origins of molecular biology: A tribute to Jacques Monod.* ASM Press, Washington, DC.

Vincent WS, Miller OL Jr, eds. 1966. *International symposium on the nucleolus—Its structure and function.* National Cancer Institute Monograph 23. US Government Printing Office, Washington, DC.

Watson JD. 1968. *The double helix: A personal account of the discovery of the structure of DNA.* Atheneum, New York.

Watson JD. 2001. *Genes, girls and Gamow.* Oxford University Press, Oxford, UK.

Watson JD. 2003. *DNA: The secret of life.* Alfred A. Knopf, New York.

Witkowski J, Gann A, Sambrook J, eds. 2008. *Life illuminated, selected papers from Cold Spring Harbor,* Vol. 2, 1972–1994. Cold Spring Harbor Laboratory Press, Cold Spring Harbor, NY.

I owe a special debt to Professor Joseph Fruton's historical account in *Proteins, Enzymes, Genes: The Interplay of Chemistry and Biology* for many references and particularly for an enlightened discussion of early protein studies, and to Professor Richard E. Dickerson, whose enlightening and entertaining account *Present at the Flood: How Structural Molecular Biology Came About* is a wonderful guide to the history of crystallography for the nonexpert.

Index

Page references followed by f denote figures.